W9-BJK-643

Elementary Statistics and Decision Making

Sidney J. Armore
The George Washington University

Charles E. Merrill Publishing Company
A Bell & Howell Company
Columbus, Ohio

Merrill Mathematics Series

Erwin Kleinfeld, *Editor*

Published by
Charles E. Merrill Publishing Co.
A Bell & Howell Company
Columbus, Ohio 43216

International Standard Book Number: 0-675-08978-6

Library of Congress Catalog Card Number: 72-95499

2 3 4 5 6 7 8 9 — 79 78 77 76 75

To my wife

Ethel

and to
the memory of her parents

Meyer and Celia Honig

to whom education and learning
were important life goals

Preface

Every effort has been made to construct this book to meet the needs of the beginner in statistics and the varied requirements of statistics courses as taught in colleges today. The required background for this text is a knowledge of the basic part of elementary algebra. The serious reader, no matter how limited his mathematical background, should encounter no difficulty in learning statistics from this text.

Actually, there is no such thing as a "standard course" in elementary statistics. Course content and level vary from college to college and even from instructor to instructor. This reflects the phenomenal development of statistical methods in recent years and the wide choice of material suitable for inclusion in an introductory statistics course.

A textbook must be made to serve various study requirements and readers of diverse backgrounds. This text attempts to accomplish these objectives by:

1. Including a wide selection of topics, each with sufficiently enriched treatment
2. Identifying various sections and whole chapters as optional to permit topics to be included or omitted as desired with no loss of continuity
3. Including sections on Algebraic Notes (all marked optional) to permit a somewhat more "mathematical" study of elementary statistics if desired
4. Adding Supplementary Study Problems after the Study Problems at the end of some chapters to provide further enrichment in the study of introductory statistics
5. Including brief, but useful treatment of additional topics in Appendix A, some of which will appeal to readers with interest in special areas, and adding appropriate study problems pertaining to this appendix

Consequently, the organization of this text very conveniently permits the reader or instructor to structure the study of statistics according to his own philosophy and objectives.

The number of study problems provided is exceptionally large (595 problems) to provide a sufficient variety as well as a sufficient number of problems for practice and review. These problems are mostly concrete "real world" problems to interest the reader, make him probe the text more carefully, and gain greater insight into the meaning and application of statistical methods.

I am indebted to the students to whom I had the pleasure to teach introductory statistics for helping me understand the problems they faced in learning statistics. I am indebted to the many associates with whom I worked as a practicing and consulting statistician in various areas of business and economics, and in education and psychology. I gained particularly from my associations with the Bureau of Agricultural Economics of the U.S. Department of Agriculture and with the Personnel Research Branch of the Department of the Army where I worked with well-trained economists and psychologists who were highly knowledgable in statistics as well.

I am also indebted to my colleagues on the statistics faculty of The George Washington University, whose comments on features essential for an acceptable statistics textbook were very helpful to me in planning this book. I would like to mention particularly Professor Harold F. Bright, formerly Chairman of the Department of Statistics and now University Provost and Vice-President for Academic Affairs, who offered me my first opportunity in college teaching; Professor Solomon Kullback, Chairman of the Department of Statistics; and Professors Arthur D. Kirsch and Jackson K. Kern.

In a very special way, I am indebted to my wife, Ethel Armore, for her understanding and assistance during the long "after work" hours I spent writing this book. She contributed in many important ways to make my work easier, including valuable suggestions relating to the production phases of textbook writing, typing the manuscript, and proofreading. My son, Jerome L. Armore, and my daughter, Susan V. Armore, assisted in problem computation and proofing. I am indeed grateful for this family effort and cooperation without which this project would have been immeasurably more difficult.

Acknowledgments

I wish to thank John Wiley & Sons, Inc., New York, N.Y. for permission to reprint Table 3.2.5 from *Introduction to Statistical Analysis and Inference for Psychology and Education*, by Sidney J. Armore, 1966 (Table B.1), and the American Statistical Association for permission to use *The Choice of a Class Interval*, by H. A. Sturges, *Journal of the American Statistical Association*, Vol. 21, 1926, as the basis for construction of this table. I am indebted to the U.S. Interstate Commerce Commission for the use of page 1 of *Table of 105,000 Random Decimal Digits*, Statement No. 4914, File No. 261-A-1, May, 1949, Washington, D.C. (Table B.3), and to the U.S. National Bureau of Standards for the use of *Tables of Normal Probability Functions*, *Applied Mathematics Series 23*, Washington, D.C. (from which I derived Table B.4).

I am indebted to the Literary Executor of the late Sir Ronald A. Fisher, F.R.S., to Dr. Frank Yates, F.R.S., and to Oliver & Boyd, Edinburgh, for permission to reprint Table III from their book *Statistical Tables for Biological, Agricultural and Medical Research* (Table B.5), and to the *Biometrika* trustees for permission to use Table 8 of *Biometrika Tables for Statisticians*, Volume I (Table B.6) and "Tables of percentage points of the inverted beta (F) distribution," *Biometrika*, Vol. 33 (1943), by M. Merrington and C. M. Thompson (Tables B.7 and B.8).

I am indebted to the U.S. Bureau of the Census for the use of *Bureau of the*

Census Manual of Tabular Presentation, by Bruce L. Jenkinson as the basis for the discussion in Section A.6 and construction of Figures A.6.1 and A.6.2.

To the Instructor

This text was prepared to fit in with the structure of introductory statistics courses as taught in most colleges today. Descriptive statistics is covered in a brief, yet adequate manner. Probability is presented in a separate chapter and is set oriented; however, it has been presented to appeal also to those who still lean toward the traditional approach.

The newer topics, such as Bayesian inference and decision theory, are included, as well as chi-square tests, the analysis of variance, nonparametric statistics, and the usual topics in statistical inference. The careful, modern, and broad treatment of regression and correlation will surprise and, I believe, please many. An appendix contains various additional topics for special areas. Some of the special features of this text may be noted as follows:

Algebraic Notes appear at the end of some chapters to provide the reader with the opportunity to include proofs and derivations if desired. These notes are presented in a simple manner to satisfy both nonmathematical and other readers.

Chapter 4 contains a coordinated comparison of the different averages (Section 4.5) and a specially prepared discussion of averaging rates of change (Section 4.3). This discussion compares the applicability of the arithmetic and geometric means and makes the geometric mean more meaningful as an average for the reader.

Chapter 5 discusses the meaning as well as the computation of the standard deviation, the most important measure in statistics.

Chapter 6 presents a carefully arranged, introductory, yet comprehensive treatment of probability from the modern point of view. Section 6.4 on how to solve probability problems should be of considerable help to many readers and welcomed by many instructors.

Chapter 7 presents an unusually comprehensive, yet introductory presentation of random sampling, including sampling with and without replacement and from finite and infinite populations. The sampling distribution concept is given unusually enriched treatment (Section 7.4). Theoretical sampling models are introduced (Section 7.5) to unify the statistical theory pertinent to specific areas in statistical inference. This is an important feature of this text and such models are presented throughout the text.

Chapter 8 includes separate treatment of the population of 1's and 0's, binomial probability distribution, and the normal approximation to binomial probabilities.

Chapters 9 and 10 discuss estimation and confidence intervals for means, proportions, and differences, as well as for total amounts and total number. The theoretical basis of confidence intervals is clearly developed.

Chapters 11 and 12 present the theoretical basis of tests of hypotheses in a carefully defined, adequately illustrated, yet introductory manner. Included are tests for means, proportions, variances, and differences. Section 11.5 on controlling decision risks contains a synthesis of tests of hypotheses concepts and relationships as well as a discussion of how to determine sample size.

Chapter 13 presents the important chi-square tests relating to frequencies.

Chapter 14 discusses the F test and one-way and two-way analysis of variance.

Chapter 15 contains a very adequate selection of nonparametric procedures.

Chapters 17 through 20 present an introductory, yet modern and unusually comprehensive treatment of regression and correlation. Included are very adequate discussions of simple and multiple regression and correlation with appropriate theoretical models, linear and curvilinear methods, confidence and prediction intervals, tests of significance, and correlation of ranks.

Appendix A presents a variety of special topics, any part of which could be included or excluded, as desired. Standard test scores (Section A.1) will appeal to certain social science readers and time series methods and index numbers (Sections A.2 through A.4) will appeal to business and economics readers. The topics in the balance of this appendix will appeal to readers in all subject areas since they discuss significant figures and rounding and methods for constructing statistical tables and charts.

A large number of study problems are included at the end of each chapter and in Appendix A. The 595 problems fall into three categories: (1) discussion and comment problems, (2) methods application problems, and (3) course enrichment problems.

The first category of problems requires evaluations, brief discussions, and comments. These problems are intended to motivate the reader to probe the text and perhaps study more carefully. The second category of problems presents real-life type situations requiring the application of statistical methods and decision making. These problems range over a broad area of subject fields, thus, providing reader enrichment in the application of the statistical methods studied. The third category of problems presents additional methods related to the topics included in a chapter. These are the supplementary study problems.

The instructor may wish to expand the material covered in a semester by assigning portions for the students to study mostly on their own. The careful instructive style of presentation makes this possible.

Each instructor, of course, will wish to decide for himself which topics to include; however, it may be useful to indicate how a half-year statistics course could be structured. Generally speaking, the Algebraic Notes could be included or excluded, depending on how "mathematical" it is desired to structure the course. Additional course material in this area is provided in the Supplementary Study Problems which appear after the Study Problems at the end of a chapter, as appropriate.

A basic statistics course could consist of Chapters 1 through 9, Chapter 11, and Chapter 17, with all optional sections omitted. In Chapter 6, perhaps only the first two sections should be included. Some instructors may wish to include only the part on probability Definition 1 in Section 6.2 and omit the balance of Chapter 6. Parts of Chapter 15 and Appendix A could be included as desired.

A more enriched statistics course could be structured in various ways. For example, Chapters 1 through 9, Chapter 11, and Chapter 17 could be included (with the Algebraic Notes included or excluded, as desired), as well as portions of Chapters 13, 14 and 15, and selected parts of other chapters in the text and in Appendix A. Instructors who wish to emphasize correlation and regression could include Chapters 17 through 20 and omit other portions of the text.

To the Reader

This is a "nonmathematical" text on elementary statistics; however, this text specifically takes into consideration that elementary statistics is studied by two types of readers, *nonmathematical readers* and *other readers*, and attempts to meet the needs of both types.

The *nonmathematical reader* is typically very much afraid of anything that looks like mathematics or feels very "rusty" in this area. This text attempts to meet the needs of such readers by careful explanations of statistical methods and applications in a clear, nonmathematical style. Each method and application is lucidly illustrated by appropriate concrete problems and solutions clearly presented. The nonmathematical reader should find this text highly suitable and should have no difficulty with any part of it.

Other readers are those who feel quite comfortable studying an area of applied mathematics, either because they enjoy thinking in terms of mathematics or are fairly knowledgable in introductory mathematics. Such readers typically find long-winded discussions tedious and distracting. This text attempts to meet the needs of such readers by sensibly brief and to-the-point discussions and by including optional material for enrichment in their study of statistics.

The presentation in this text is oriented toward all subject areas. There is considerable advantage in studying introductory statistics in terms of a variety of subject areas, since it indicates the general applicability of statistical methods, provides enrichment for the reader, and may be expected to contribute to a better understanding of statistical methods and their application.

The treatment of statistical methods in this text is modern in every way. The exposition is instructive and careful, the scope is broad, yet the presentation was restricted in length to provide a useful and practical text. Even though the scope of topics included is broad, no topic is presented without adequate treatment. The presentation avoids too much brevity as well as too much discussion. Every effort was made to say it "just right" and "just enough."

Certain sections and whole chapters have been marked optional to identify those parts of the text which may be omitted with no loss of continuity. Readers and instructors may choose which of these parts to include on a first or second reading or to exclude altogether. In addition, supplementary study problems are included as appropriate after the study problems presented at the end of a chapter. This provides additional topics for study, if desired.

This text avoids the pitfalls of making introductory statistics appear "easy to learn" by leaving out many of the necessary explanations of the underlying statistical theory involved. This text provides appropriate explanations of the underlying statistical theory and the "why" of the statistical methods presented. The explanations and applications are always presented in a satisfying and acceptable manner.

This text has been organized and presented in a manner which should appeal to most readers. The serious reader should have no difficulty in following this text and learning elementary statistics. The large number of study problems provided (595 problems) provides sufficient material for practice and review.

The numbering of sections, tables, figures, and equations has been arranged for reader convenience. For example, Section 5.2 refers to the second section of Chapter 5. Table 7.4.1 refers to the first table in the fourth section of Chapter 7. Figures and equations are numbered in the same way as tables.

Contents

Part III   Statistical Basis for Decision Making

Part IV   Statistical Decision Making

Part V   Statistical Decision Making—Further Topics

Part VI   Association and Prediction

Appendix A   Additional Topics

Appendix B   Tables

Appendix C   Answers to Even-Numbered Problems

Index

Symbols used in this text

Symbol	Explanation
$a < b$	a is less than b
$a > b$	a is greater than b
$a \leq b$	a is less than or equal to b
$a \geq b$	a is greater than or equal to b
$a \pm b$	a plus or minus b
$a \neq b$	a is not equal to b
∞	infinity

Part I
Statistics:
A General View

Chapter 1

Statistics as a
Field of Study

1.1 A Tool for Decision Making

The popular meaning of _statistics_ is a set of numerical data, such as the number of traffic accidents per month or the number of students per class. More technically, statistics means methods for collecting, organizing, interpreting, describing, and presenting numerical data. Statistical methods are used in nearly every professional field and in nearly every area of business and production, such as in business and economics, public administration, education, anthropology, psychology, sociology, medicine, engineering. The same basic statistical methodology is used in all fields.

Generally speaking, there is no restriction to the kind and quantity of numerical data to which statistical methods can be applied. When the need to use statistics arises, it is always because there is a problem to be solved, a decision to be made, or a question to be answered. _Generally, a most important part of the_ *application of statistical methods is to carefully and precisely identify and define the problem to be solved, the decision to be made, or the question to be answered.*

Until recent times, statistical methods have been used for data collection and data description, to describe events of the past, and to provide a basis for thinking and planning ahead, usually on an intuitive and judgment basis. In recent times, however, the major use of statistics is in the area of *decision making*. This is why statistical methods have found such whole-hearted and widespread acceptance. Today the major emphasis and usefulness of statistics is in sample survey design, estimation, hypothesis testing, forecasting and prediction, and quality control. An understanding of statistical methods, at least at the introductory level, is a necessity for the progressive man and woman in a profession or in the business world.

1.2   Is Statistics Difficult?

Many readers approach the study of introductory statistics with great fear. Such fear is needless. Only a knowledge of arithmetic and the bare essentials of elementary algebra is needed to follow the discussions in this text.

Generally speaking, it is necessary to study each chapter thoroughly before going on to later chapters, since there is a continuity of development of statistical methods from chapter to chapter. It is absolutely essential that you test and strengthen your learning by carefully working out the practice problems in each chapter.

Chapter 2

Basic Concepts

2.1 Populations and Descriptive Statistics

Statistical methods deal with two types of data, *population* data and *sample* data. *The complete set of data relating to an area of interest is called the population or the universe.* The annual incomes of heads of families in Michigan make up a population of incomes and the IQ scores for college freshmen make up a population of IQ scores. If we are interested in the number of errors made by each of the 25 typists in a large firm during a certain day, we are concerned with a population of 25 error counts; however, as a matter of convenience, we may refer to this as a population of 25 typists.

Populations encountered in practice are *finite* (limited in size); however, when a finite population is very large, it is general practice to treat it as if it were *infinite* (unlimited in size) when applying statistical procedures. Infinite popu-

5

lations, such as the population of heads and tails obtained when tossing a coin an unlimited number of times, is a theoretical concept.

Statistical methods appropriate for population data are called *descriptive statistics* and comprise the methods for organization, summarization, description, and analysis. Suppose a large firm has collected detailed information relating to its 248 employees—length of employment with the firm, salary, date of last promotion, age, level of education attained, etc. It is not possible to learn much from such a large quantity of data without some form of organization and summarization. Descriptive statistics provide the methods to organize, summarize, and evaluate such data.

2.2　Inferring from Samples

A set of data represents a sample if it contains only part of the total population, such as 50 grades selected from a population of 325 grades or 20 mechanics selected from a population of 175 mechanics. Usually, we deal with sample data, not population data. This is not surprising when it is realized that collection and analysis of data for an entire population is usually expensive and time consuming. Generally, sample data are collected when it is desired to estimate population characteristics.

Use of sample data to estimate population characteristics is a form of *inductive inference* since it involves reasoning from the particular (sample) to the general (population). Statisticians speak of this as *generalizing from the sample to the population*. If a sample of 30 retail stores shows that 10 percent are less than two years old, we may generalize to the population of retail stores by inferring that 10 percent of all the stores in the population are less than two years old. Statistical methods used to generalize from a sample to the population are called *inductive statistics*.

In statistics, inductive inference is called *statistical inference*. Risk is always present when an inference is made. Suppose it is inferred from a sample of 100 Cleveland housewives, that 30 percent of the housewives in Cleveland prefer soap A. How accurate is this estimate (inference)? Is it just about right or is it way off? Inductive statistics, based on the laws of probability, provide the methods for making statistical inferences from sample data and for evaluating the risks involved.

Notice that the population concept provides the frame of reference for all statistical methods. Descriptive methods are used to describe a population based on population data. Inductive statistics are used to describe a population based on sample data.

2.3　Variables: Discrete and Continuous

A quantity which can have only a single value is a *constant*, such as 3, 6, or the number of days in a week. On the other hand, a quantity which takes on different values is a *variable*, such as the number of days in a month, the grade per student on an exam, or the time required per person to do a job.

A variable is *discrete* if it can take on only specified values, such as the variable "number of children in a family" can take on only whole number values. A value between, for instance, 3 and 4 cannot occur for this variable, indicating that discrete variables contain gaps. Discrete variables result from counting. Such data are often called *enumeration data*.

Variables which result from measuring (other than counting) are *continuous* variables, such as height, length, temperature. Continuous variables do not contain gaps since they can take on any value between specified limits. For example, the length of an object may be measured to any desired degree of precision, such as 12.8 feet or 12.769 feet, etc., without any restriction, since the continuous variable "length" does not contain any gaps. Such data are often called *measurement data*.

2.4 Symbols and Summation

The study of statistics involves the use of a specialized form of communicating, *symbolic notation*. It is the needless worry about working with symbols which evokes much of the fear accompanying the study of statistics. In introductory statistics, we use relatively simple symbolic notation.

Consider the set of weights (in pounds): 21, 10, 15, 6, etc. The variable in this set is weight (in pounds). Let X = weight (in pounds) and let X_1, X_2, X_3, etc., denote the individual weights. The symbol X_1 is read "X sub 1" or more simply, "X one." Then, $X_1 = 21$ lbs, $X_2 = 10$ lbs, etc. Notice that we use a letter (X) to denote the variable (weight). Then we attach a *subscript* to this letter (1, 2, etc.) to create a specific symbol for the individual weights, such as X_3 represents the third weight. Sometimes we wish to refer to a single weight in the set in a general way, without specifying any particular one of the weights. We use the symbol X_i for this purpose, where i is a *variable subscript* which can take on each of the values 1, 2, etc., as required. If i is put equal to 5, then $X_i = X_5$, which denotes the fifth weight in the set.

Of course, it does not matter which symbols are used for the variable and the subscript, just as long as we know for what each symbol stands. We may use j for the subscript and write X_j; or we may use Y for the variable and write Y_i or Y_j depending on whether we choose i or j for the subscript symbol.

We will follow conventional usage of symbols as far as these are available. We will use X to denote the variable if one variable is involved in a problem (the typical case for us). We will use Y to denote a second variable and Z in the unusual case where we deal with a third variable. The lower case letters a, b, c, d, k, g are often used to represent constants (sometimes their upper case counterparts are used). Usually i is used as the subscript. If a second subscript is needed, j is usually used.

Statistical procedures nearly always involve the addition (summation) of sets of data, called the *summation operation*. Let X represent the number of hospital admissions per day, so that X_1, X_2, X_3, X_4 denote admissions for four days. We may represent the total (summation) of admissions for the four days as $\sum_{i=1}^{4} X_i$. $\sum$ is the capital Greek letter sigma and stands for "the sum of." The full symbol $\sum_{i=1}^{4} X_i$ is read as "the sum of X_i, i varying from 1 through 4." We may write

$$\sum_{i=1}^{4} X_i = X_1 + X_2 + X_3 + X_4$$

If it is given that $X_1 = 8$, $X_2 = 12$, $X_3 = 6$, and $X_4 = 4$, we have

$$\sum_{i=1}^{4} X_i = 8 + 12 + 6 + 4 = 30$$

Or
$$\sum_{i=1}^{4} X_i = 30$$

The letter beneath $\sum$, i, is called the *summation index*, which varies from 1 ($i = 1$ is shown below $\sum$) to 4 (as shown just above $\sum$). We may want to add only some of the X_i values. We indicate this as follows:

$$\sum_{i=1}^{3} X_i = X_1 + X_2 + X_3$$

or
$$\sum_{i=3}^{4} X_i = X_3 + X_4$$

More generally, if we want to add N quantities, we may write

$$\sum_{i=1}^{N} X_i = X_1 + X_2 + X_3 + \ldots + X_N \tag{2.4.1}$$

In practice, the summation notation is simplified as much as possible. We may write $\sum X_i$, which is read "sum of X_i," if it is clear which X_i values are to be added. More often, we write $\sum X$, which is read "sum of X." Of course, abbreviated notations are to be used only when it is absolutely clear which X values are to be added. Fortunately, this is often the case. In our study of statistics, we will encounter summation types such as

$$\sum_{i=1}^{N} X_i Y_i = X_1 Y_1 + X_2 Y_2 + X_3 Y_3 + \ldots + X_N Y_N \tag{2.4.2}$$

$$\sum_{i=1}^{N} k X_i = k X_1 + k X_2 + k X_3 + \ldots + k X_N \tag{2.4.3}$$

$$\sum_{i=1}^{N} g(X_i - k) = g(X_1 - k) + g(X_2 - k) + g(X_3 - k) + \ldots \\ + g(X_N - k) \tag{2.4.4}$$

Notice that in Equations 2.4.3 and 2.4.4, k and g denote constants and so do not need subscripts. The foregoing equations are similar to Equation 2.4.1 in that they all represent summations. In Equation 2.4.1 individual X_i terms are to be added; whereas, in Equation 2.4.2 individual $X_i Y_i$ products are to be added. In Equation 2.4.4, the individual terms to be added are the $g(X_i - k)$ products.

Three simple *summation rules,* very useful in the development and application of statistical procedures, are:

Rule 1: $$\sum_{i=1}^{N} (X_i \pm Y_i) = \sum_{i=1}^{N} X_i \pm \sum_{i=1}^{N} Y_i$$

This rule states that *the summation of a sum (or difference) of terms is equal to the sum (or difference) of the individual summations of each term.* Suppose $N = 3$, so that we have three terms to be added, and $X_1 = 10$, $X_2 = 20$, $X_3 = 30$ and $Y_1 = 2$, $Y_2 = 4$, $Y_3 = 6$. Then, $\sum X = 60$, $\sum Y = 12$, $\sum (X + Y) = (10 + 2) + (20 + 4) + (30 + 6) = 72$, and $\sum (X - Y) = (10 - 2) + (20 - 4) + (30 - 6) = 48$. We may verify Rule 1 as follows:

$$\sum (X + Y) = 60 + 12 = 72$$
$$\sum (X - Y) = 60 - 12 = 48$$

A simple algebraic proof of this rule is presented in Section 2.5, Note 2A.

Rule 2: $$\sum_{i=1}^{N} kX_i = k \sum_{i=1}^{N} X_i$$

This rule states that *the summation of a constant times a variable is equal to the constant times the summation of the variable.* Suppose $N = 3$, $k = 10$, $X_1 = 3$, $X_2 = 6$, $X_3 = 9$. Then, $\sum X = 18$ and $\sum kX = (10)(3) + (10)(6) + (10)(9) = 180$. We may verify Rule 2 as follows:

$$\sum kX = (10)(18) = 180$$

Note 2B, Section 2.5 presents a simple algebraic proof of this rule.

Rule 3: $$\sum_{i=1}^{N} k = Nk$$

This rule states that *the summation of a constant is equal to the constant times the number of times it appears in the summation.* Obviously, if the constant 5 is to be added three times, the total will equal $(3)(5) = 15$. An algebraic proof appears in Section 2.5, Note 2C.

The foregoing discussion was primarily intended to introduce the reader to some of the symbols we will use. Other symbols will be introduced as needed.

2.5 Optional: Algebraic Notes

In each algebraic note, the equation which is the subject of the proof is identified. Equations which have already appeared in the text are identified by the equation numbers originally assigned. New equations are identified by equation numbers keyed to this section.

Note 2A: To prove

Rule 1: $$\sum_{i=1}^{N} (X_i \pm Y_i) = \sum_{i=1}^{N} X_i \pm \sum_{i=1}^{N} Y_i$$

Expand the summation of the sum, $\sum (X + Y)$, to obtain

$$\sum_{i=1}^{N} (X_i + Y_i) = (X_1 + Y_1) + (X_2 + Y_2) + (X_3 + Y_3) + \cdots \\ + (X_N + Y_N) \tag{2.5.1}$$

Remove the parentheses on the right-hand side of Equation 2.5.1 and collect X and Y terms separately to obtain

$$\sum_{i=1}^{N} (X_i + Y_i) = X_1 + X_2 + X_3 + \cdots + X_N + Y_1 + Y_2 + Y_3 \\ + \cdots + Y_N \tag{2.5.2}$$

Then, using the summation notation to express the sum of the X's and the Y's on the right-hand side, we complete the proof as follows:

$$\sum_{i=1}^{N} (X_i + Y_i) = \sum_{i=1}^{N} X_i + \sum_{i=1}^{N} Y_i$$

We can prove Rule 1 for the summation of a difference, $\sum (X_i - Y_i)$, in the same way.

Note 2B: To prove

Rule 2:
$$\sum_{i=1}^{N} kX_i = k \sum_{i=1}^{N} X_i$$

Expand the summation on the left-hand side to obtain

$$\sum_{i=1}^{N} kX_i = kX_1 + kX_2 + kX_3 + \cdots + kX_N$$
$$= k(X_1 + X_2 + X_3 + \cdots + X_N)$$
$$= k \sum_{i=1}^{N} X_i$$

Note 2C: To prove

Rule 3:
$$\sum_{i=1}^{N} k = Nk$$

Since the summation on the left-hand side states that we are to add the constant k N times, we may write

$$\sum_{i=1}^{N} k = k + k + k + \cdots + k \qquad (N \ k\text{'s})$$
$$= Nk$$

Study Problems

1.   Define and illustrate:

a. Sample	b. Population
c. Infinite population	d. Finite population
e. Universe	f. Inductive inference
g. Statistical inference	

2.   Explain briefly and illustrate the difference between descriptive and inductive statistics.

3.   A lawmaker is interested in how many times each of the 500 members of the legislative body abstained from voting on major legislation during the past two years. Is he interested in a population of lawmakers or a population of counts? Comment briefly.

4.   Identify the specific population involved in each case:
      a.   A newspaper is compiling data on date of first subscription for its suburban subscribers.
      b.   An anthropologist is interested in studying a certain skull measurement of ten-year-old male American Indians.
      c.   A parking lot attendant records "length of time parked" for a sample of 22 automobiles selected from cars parked on a given day.
      d.   A teacher reviews the IQ scores for a sample of seventh-graders who were not up to grade in their work.
      e.   A political analyst asks his assistant to collect data on attitude toward a certain candidate from a sample of high-income voters.

5.   Indicate whether inductive or descriptive statistics is involved (defend your choice):
      a.   A psychologist studies the clerical aptitude scores for last year's high school graduates in a city.
      b.   A medical technician counts the red blood cells in a blood sample.
      c.   A sociologist studies the attitudes of pre-school children toward children of the opposite sex based on an analysis of sample data.
      d.   A department store manager studies the daily number of merchandise returns for the last six months, by department.
      e.   A political scientist studies voter reaction to a new law based on data collected from a sample of cities.

6.   Compare and illustrate:
      a.   Variable, constant
      b.   Discrete variable, continuous variable

7.   Indicate for each illustration whether it represents a constant, discrete variable, or continuous variable:
      a.   Number of days per month
      b.   Plant height after a week of growth
      c.   Number of failing students per class
      d.   Daily temperature at 3 P.M. in Boston
      e.   Number of days per week
      f.   Weight of brass fittings produced by a machine
      g.   Time required to do a certain job
      h.   Corporate taxes paid, per company

8.   Given:

$a = 5$	$X_1 = -1$	$Y_1 = 4$	$Z_1 = 2$
$k = 2$	$X_2 = 5$	$Y_2 = -3$	$Z_2 = 4$
$g = 10$	$X_3 = 3$	$Y_3 = 0$	$Z_3 = 8$
	$X_4 = -2$	$Y_4 = 2$	$Z_4 = 6$

Determine the value:

a. $\sum_{i=2}^{4} X_i$ b. $\sum_{i=1}^{3} k(X_i Z_i)$

c. $\sum_{i=1}^{5} g$ d. $g \sum_{i=1}^{2} (Y_i + X_i)$

e. $\sum_{i=2}^{4} (aY_i - kZ_i)^2$ f. $(X_3 - Y_4 + Z_1)ak + 100$

g. $20 \sum_{i=3}^{4} \dfrac{X_i}{k}$ h. $\sum_{i=1}^{10} 1{,}000 - \sum_{i=2}^{3} kX_i Y_i$

9. Given:

$N_1 = 0$	$M_1 = 3$	$L_1 = 2$	$G = 2$
$N_2 = 3$	$M_2 = 2$	$L_2 = 4$	$B = -1$
$N_3 = 7$	$M_3 = 1$	$L_3 = 6$	

Determine the value:

a. $\sum_{i=1}^{3} (M_i - 1)$ b. $\sum_{i=1}^{3} B(N_i - 2)$

c. $\sum_{i=1}^{3} \dfrac{1}{10} (N_i - M_i)$ d. $\dfrac{1}{G} \sum_{i=1}^{3} (L_i - N_i)^2$

e. $GB(N_2 - M_1^2 + BL_3)$ f. $B \sum_{i=1}^{5} G$

10. Write in condensed summation form:
   a. $X_1 + X_2 + X_3 + X_4 + X_5$
   b. $Y_{11} + Y_{12} + Y_{13} + Y_{14}$
   c. $gZ_{12} + gZ_{13} + gZ_{14} + gZ_{15}$
   d. $\dfrac{X_1 + Y_1}{k} + \dfrac{X_2 + Y_2}{k} + \dfrac{X_3 + Y_3}{k}$
   e. $g(Y_2 - L_2) + g(Y_3 - L_3) + g(Y_4 - L_4)$
   f. $(X_1 - G)^2 + (X_2 - G)^2 + \ldots + (X_N - G)^2$
   g. $K + K + K + K + K$
   h. $\dfrac{2X_1 - kY_1}{M_1} + \dfrac{2X_2 - kY_2}{M_2} + \ldots + \dfrac{2X_m - kY_m}{M_m}$
   i. $135 - \dfrac{Y_1}{2C} - \dfrac{Y_2}{2C} - \dfrac{Y_3}{2C} - \ldots - \dfrac{Y_j}{2C}$

11. Write in expanded form:

a. $\sum_{j=1}^{3} X_j$ b. $\sum_{i=4}^{6} (X_i - Y_i)$

c. $\sum_{i=1}^{n} (X_i - K)^2$ d. $\sum_{i=7}^{9} \dfrac{L_i - BM_i}{aT_i}$

e. $\sum_{i=1}^{3} G$

Supplementary Study Problems

12. Prove the equalities, indicating the summation rules used:

a. $\sum_{i=1}^{n} (gX_i - Y_i) = g \sum_{i=1}^{n} X_i - \sum_{i=1}^{n} Y_i$

b. $\displaystyle\sum_{i=1}^{N} \frac{X_i + Y_i}{k^2} = \frac{1}{k^2}\left(\sum_{i=1}^{N} X_i + \sum_{i=1}^{N} Y_i\right)$

c. $\displaystyle\sum_{i=2}^{k} \frac{aL_i - T_i^2}{agM_i} = \frac{1}{g}\sum_{i=2}^{k} \frac{L_i}{M_i} - \frac{1}{ag}\sum_{i=2}^{k} \frac{T_i^2}{M_i}$

d. $\displaystyle\sum_{i=4}^{8} \left(\frac{Y_i + gT_i}{g} - k\right) = \frac{1}{g}\sum_{i=4}^{8} Y_i + \sum_{i=4}^{8} T_i - 5k$

e. $\displaystyle\sum_{i=1}^{M} (X_i - 10g)a = a\sum_{i=1}^{M} X_i - 10Mga$

f. $\displaystyle\sum_{i=1}^{N} \frac{(X_i - k)^2}{n} = \frac{1}{n}\sum_{i=1}^{N} X_i^2 - \frac{2k}{n}\sum_{i=1}^{N} X_i + \frac{Nk^2}{n}$

Part II
Descriptive Statistics and Analysis

Chapter 3

Data Organization
and Distributions

3.1 Frequency Distributions

Part II presents the methods of data organization, summarization, and description applicable to population data; however, these methods, sometimes with a little adjustment, are applicable to sample data as well. These adjustments will be indicated as appropriate.

When faced with a large quantity of *raw (unorganized) data*, analysts and administrators find it difficult (often impossible) to gain any understanding of the distribution of the data or of the relationships implied by the data. Table 3.1.1 presents the home-to-work travel times for 75 workers. Clearly, these travel times must somehow be organized and summarized if an understanding of these data is to be obtained. A simple type of organization is to *rank* the data, for instance, from low to high, however, ranking provides little aid in comprehending the data. It is necessary to summarize as well as organize the data to facilitate understanding.

TABLE 3.1.1

Home-to-work travel time for 75 workers (minutes).

4.8	9.3	8.2	6.7	12.0	3.9	11.4	6.2
6.4	9.3	14.0	10.4	4.2	14.9	17.7	15.3
12.9	15.7	3.4	18.1	20.4	10.7	16.7	9.7
18.4	14.3	7.3	8.8	13.7	8.5	12.6	14.0
10.1	19.6	5.7	9.0	3.6	10.7	21.4	4.9
23.7	3.9	17.7	11.2	10.7	7.3	11.3	
10.7	15.4	12.6	6.4	16.4	7.3	16.6	
7.5	20.9	9.6	15.7	5.5	11.8	11.9	
14.4	22.7	22.0	14.6	13.0	18.4	12.7	
19.0	13.4	6.5	10.1	8.7	12.4	11.3	

A widely used form of data organization and summarization is the *frequency table*. Such a table is constructed by dividing the *range* of the data (travel times) into *sub-ranges* or *class intervals* and showing the *frequency* or number of travel times which fall into each *interval*. We see from Table 3.1.1 that the least travel time is 3.4 minutes and the most is 23.7 minutes so that the range of the data is 23.7 − 3.4 or 20.3 minutes. Into how many intervals should we divide the range? If we use too many intervals, the resulting frequency table will be too detailed and still difficult to comprehend. On the other hand, if we use too few intervals we will over-summarize the data and lose too much information. Table B.1 in Appendix B provides a guide as to the number of intervals (or classes) to use. According to this table, we should use about seven intervals when summarizing 75 quantities.

Dividing the range 20.3 minutes by the suggested number of intervals, seven, we obtain 20.3/7 = 2.9 minutes, which represents the *length* of each of the seven intervals. It is good practice to select a convenient number for length of class interval. Therefore, let us round 2.9 minutes to 3.0 minutes. Each class interval is identified by a pair of *class limits*, a *lower limit* and an *upper limit*. It is also good practice for the lower limits to be convenient numbers. Since the lowest value in the set of 75 travel times is 3.4 minutes, let us round this to 3.0 minutes and use it as the lower limit of the first class interval, as shown in Table 3.1.2. Then, add the selected length of class interval (3.0) successively to obtain the lower limits for the seven intervals (3.0 + 3.0 = 6.0, 6.0 + 3.0 = 9.0, etc.).

Class intervals must be determined so that they do not overlap. Therefore, if the lower limit of the second interval is 6.0 minutes, the upper limit of the first interval must be 5.9 minutes, as shown in Table 3.1.2. In other words, the first interval includes travel times as low as 3.0 minutes and as high as 5.9 minutes; whereas, the second interval includes travel times from 6.0 minutes through 8.9 minutes. The seven intervals are shown in Table 3.1.2. The number of travel times which fall into each interval (*class frequency*) is determined by tallying, as shown in the table. As an illustration, the first travel time listed in Table 3.1.1 is 4.8 minutes and this is tallied by placing a vertical stroke in the interval in which it falls (3.0–5.9). Each fifth stroke is made diagonally to tie

TABLE 3.1.2

Home-to-work travel time for 75 workers (minutes): frequency distribution tally table.

Class interval	Tally	Class frequency
3.0–5.9	ⅢⅢ ////	9
6.0–8.9	ⅢⅢ ⅢⅢ ///	13
9.0–11.9	ⅢⅢ ⅢⅢ ⅢⅢ ///	18
12.0–14.9	ⅢⅢ ⅢⅢ ⅢⅢ	15
15.0–17.9	ⅢⅢ ////	9
18.0–20.9	ⅢⅢ //	7
21.0–23.9	////	4

the five strokes into a bundle and facilitate counting. The tallies in each class are counted and this frequency is placed in the class frequency column.

Examine the two columns of information in Table 3.1.2, "class interval" and "class frequency." The understanding obtainable from these data is certainly considerable compared with the understanding obtainable from the raw data in Table 3.1.1. It can easily be seen that most of the travel times are between 6.0 minutes and 14.9 minutes, with the largest number falling in the 9.0–11.9 minute class interval. You pay a small penalty for this greater comprehension of the data, since a frequency table does not show the exact travel time for each of the 75 workers. The class frequency column does, however, clearly show the *frequency distribution* of travel times for these workers.

The class interval and class frequency columns make up the basic frequency table. Various additional columns of information useful for analytical and computational purposes may be derived from these basic data. Table 3.1.3 presents such additional information. Notice that the first column is headed "discrete class limits" instead of class interval, since this is a more descriptive heading. This column heading takes into account that the variable of the data "home-to-work travel time" is a discrete variable and that this column presents the *lower* and *upper discrete class limits* to identify the class intervals. It should be noted that any *recorded variable* is discrete, even continuous variables. The travel time variable is continuous; however, it is recorded to a tenth of a minute, so that there can be no values between say 5.6 minutes and 5.7 minutes. Hence, there are gaps between consecutive tenths, making the recorded variable discrete. Notice that the discrete class limits are in the same units (tenths) as the raw data (Table 3.1.1).

The *length of class interval* can be computed from a completed frequency table by determining the difference between *successive lower limits* or between *successive upper limits*. The *class midpoint* represents the middle of a class interval and is computed as half the sum of the limits for a class. The first class midpoint in Table 3.1.3 is obtained as $(3.0 + 5.9)/2 = 4.45$. Another name for class midpoints is class marks.

Notice that there are gaps between the discrete class limits which identify adjacent intervals, such as between the interval 3.0–5.9 and the interval 6.0–8.9.

TABLE 3.1.3

Home-to-work travel time for 75 workers (minutes): frequency distribution.

Discrete class limits	Frequency	Class midpoints	Continuous class limits	Less than cumulative frequency	More than cumulative frequency
			2.95	0	75
3.0–5.9	9	4.45			
			5.95	9	66
6.0–8.9	13	7.45			
			8.95	22	53
9.0–11.9	18	10.45			
			11.95	40	35
12.0–14.9	15	13.45			
			14.95	55	20
15.0–17.9	9	16.45			
			17.95	64	11
18.0–20.9	7	19.45			
			20.95	71	4
21.0–23.9	4	22.45			
			23.95	75	0

The gap (no recorded travel time is possible) is between the upper limit 5.9 minutes and the lower limit 6.0 minutes. In the same way there is a gap between the interval 6.0–8.9 and the interval 9.0–11.9. The gap here is between the upper limit 8.9 minutes and the lower limit 9.0 minutes. The *continuous class limits* shown in Table 3.1.3 close these gaps and, in essence, convert the recorded discrete variable "travel time expressed in tenths" to a continuous variable. The continuous class limits are computed as half the sum of the upper discrete limit of one class and the lower discrete limit of the next higher class. The continuous class limit 5.95 minutes in Table 3.1.3 is computed as half the sum of 5.9 minutes and 6.0 minutes; the continuous class limit 8.95 minutes is computed as half the sum of 8.9 minutes and 9.0 minutes; etc. The difference between successive continuous class limits is 3.0 (length of class interval). The first continuous class limit in Table 3.1.3 is computed as $5.95 - 3.0 = 2.95$ minutes and the last is computed as $20.95 + 3.0 = 23.95$ minutes.

The discrete class limits mark off *discrete classes*. The continuous class limits "spread out" each discrete class to eliminate the gaps between class intervals and mark off continuous classes. More specifically, the discrete class 3.0–5.9 becomes the continuous class 2.95–5.95, the discrete class 6.0–8.9 becomes the continuous class 5.95–8.95, etc. Notice that the continuous class intervals do overlap, since the upper class limit of one class is also the lower class limit of the next higher class. For example, the upper limit of the continuous class 2.95–5.95 is also the lower limit of the next higher continuous class 5.95–8.95. This is why the continuous limits are placed between the classes in Table 3.1.3. The continuous class limits, unlike the discrete class limits, are not in the same units as the raw data. In our illustration, the continuous class limits are in hundredths; whereas, the raw data are in tenths. The two columns of information class midpoints and continuous class limits are useful for computational purposes and will be used in Chapters 4 and 5.

The last two columns in Table 3.1.3 provide useful analytical information and are also used for computational purposes. They are computed to correspond to the continuous class limits. The "less than cumulative frequency" column shows how many of the 75 travel times are less than each continuous class limit. The first "less than" frequency is 0, indicating that none of the 75 travel times are less than 2.95 minutes; the third is 22, indicating that $9 + 13$ or 22 travel times among the 75 are less than 8.95 minutes, etc. Notice that the class frequencies are *cumulated* as you move to higher and higher continuous class limits. Clearly, as you move to higher continuous class limits, the number of travel times which are less than a given limit is greater.

Similarly, the "more than cumulative frequency" column shows how many of the 75 travel times are more than each continuous class limit. The first number in this column is 75, indicating that all 75 travel times are more than 2.95 minutes; the second is 66, indicating that $75 - 9$ or 66 travel times are more than 5.95 minutes, etc. Clearly, the "more than" frequencies become smaller as you move to higher and higher continuous class limits. We may learn from the cumulative frequency columns, for example, that 22 of the 75 travel times are less than 8.95 minutes and the balance (53 travel times) are more than 8.95 minutes.

The frequency distribution exhibited by a frequency table is a useful means for organizing and summarizing a set of raw data. It should, however, be constructed thoughtfully. Sometimes, it may not be necessary to use as many class intervals as indicated in Table B.1 and sometimes you will want to use more classes. The purpose for which the table is constructed is the important consideration. Finally, even though Table B.1 suggests a certain number of classes to help you begin the construction of a frequency table, the number of classes you end up with may be one more or less, since you need to provide a sufficient number of classes to include all your data.

The frequency tables discussed show class intervals of *equal length;* however, classes of *unequal length* are used in a frequency table where the raw data contain a few quantities which are much smaller or much larger than the others. In such instances, the class interval may be, for example, of length five units for all classes except perhaps the first or last class which may be of length ten units. In some frequency tables of this type the first or last class may be an *open-end interval,* such as "10 or less" or "over 70."

3.2   Distribution Concepts and Graphic Representation

The frequency distribution exhibited by a well-constructed frequency table offers considerable aid in understanding how a set of data is distributed. It is sometimes useful, however, to display the distribution in the form of a graph.

Three types of graphs are often used for this purpose: the *frequency polygon*, the *histogram*, and the *ogive*. We will illustrate each of these using the frequency distribution of travel times in Table 3.1.3.

Figure 3.2.1 presents the *frequency polygon*. The horizontal axis contains the scale of travel times, with the class midpoints marked off. Notice that the midpoint for a class below the lowest class (1.45 minutes) and the midpoint for a class above the highest class (25.45 minutes) are also marked off. The scale of class frequencies is shown on the vertical scale. The graph is constructed by placing a dot for each class interval directly above the midpoint for the class, at a height corresponding to the class frequency, and connecting the dots by straight lines. Then, the frequency polygon is closed at both ends by connecting the dot above the lowest midpoint (4.45 minutes) to the dot (at zero frequency) for the added lower midpoint (1.45 minutes) and, similarly, connecting the dot above the highest midpoint (22.45 minutes) to the dot (at zero frequency) for the added higher midpoint (25.45 minutes). *Construction of the frequency polygon assumes that all travel times in a class are equal to the class midpoint.* The assistance afforded by this type of graph in displaying the shape of the frequency distribution is apparent.

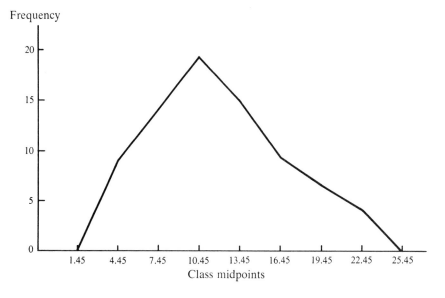

FIGURE 3.2.1

Frequency polygon.

Figure 3.2.2 presents the *histogram*. The continuous class limits are marked off on the horizontal axis and the class frequencies on the vertical axis. The frequency for each class is plotted as a horizontal line over the full length of the class interval, at a height corresponding to the class frequency. The ends of this horizontal line are connected by vertical lines to the corresponding continuous class limits. Clearly, the histogram has the appearance of a bar chart, with bars of equal width placed close together. *Construction of the histogram assumes that the items in a class are evenly distributed over the full length of a class interval.*

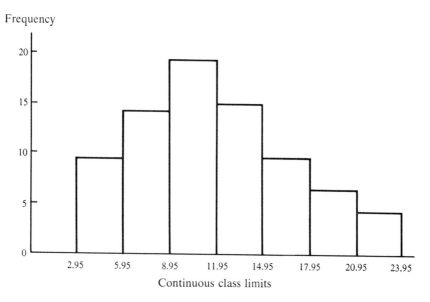

FIGURE 3.2.2

Histogram.

In the frequency polygon, the *height* of the dot above a class midpoint represents the class frequency. In a histogram, it is the *area of a bar* which represents the class frequency. In a histogram, we may also think of the combined area of adjacent bars as representing the combined frequencies of the corresponding classes. Then, the area under the *entire histogram* represents the total of all class frequencies (or the full set of data). *The use of areas under a distribution curve, such as a histogram, to represent frequencies is a fundamental notion in statistical inference*, as will become apparent in later chapters.

Figure 3.2.3 presents the *ogive*, which is a graphic presentation of the *cumulative frequencies*. Two ogives are shown, a "more than ogive" and a "less than ogive." The continuous class limits are marked off on the horizontal axis and the cumulative frequencies on the vertical axis. Each cumulative frequency is plotted above the corresponding continuous class limit at the proper height and the plotted points are connected by straight lines.

Frequency distributions may be expected to vary as to *shape*. Figure 3.2.4 presents the histograms for four types of distributions which illustrate types which may be encountered in practice. Part (A) illustrates a *symmetrical distribution with a central peak.* Part (B) illustrates a *skewed distribution*, with a "tail" trailing out to the right. This is known as a *right-skewed* or *positively-skewed distribution*. Some distributions are just the reverse, the "tail" trails out to the left and is known as a *left-skewed* or *negatively-skewed distribution*. Part

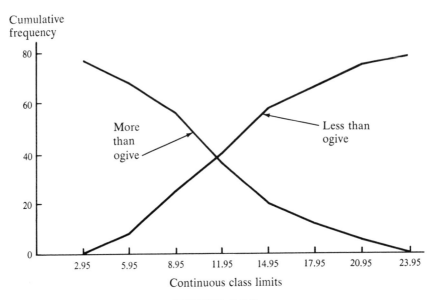

FIGURE 3.2.3

Ogives.

(C) illustrates an *exponential distribution* which has increasing class frequencies as you move out to the right. This is also called a *J-distribution*. Such a distribution shape may be reversed and show the highest class frequency in the first class and declining class frequencies as you move to the right. Part (D) shows a *U-distribution*, with the center classes containing the smallest frequencies.

Statistical methods, whether descriptive or inferential, are directed toward description of a distribution. So far we have considered frequency distributions. In later chapters, we will study *probability distributions* and especially *sampling distributions*. The statistician's interest in a set of data relates to the distribution underlying the data. The frequency distribution revealed by a frequency table represents the distribution of a set of data for a population. From a theoretical point of view, the histogram is the preferred graphic representation, since areas under the histogram represent frequencies for individual classes or adjacent classes. It is sometimes useful to approximate a histogram with a smooth, continuous curve, as in Figure 3.2.5. The continuous curve may be sketched freehand, and for many purposes, this is adequate, or it may be fitted mathematically to the data.

The distribution displayed by a histogram, in the case of *sample data*, may be considered representative of the population from which the sample was selected. It is always a question whether the observed distribution of sample data adequately represents the population distribution. Often, when dealing with sample data, a theoretical distribution of an infinite population is used to

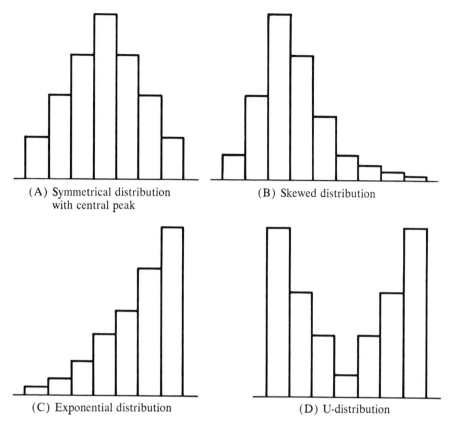

(A) Symmetrical distribution
    with central peak

(B) Skewed distribution

(C) Exponential distribution

(D) U-distribution

FIGURE 3.2.4

Four types of distributions.

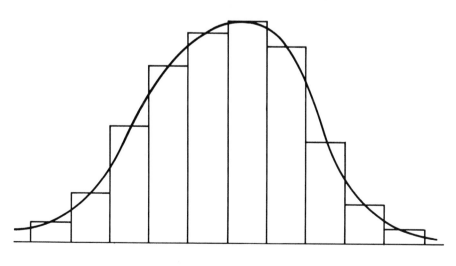

FIGURE 3.2.5

represent the population being studied. Sometimes a discrete theoretical distribution such as the *distribution* of 1's and 0's is used. Other times, a continuous theoretical distribution such as the *normal distribution* is used. These theoretical distributions are discussed in Chapter 8.

Figure 3.2.6 illustrates the process whereby a theoretical continuous distribution is developed. Part (A) presents the histogram for a set of data. Now, suppose that the length of class interval is reduced and the number of classes and the size of the set of data (number of items in the set) are increased. Then,

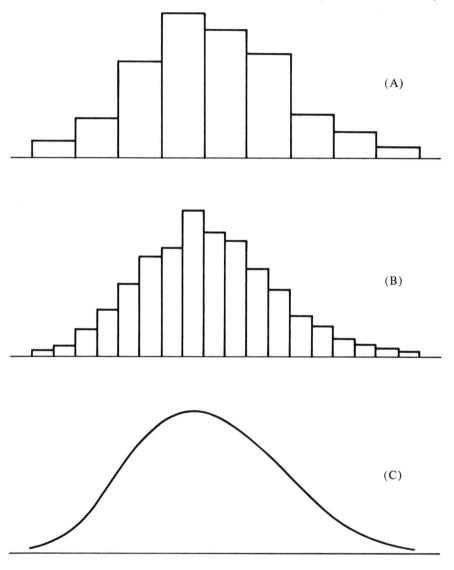

(A)

(B)

(C)

FIGURE 3.2.6

Development of a theoretical continuous distribution.

as illustrated in part (B) of the figure, the shape of the distribution represented by the histogram will become smoother. If this process is continued until the length of class interval becomes infinitesimally small and the number of classes and the set of data become infinitely large, the shape of the histogram will become a smooth, continuous curve as in part (C) of the figure.

A most important feature of a continuous distribution curve, such as just developed, is the use of the area under the curve to represent frequencies, as discussed in the case of the histogram. If each class frequency is expressed as a *proportion* of the total number of items in the set of data (often called a *relative frequency*), the sum of the proportions for all the classes of a frequency distribution is equal to unity. The bars of a histogram as well as areas under a continuous curve also represent relative frequencies. *Interpretation of areas under a continuous distribution curve as relative frequencies is of major importance in statistical inference.* More will be said about this in later chapters.

3.3   Nominal and Ordinal Scale Data

Analysts and administrators in all fields deal with *qualitative data* as well as *quantitative data.* We may divide qualitative data into two types, *nominal scale data* and *ordinal scale data. Nominal scale data* represent merely category information. For example, the numbers of men in a large population by state of birth are nominal scale data, the nominal scale (categories) being "state of birth." The quantities of shoes sold, by color, are nominal scale data, the nominal scale (categories) being "color." On the other hand, *ordinal scale data* represent categories which may be *ranked* in some order of "greater than" or "less than." For example, the numbers of college faculty members by academic rank (professor, associate professor, etc.) are ordinal scale data, the ordinal scale being "academic rank." Or the numbers of voters by attitude toward a new tax (opposed, neutral, in favor) are ordinal scale data, the ordinal scale being "attitude."

Suppose it is desired to show the frequency distribution of 83 three-car families, by county. This is most easily accomplished by preparing a table such as Table 3.3.1 and tallying. These data are nominal scale data, with "county"

TABLE 3.3.1

Frequency distribution of 83 three-car families, by county.

County	Tally	Number of three-car families
Arbor	ℕ ℕ	10
Buck	ℕ ℕ ℕ	15
Core	ℕ ///	8
Dill	ℕ ℕ //	12
Dover	ℕ ℕ ℕ ///	18
Ethel	ℕ ℕ ℕ ℕ	20

Number of
three-car
families

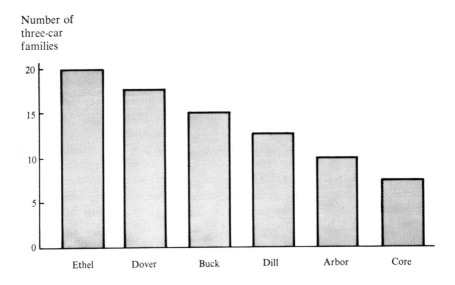

FIGURE 3.3.1

Bar chart: distribution of 83 three-car families, by county.

representing the nominal (category) scale. Nominal and ordinal scale data may be graphically presented by a bar chart, as in Figure 3.3.1. Notice that in the frequency table (Table 3.3.1) it is best to show the counties in alphabetical order to facilitate finding a county; however, the bar chart is best constructed by showing the counties in order of county frequency (number of three-car families in a county).

Study Problems

1. List four ways to compute the length of class interval for a frequency distribution.

2. List two ways to compute the class midpoints for a frequency distribution.

3. About how many classes should be used when constructing a frequency table for the following quantities of raw data: (a) 45, (b) 175, (c) 400, (d) 200,000, and (e) 1,000,000?

4. The weekly number of traffic accidents in a city for a number of years is to be organized into a frequency table of eight classes. If the lower discrete class limit of the first class interval is 12 and the class length is 3, determine (a) the 8 discrete class intervals and (b) the class midpoints.

5. Given discrete class limits 20–23, 24–27, 28–31, 32–35, and 36–39, determine the class midpoints, continuous class limits, and length of class interval.

6. Given discrete class limits .010–.019, .020–.029, .030–.039, .040–.049, .050–.059, and .060–.069. Determine the class midpoints, continuous class limits, and length of class interval.

7.   Scores on a sociability scale were organized into a frequency distribution, as follows:

Score	Class frequency
15.0–17.9	4
18.0–20.9	12
21.0–23.9	18
24.0–26.9	10
27.0–29.9	6
	50

Determine: (a) class midpoints, (b) continuous class limits, (c) more than cumulative frequencies, (d) less than cumulative frequencies, and (e) length of class interval.

8.   Price quotations for a commodity were determined for a sample of 30 cities and organized into a frequency distribution, as follows:

Price	Number of cities in sample
$2.50–$2.74	2
2.75– 2.99	4
3.00– 3.24	5
3.25– 3.49	6
3.50– 3.74	8
3.75– 3.99	4
4.00– 4.24	1

Determine (a) through (e) as in Problem 7.

9.   Information is lost when raw data are organized into a frequency distribution. Explain.

10.   List four features in the following frequency distribution which are not acceptable:

Discrete class limits	Frequency
20.17–21.29	23
21.29–22.41	216
22.41–23.53	312
23.53–24.65	17
	568

11.   Inches of rainfall for an area were collected for a sample of 70 days as follows:

24	37	24	19	38	23	26
29	27	34	30	28	22	24
20	24	28	22	29	21	26
27	24	30	25	25	18	23

Descriptive Statistics and Analysis

26	26	29	27	28	21	30
23	25	25	28	24	23	28
29	31	31	31	26	32	33
24	22	25	22	33	28	24
27	24	19	31	34	28	34
27	25	36	30	25	19	23

    a. Organize the data into a frequency distribution (show discrete class limits, tally table, and class frequencies).
    b. Determine the class midpoints.
    c. Determine the continuous class limits.
    d. Determine the less than cumulative frequencies.
    e. Determine the more than cumulative frequencies.

12. Scores on computer science aptitude were obtained for 53 students as follows:

161	177	164	168	156	151	150	160	151
161	166	164	164	156	166	160	155	157
151	158	152	178	161	152	158	167	153
152	174	151	155	160	163	152	166	168
155	157	158	158	155	152	153	156	150
167	150	173	178	174	153	156	151	

Determine (a) through (e) as for Problem 11.

13. Marriage rates per 10,000 population for 49 large cities were as follows:

16.45	16.24	16.42	16.45	16.69	15.27	16.70
15.54	15.75	16.11	16.61	16.22	16.61	15.89
15.53	16.12	16.73	16.31	16.06	15.81	15.99
16.27	15.31	16.70	16.58	15.48	15.48	15.34
15.34	16.68	16.68	16.61	15.73	16.58	15.26
16.38	15.32	15.27	15.34	15.61	16.45	15.91
15.74	16.42	16.61	16.02	16.23	15.60	15.51

Determine (a) through (e) as in Problem 11.

14. In constructing a frequency polygon, what assumption is made concerning the values of the items in a class?

15. In constructing a histogram, what assumption is made concerning the values of the items in a class?

16. Name two kinds of distributions of interest in statistics, other than frequency distributions.

17. Name three different distribution types (shapes).

18. Name a theoretical distribution which is (a) continuous and (b) discrete.

19. If a frequency distribution is constructed several times for a given set of data, each succeeding time with smaller class length, how would you expect the pattern of class frequencies to be affected?

20. Construct the histogram and "more than" ogive for the following problems: (a) Problem 7, (b) Problem 8, and (c) Problem 11.

21. Construct the frequency polygon and "less than" ogive for the following problems: (a) Problem 12 and (b) Problem 13.

22. Compute the relative frequencies (proportions) of scores in the five class intervals for the frequency table in Problem 7.

23. Name, define, and give three illustrations for each of the two types of qualitative data.

24. For each item below indicate whether it represents quantitative or qualitative data and, if qualitative, specify which type:

    a. Corporate executives identified only by position title (president, vice president, etc.).

    b. Flowers identified by species.

    c. Club members identified by length of membership.

    d. Women identified by marital status.

    e. Students identified as freshmen, sophomore, etc.

25. Twenty alumni attending a reunion were asked their state of residence and gave the following answers: New York, Georgia, New York, California, New York, New York, California, California, Vermont, New York, California, Georgia, New York, California, Illinois, Illinois, California, Illinois, New York, and Illinois. Construct a frequency table and a bar chart for these data.

26. Fifteen experts were asked how they thought the stock market would behave during the next week or two. They gave the following answers: small uptrend, mild downtrend, retain present position, small uptrend, small uptrend, sharp downtrend, sharp uptrend, sharp uptrend, mild downtrend, small uptrend, small uptrend, retain present position, small uptrend, retain present position, and sharp uptrend. Construct a frequency table and a bar chart for these opinions. (Arrange the opinions in a logical sequence in the table and chart.)

Supplementary Study Problems

27. Obtaining the cumulative frequencies ("more than" and "less than") as in Table 3.1.3 is useful for certain computational purposes (as discussed in Chapters 4 and 5). Other ways of presenting cumulative frequencies are sometimes preferred for other purposes, such as the following for "expenditures for bowling per family":

	Expenditures per family	Cumulative number of families
(1)	Less than $10	0
	Less than $20	8
	Less than $30	19
(2)	$10 or less	2
	$20 or less	12
	$30 or less	24
(3)	$10 or more	40
	$20 or more	32
	$30 or more	21

$$(4) \left\{ \begin{array}{ll} \text{More than \$10} & 38 \\ \text{More than \$20} & 29 \\ \text{More than \$30} & 15 \end{array} \right.$$

Using the frequency distribution in Problem 7, construct (as just illustrated):

      a.   A "less than" cumulative frequency table as in form (1) (less than 15.0, less than 18.0, etc.)

      b.   An "or more" cumulative frequency table as in form (3) (15.0 or more, 18.0 or more, etc.)

28.   We have been considering frequency distributions with classes of equal length. Sometimes classes of unequal length are preferred, especially if the frequencies associated with high or low values (or both) are very small. The following frequency distribution of the number of visitors per day at an art exhibit for 25 days has unequal class intervals:

Number of visitors	Number of days
10–29	5
30–39	20
40–49	15
50–89	9

The differences in length of class interval must be taken into consideration when constructing a histogram. If a class interval is twice as long, the corresponding bar in the histogram should be only half as high. That is, divide the class frequency for that class by two before constructing the bar. If the class interval is 2.5 times as long, then divide the class frequency by 2.5 before constructing the bar, etc. Construct a histogram for the frequency distribution in this problem.

Chapter 4

Central Tendency of
a Distribution

4.1 What Is an Average?

A most often used descriptive measure of a distribution is a measure of its *central tendency*, called an *average*. We speak about average price, average grade, average speed, etc. What is meant by "average"? Usually, we mean some quantity which is typical of a set of quantities or, in some way, located in the center of the set. *Generally, an "average" means a single value which, in some sense, summarizes a set of data.* If a retail store states that customer purchase returns average $150 a week, this single amount gives a general understanding of the weekly amount of returns, even though we know it may sometimes be much more or much less.

Averages may be computed in different ways. We will study the *arithmetic mean*, which is the most frequently used and the most important average. We will also discuss the *median* and say a little about the *mode*. In our discussion of

33

averaging rates of change, we will discuss the *geometric mean* which is a specialized type of average. Each of these averages measures the central tendency of a
distribution in a different way. Although each average may represent a *typical
value* of a set of data, this is not always true for any average. Consider the four
weights (in pounds) 2, 14, 80, 190. It is possible, of course, to find a *central value*
for these four weights; however, it is clearly impossible for any single weight to
be *typical* of the set.

The widespread use of computers, which has been enhanced by increased
availability of time-sharing facilities, reduces the need for manual and desk
calculator computations. Nevertheless, such computations are still required
where computer facilities are not available and, even where available, demands
for their use often exceed computer capacity. It will be observed in this and
subsequent chapters, that the computational procedures presented and illustrated are of the manual and desk calculator variety. This is done for several
reasons. Primarily, the objective in all instances is to teach statistical concepts
and methods, and the computational procedures are presented for this purpose.
Certain shortcut or alternative computational procedures presented are often
directed toward this objective as well as to serve the needs of those readers who
do not have the use of computer facilities. Of course, the methods and procedures
presented can be used to prepare computer programs as desired.

4.2 Arithmetic Mean

When we speak of an average, we nearly always mean the *arithmetic mean*,
usually referred to as the *mean*. Consider the numbers of errors made by each
of four clerks: 3, 4, 2, 5. We compute the mean by dividing the sum of these
error counts (14) by the number of error counts in the set (4) to obtain $14/4 =$
3.4 errors per clerk. *The arithmetic mean of a set of quantities is computed by
dividing the sum of the quantities by the number of quantities in the set.*

It is useful in statistics to express computational procedures in symbolic
notation. This facilitates comprehension of the procedure and its application to
problems encountered in practice. We will use μ (the Greek letter mu) to denote
a *population mean* and $\overline{X}$ (read "X bar") to denote a *sample mean*. It may appear
unnecessary, at this time, to have two symbols for the mean, however, the
usefulness of two symbols will become apparent later on. We will use N to
represent *population size* and n for *sample size*. We may express the computation
of the arithmetic mean for a set of X_i quantities as follows:

$$\text{\textit{Population}} \quad \mu = \frac{X_1 + X_2 + X_3 + \ldots + X_N}{N} = \frac{\sum X_i}{N} \tag{4.2.1}$$

$$\text{\textit{Sample}} \quad \overline{X} = \frac{X_1 + X_2 + X_3 + \ldots + X_n}{n} = \frac{\sum X_i}{n} \tag{4.2.2}$$

Equation 4.2.1 for population data and Equation 4.2.2 for sample data
express the computation of the mean as previously illustrated. The far right
form in each equation expresses the same procedure, using the summation

notation in the numerator. If a population of 100 lengths totals 2,375 feet, we may compute the arithmetic mean of the lengths using Equation 4.2.1 and obtain $\mu = 2,375/100 = 23.75$ feet.

Notice that, even though there are differences in the symbols used to express the population and sample means, the computational procedures are exactly the same. There is, however, an important difference in the interpretation of μ and $\overline{X}$. In the case of population data, μ is the population mean and represents a measure of central tendency for the population distribution. The sample mean $\overline{X}$ may be considered as the arithmetic mean for the sample data. This interpretation is similar to the interpretation of μ, however, when sample data are collected, it is never because there is any particular interest in the sample quantities themselves. *The interest in a sample is only for the information it can provide about the population from which it was selected. It has been shown by mathematical statisticians that the sample mean $\overline{X}$ can be interpreted as an estimate of the population mean μ.* For example, if we compute $\overline{X} = \$85.25$ for a sample of typists' weekly salaries selected from all typists' salaries in Cleveland, we may estimate that the mean weekly salary paid to typists in Cleveland is $85.25. This is the interpretation of interest when $\overline{X}$ is computed.

Sometimes the means are known for two or more samples and it is desired to compute the mean for all samples combined. Table 4.2.1 presents the mean weights for three samples of fittings produced by a certain machine. Of course, it is to be expected that the sample mean will vary from sample to sample. The larger the sample size, the more confidence we have in using the sample mean as an estimate of the population mean. (Sampling will be considered more fully later in the text.) The mean of the samples combined, called the *general mean* $\overline{X}$ for the three samples, is computed as follows ($\overline{X}_i$ denotes the mean and n_i the sample size for a particular sample):

$$\overline{X} = \frac{\sum \overline{X}_i n_i}{\sum n_i} \tag{4.2.3}$$

(The proof for this equation is shown in Note 4A, Section 4.6.)

TABLE 4.2.1

Mean weight (ounces) for three samples of fittings: computation of the general mean.

Sample	Sample size, n_i	Sample mean weight, $\overline{X}_i$	$\overline{X}_i n_i$
1	30	4.3	129.0
2	40	3.9	156.0
3	25	4.0	100.0
Total	95		385.0

The numerator of the foregoing equation is computed by multiplying each sample mean $\overline{X}_i$ by its sample size n_i and adding these products for the samples to be combined. The denominator is the sum of the sample sizes. We may compute the general mean for the three samples in Table 4.2.1, where $\sum \overline{X}_i n_i = 385.0$ and $\sum n_i = 95$, as $\overline{X} = 385.0/95 = 4.1$ oz. per fitting. Table 4.2.1 shows a convenient way to organize the data and the computations.

We may compute a general mean for two or more populations combined. Suppose we know the mean number of shirts sold per week for each of three stores operated by a company and wish to compute the general mean number of shirts sold per week for the three stores together. We follow the same procedure as indicated in Equation 4.2.3. However, this equation should be rewritten for populations by substituting μ for $\overline{X}$, μ_i for $\overline{X}_i$, and N_i for n_i.

Sometimes the quantities to be averaged are considered to have different degrees of importance. As an illustration, Table 4.2.2 presents a student's grades on two midterm examinations and a final examination. The teacher wishes to compute an average of these grades to determine a semester grade for the student. However, he does not wish to give each grade the same weight in determining the average. The weights (w_i) chosen by the teacher are shown in the table. The *weighted mean* μ_w is computed as follows:

$$\mu_w = \frac{\sum X_i w_i}{\sum w_i} \tag{4.2.4}$$

Applying this equation to our illustration, we compute the student's weighted mean grade as $825/10 = 82.5$. Table 4.2.2 shows a convenient way to organize the data and the computations.

TABLE 4.2.2

Grades on two midterms and a final examination for a student: computation of the weighted mean.

	Grade X_i	Weight w_i	$X_i w_i$
Midterm I	80	3	240
Midterm II	95	3	285
Final	75	4	300
Total		10	825

Table 4.2.3 presents a frequency distribution for a sample of 50 prices. How do we go about computing the arithmetic mean for a frequency distribution when we do not know the individual quantities making up the raw data? *We make the assumption that the items in a class interval are equal to the class midpoint.* Applying this assumption to the data in Table 4.2.3, we assume that the nine prices in the $10–$14 class are each $12, the ten prices in the $15–$19 class are each $17, etc. Then, we compute the mean by adding the 50 assumed prices and dividing by 50. However, notice that nine of these assumed prices are $12, ten are $17, etc. Consequently, the sum of the 50 prices may be obtained by adding $(9)(\$12) + (10)(\$17) + \ldots$ and dividing by 50 (sum of the class frequencies). This computational procedure may be expressed as

$$\overline{X} = \frac{\sum X_i f_i}{\sum f_i} \tag{4.2.5}$$

TABLE 4.2.3

*Frequency distribution of a sample of 50 prices: compu-
tation of the arithmetic mean.*

Discrete class limits	Midpoint X_i	Frequency f_i	$X_i f_i$
$10–$14	$12	9	$ 108
15– 19	17	10	170
20– 24	22	15	330
25– 29	27	9	243
30– 34	32	7	224
Total		50	$1,075

Referring to the sample of 50 prices in Table 4.2.3, we compute the mean as
$1,075/50 = $21.50 in accordance with Equation 4.2.5. Table 4.2.3 shows a
convenient way to organize the data and the computations. Of course, the
assumption that all items in a class are equal to the class midpoint will usually
lead to some discrepancy in the mean computed for a frequency distribution
compared to the true mean, which would be obtained if the actual item values
were used in the computations. However, for a well-constructed frequency
distribution, the discrepancy is usually not important.

Equation 4.2.5 may be rewritten for computing a mean for population data
organized into a frequency distribution by substituting μ for $\overline{X}$. Note 4B in
Section 4.6 presents a shortcut method for computing the mean for a frequency
distribution.

4.3 Optional: Averaging Rates of Change

The geometric mean G is used for averaging rates of change. The geometric
mean of two quantities is the square root of their product. For example, the
geometric mean of 20 and 5 is $\sqrt{20 \cdot 5} = \sqrt{100} = 10$. The geometric mean of
three quantities is the cube root of their product. For example, the geometric
mean of 8, 4, and 2 is $\sqrt[3]{8 \cdot 4 \cdot 2} = \sqrt[3]{64} = 4$. Generally, for N quantities, the
geometric mean is

$$G = \sqrt[N]{X_1 X_2 X_3 \ldots X_N} \qquad (4.3.1)$$

The geometric mean is more easily computed using logarithms:

$$\text{Log } G = \frac{\Sigma \text{ Log } X_i}{N} \qquad (4.3.2)$$

The geometric mean is the appropriate measure to use when the *average rate
of change or average rate of growth* is to be computed. Table 4.3.1 presents the
population of a community for a four-year period. The average yearly rate of
population growth is obtained by computing the geometric mean of link relatives. The link relative column in the table, X_i, expresses the population for a

year as a proportion of the population the year before. If Y_i denotes the population for a year and Y_{i-1} the population of the year before, then we compute the *link relative* X_i as

$$X_i = \frac{Y_i}{Y_{i-1}} \tag{4.3.3}$$

TABLE 4.3.1

Population of a community during a four-year period: computation of the average yearly rate of growth.

| | | Link relative | |
Year	Population Y_i	$X_i = \dfrac{Y_i}{Y_{i-1}}$	Log X_i
1	1,000	—	—
2	1,100	1.10	0.041
3	1,600	1.45	0.161
4	3,500	2.19	0.340
Total		4.74	0.542

The first link relative in Table 4.3.1, 1.10, indicates that the population in year 2 was 1.10 or 110 percent of the population in year 1 so that the rate of growth from year 1 to year 2 was .10 or 10 percent. We must determine the logarithm of each link relative to apply Equation 4.3.2. Using $\sum \text{Log } X_i$ as shown in the table and Equation 4.3.2, we compute Log $G = 0.542/3$ or 0.18067. Then $G = $ antilog 0.18067 or 1.52. This indicates that, on the average, the population for a year was 1.52 or 152 percent of the population the year before. In other words, the average rate of population growth was .52 or 52 percent per year.

Let us test this result by increasing the year 1 population (1,000) by 52 percent a year for three years and compare the computed year 4 population with the actual (3,500). Carrying out this computation we obtain

$$(1,000)(1.52)(1.52)(1.52) = 3,512$$

Notice that we multiply 1,000 by 1.52, not .52, and that we multiply by 1.52 three times to account for the average growth over the three years following year 1. The result, 3,512, differs from 3,500 only due to rounding errors.

It is incorrect to use the arithmetic mean of link relatives as a measure of the average rate of growth. Using the data in Table 4.3.1, the arithmetic mean of the link relatives (X_i) is 1.58, indicating an average yearly rate of population growth of .58 or 58 percent. If we increase the year 1 population by this average rate for three years, we obtain

$$(1000)(1.58)(1.58)(1.58) = 3,944$$

Clearly, the computed year 4 population, 3,944, is too large.

If we write the link relatives in fraction form (Equation 4.3.3), we may compute the geometric mean for the link relatives in Table 4.3.1 using Equation 4.3.1 as follows:

$$\sqrt[3]{\frac{1,100}{1,000} \cdot \frac{1,600}{1,100} \cdot \frac{3,500}{1,600}} = \sqrt[3]{\frac{3,500}{1,000}} = 1.52$$

Notice that after canceling the geometric mean depends only on the year 1 and year 4 population figures. When the geometric mean is to be used as a measure of the average rate of growth, the equation

$$\text{Log } G = \frac{1}{N-1} \text{ Log } \frac{Y_N}{Y_1} \tag{4.3.4}$$

is most convenient, where:

Y_1 = first amount in a set
Y_N = last amount in the set
N = number of amounts in the set

Suppose the number of viewers of a TV show was estimated at 2,000 for the first performance and 10,000 for the tenth performance. We may use Equation 4.3.4 to compute the average rate of increase per performance in the number of viewers as follows:

$$\text{Log } G = \frac{1}{10-1} \text{ Log } \frac{10,000}{2,000} = .07767$$

Then, $G = 1.20$ indicates an average rate of increase in viewers of 20 percent from performance to performance.

4.4   Median, Position Measures, Mode

The median of a set of quantities is the value centrally located so that the number of quantities in the set with a lower value is equal to the number with a higher value. We use M to denote the median for sample data as well as population data. The median is determined in the same way for both types of data. (In this section, we will use N to denote both population size and sample size.)

Given the IQ scores 120, 100, 96, 118, 110. The first step in determining the median is to rank the quantities, for instance from low to high: 96, 100, 110, 118, 120. Then, determine the *median position* by computing $(N + 1)/2$. In our illustration, $N = 5$ and we compute the median position as $(5 + 1)/2 = 3$. This indicates that the third quantity *in the ranked sequence* is the median. In our example, the third IQ score, counting from *either direction*, is 110. Then, 110 is the median. Suppose we wish to determine the median for an even number of quantities, such as the four ages 15, 13, 20, and 18. Ranking, we obtain 13, 15, 18, 20. The median position is $(4 + 1)/2 = 2.5$. This indicates that the median is between the second and third ages in the ranked sequence—between 15 and 18. *Then, the median is computed as the arithmetic mean of these two middle quantities.* In our problem, the median age is $(15 + 18)/2 = 16.5$.

If a set of quantities contains ties, how do we determine the median? Given the six prices (in cents): 3, 2, 2, 4, 3, 2. Ranking, we obtain 2, 2, 2, 3, 3, 4. The median position is $(6 + 1)/2 = 3.5$, so that the median is between the third and fourth prices in the ranked sequence—between 2 cents and 3 cents. Computing the arithmetic mean, the median is 2.5 cents.

Table 4.4.1 presents a frequency distribution of the number of days in a hospital (day counts) for a sample of 150 men. How do we determine the median for a frequency distribution? Let us examine the histogram (Figure 4.4.1) for these data. It will be recalled that the area under a histogram represents the number of items summarized in the frequency table. The area under the histogram in Figure 4.4.1 represents the hospital day counts for the 150 men. The vertical dotted line in Figure 4.4.1 has been constructed to divide the area under the histogram into two equal parts, so that each part represents $N/2 = 150/2$ or 75 day counts. Clearly then, the point on the horizontal scale where this vertical dotted line is constructed is the median, as noted in the figure. *The median for a set of quantities organized into a frequency distribution is that value (M) where half the quantities (N/2) have a lower value and half have a higher value.*

TABLE 4.4.1

*Frequency distribution of the number of days in a hospital
for a sample of 150 men: computation of the median.*

Discrete class limits	Frequency f_i	Continuous class limits	Less than cum f_i
		−.5	0
0–2	65		
		2.5	65
3–5	(30)		
		5.5	95
6–8	25		
		8.5	120
9–11	10		
		11.5	130
12–14	8		
		14.5	138
15–17	7		
		17.5	145
18–20	5		
		20.5	150

The median may be computed using the continuous class limits and the associated "less than cumulative frequencies" (less than cum f_i). First, compute $N/2$, which in our illustration is 75. This tells us how many of the items in the frequency distribution make up half the items included. Inspect the less than cum f_i column in Table 4.4.1. Notice that 75 falls between the two cumulative frequencies 65 and 95 which are associated with the continuous class limits 2.5 and 5.5, respectively. Clearly, the median number of day counts (which is associated with 75 as the less than cum f_i) must lie between 2.5 days and 5.5 days.

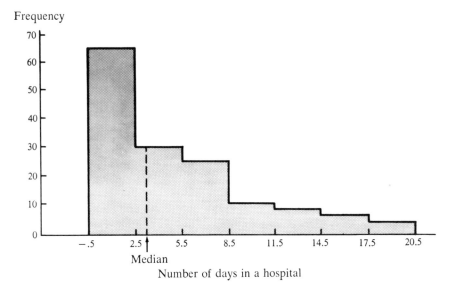

FIGURE 4.4.1

Determination of the median for a frequency distribution.

The class interval defined by these two continuous class limits, 2.5–5.5, is called the *median class,* since it contains the median.

Finally, *the median is computed by the method of linear interpolation based on the assumption that the items in the median class are evenly distributed over the class interval.* Let us write the continuous class limits of the median class and the associated less than cum f_i as follows:

$$\begin{Bmatrix} 2.5 \\ M \\ 5.5 \end{Bmatrix} \qquad \begin{Bmatrix} 65 \\ 75 \\ 95 \end{Bmatrix}$$

Notice the four braces and also that M is written between the class limits and $75 \ (= N/2)$ is written between the cumulative frequencies. Linear interpolation requires equating ratios of differences as follows:

$$\frac{M - 2.5}{5.5 - 2.5} = \frac{75 - 65}{95 - 65}$$

Note that the two short braces identify quantities related to the two *numerator differences* and the two long braces identify quantities related to the two *denominator differences.* Carrying out the subtractions as far as possible, we obtain

$$\frac{M - 2.5}{3} = \frac{10}{30}$$

Notice that the denominator on the left side is the length of class interval (3) and the denominator on the right is the frequency in the median class (30). Solving for M, we obtain the median day count as

$$M = 2.5 + (3)\frac{10}{30} = 3.5$$

Based on the foregoing result, we may write the equation for the median computed for a frequency distribution as follows:

$$M = L + g\frac{d}{f_M} \tag{4.4.1}$$

where: L = lower continuous class limit of the median class
   g = length of class interval
   d = difference between $N/2$ and the less than cum f_i associated with L
   f_M = frequency of the median class

In our example, $L = 2.5$, $g = 3$, $d = 75 - 65 = 10$, and $f_M = 30$. As an aid to computation, it is a good idea to bracket the continuous class limits of the median class and also the associated less than cumulative frequencies and place parentheses around the median class frequency, as shown in Table 4.4.1. We may conveniently identify the data to be substituted into Equation 4.4.1 as follows:

   L = smaller of the two limits in the bracket
   g = difference between the two limits in the bracket
   d = $N/2$ − (less than cum f_i associated with L)
   f_M = class frequency in parentheses

The median marks off the halfway or 50 percent point for a set of quantities, since half the quantities are less than the median and half are more. Consequently, the median is also called the *50th percentile* and denoted as P_{50}. Generally, we speak of P_i as the *ith percentile* which is *a value below which i percent of the items in a set fall.* For example, P_{10} is the 10th percentile and indicates a value below which 10 percent of the items in a set fall. *Percentiles,* which are called *measures of position,* may be computed by an equation similar to Equation 4.4.1, as follows:

$$P_i = L + g\frac{d}{f_i} \tag{4.4.2}$$

where: L = lower continuous class limit of the *i*th percentile class
   g = length of the *i*th percentile class
   d = difference between *i* percent of the quantities in a set and the less than cum f_i associated with L
   f_i = frequency of the *i*th percentile class

Let us compute P_{75}, the 75th percentile, for the 150 hospital day counts in Table 4.4.1. First, $(.75)(150) = 112.50$, the number of day counts less than P_{75}. Note that 112.50 falls between the less than cumulative frequencies 95 and 120, which are associated with the continuous class limits 5.5 and 8.5, respectively. The 75th percentile class is 5.5–8.5, with a class frequency of 25. Applying Equation 4.4.2 we obtain

$$P_{75} = 5.5 + (3)\frac{112.50 - 95}{25} = 7.6$$

Then, 75 percent of the hospital day counts for the 150 men are less than 7.6 days.

The percentiles P_{25}, P_{50}, and P_{75} are usually called *quartiles*. The 25th percentile is the first or lower quartile, the 50th percentile is the second quartile or the median, and the 75th percentile is the third or upper quartile. Symbols often used are Q_1 for P_{25}, Q_2 or M for P_{50}, and Q_3 for P_{75}. The nine percentiles P_{10}, P_{20}, P_{30}, . . . , P_{90} are usually called *deciles* where D_1 or P_{10} is the first decile, D_2 or P_{20} is the second decile, etc.

The *mode* is the easiest type of average to compute and to understand. *In a set of quantities, the mode is that quantity which appears most often.* In the set of plant heights (in inches) 12, 16, 14, 16, 14, 14, 12, 11, 14, 14, 13, 14, we find that the height 14 inches appears most often (six times). Hence, the *mode* for this set, the *modal height*, is 14 inches. In a set of data, where no quantity appears most often, there is no mode (it does not exist).

4.5 Properties and Comparisons

The geometric mean is a special purpose average; however, there is sometimes a question as to when to use the arithmetic mean, the median, and the mode. We will examine some of the properties of these three averages, compare them, and indicate which one is to be used in a given problem.

The arithmetic mean is determined by every item in a set of quantities. Therefore, the mean is a very sensitive average. In the set 5, 3, 7, the mean is 5. If one of these quantities is changed, the mean will change. For example, if 5 is changed to 20, the mean becomes 10 instead of 5. This is an important and usually desirable property of the mean. This property, however, makes the mean less desirable when an average is required for a set of data which contains extreme values. In the set 4, 7, 100, the mean (37) is strongly influenced by the extreme value 100 and is not representative of any of the quantities in the set.

If the mean, for example μ, is subtracted from a quantity in a set (X_i), we obtain the *deviation from the mean* ($X_i - \mu$), a very important concept in statistics. We will denote this deviation by the lower case letter x_i, so that $x_i = X_i - \mu$. *The sum of the deviations from the mean of a set of N quantities is equal to zero.* We may express this as

$$\Sigma x_i = \Sigma (X_i - \mu) = 0 \tag{4.5.1}$$

(The proof is left as an exercise for the reader.)

Given the set 2, 6, 4, 12, with mean 6. The deviations from the mean are: $2 - 6 = -4$, $6 - 6 = 0$, $4 - 6 = -2$, $12 - 6 = 6$. The algebraic sum of these deviations (the sum which takes into account the plus and minus signs) is $(-4) + (0) + (-2) + (6) = 0$. Equation 4.5.1 may be rewritten for sample data by substituting $\overline{X}$ for μ.

In a set of N quantities, we may compute the deviation of each quantity X_i from a constant k to obtain the deviation ($X_i - k$). Then, squaring each deviation we obtain $(X_i - k)^2$ and adding the N squared deviations we obtain $\sum (X_i - k)^2$. *This sum of the squared deviations from a constant k is a minimum if the value chosen for k is the mean of the set.* This is called the *least squares property of the mean* and may be expressed as

$$\sum (X_i - k)^2 \quad \text{is a minimum if } k = \mu \quad \text{(or } \overline{X}) \qquad \text{(4.5.2)}$$

The reader should verify this property of the mean by computing the sum of the squared deviations of the set 10, 14, 13, 3 from its mean and also from $k = 9$.

There is one and only one mean for a set of quantities. If the individual quantities in a set are not known, the mean can be computed if the total of the items in the set $(\sum X_i)$ is known as well as the number of items in the set. Why? (Hint: Examine the equations for the mean in Section 4.2, Equations 4.2.1 and 4.2.2.)

The median is considerably less sensitive to the particular quantities in a set than the mean, because the median is an average of position (centrally located in the set) and not a computed average like the mean. In the set 3, 8, 10, 11, 14, the median is 10 (the quantity in the center). If 14 is changed to 1,400, the median remains 10 (which is still the quantity in the center).

In our previous discussion of deviations such as $(X_i - k)$, it was noted that some deviations are positive and some negative. Sometimes we are interested in the *magnitude* of the deviations, not the direction (plus or minus). In such instances, we are interested in the *absolute deviations*. The deviations of 7 and 10 from 9 are $7 - 9 = -2$ and $10 - 9 = 1$. The absolute deviations are 2 and 1, respectively. Clearly, to obtain absolute deviations, we must consider all deviations as positive. The sum of the absolute deviations of a set of N quantities (X_i) from a constant k is a minimum if k is equal to the median for the set. We may express this as

$$\sum |X_i - k| \quad \text{is a minimum if } k = M \qquad \text{(4.5.3)}$$

Notice the symbol $|X_i - k|$ which denotes the *absolute deviation*. The reader should verify this property of the median by computing the sum of the absolute deviations of the set 3, 5, 8, 11, 14 from its median and also from $k = 9$.

The *mode* is an average of limited value and is used infrequently. The mode is practically insensitive to the particular values included in a set. Changes in the quantities included in a set have no effect on the mode, except if the value with the maximum frequency is affected.

In a symmetrical distribution with a central peak, as in Figure 4.5.1, part (A), the mean, median, and mode are equal, so that there is no problem as to which to use. In a negatively skewed distribution, as in part (B) of the figure, the items are concentrated at the higher values and some items are strung out in a tail at the lower end of the scale of values. These extreme (tail) values influence the mean, pulling it down in value, but have no influence on the median and mode. Therefore, in such distributions, the mean has the lowest value, the mode has the highest value, and the median lies somewhere between the two (as shown in the figure). In a positively skewed distribution, as illustrated in part (C) of the figure, the relationship among these averages is reversed.

Generally, the arithmetic mean is preferred over the median or mode, because it is more rigorously defined, it can be treated algebraically, and because it tends to have less *sampling variability*. Sampling variability will be discussed in later

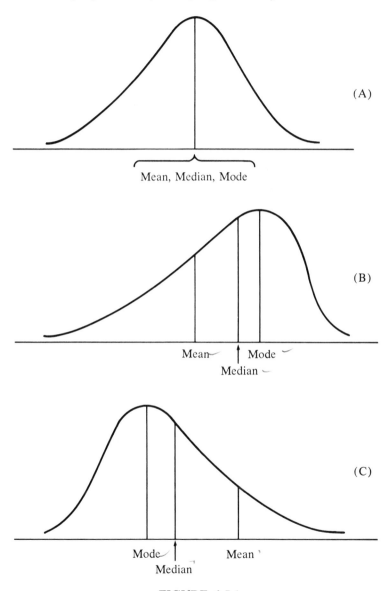

(A)

Mean, Median, Mode

(B)

Mean⌐   ↑ Mode ⌐
         Median ⌐

(C)

Mode⌐ ↑   Mean ⌐
    Median⌐

FIGURE 4.5.1

Relative positions of the mean, median and mode for three types of distributions.

chapters. It may be noted now that the theory underlying statistical inferences based on sample data is based on the notion of *repeated sampling*, a fundamental notion in statistical theory.

For example, a sample of 30 bolts produced by a machine may show an average diameter of .52 cms, and based on this information, an inference is made as to the average diameter for all bolts produced by this machine. Only one sample of 30 bolts is selected; however, the statistical theory involved in making the inference is based on the notion of selecting a large number of samples of 30 bolts each, determining the average for each sample, and taking into consideration how this average varies from sample to sample. The variation from sample to sample is called *sampling variation*. According to statistical theory to be discussed later, the less the variation in the sample average from sample to sample, the more likely is it that the average computed for the particular sample selected will be a good estimate of the population average.

In the typical sampling situation, *sampling variability of an average is less if the average used is the arithmetic mean.* This is a critically important advantage of the mean when samples are used to generalize a population.

Generally, the median is the preferred average to describe data which include extreme values, such as data with highly skewed distributions. The mean is too much affected by the extreme values and the mode merely reflects the value with the maximum frequency. In such situations, the median is most representative of the data.

The mode is the preferred average where interest centers on the item with the maximum frequency. For example, a retailer may ask a sample of customers which color of luggage is preferred because he wants to decide which color to purchase for his small inventory of luggage. This retailer is interested in the color chosen most often by the sample of cutomers, the *modal color*.

The mean can be computed only for quantitative data, whereas, the median and mode can be determined for qualitative data as well. The mode can be determined for any type of qualitative data, whether nominal or ordinal scale data. The median, on the other hand, can be determined only for ordinal scale qualitative data. For example, the only average which can be computed for the color choices of luggage of the sample of customers previously mentioned (nominal scale data) is the mode.

Table 4.5.1 presents the distribution of a sample of voter attitudes toward an issue. The attitude scale represents an ordinal scale. Clearly, it is not possible to compute an arithmetic mean for these data. The modal attitude is easily determined as "very much in favor," since it shows the largest frequency (60) in the sample of 140 voters. We may compute the median attitude by first determining the position of the median as $(140 + 1)/2 = 70.5$. Compute the cum f_i column, as shown in the table. This column indicates that 60 sample voters are "very much in favor"; $60 + 30$ or 90 voters are "in favor" or better; etc. Looking down the cum f_i column, notice that the 70th and 71st voters' attitudes are included in the 90 who are at least in favor of the issue. Hence, the median attitude is "in favor." In this illustration, the median would usually be preferred to the mode, since it is more representative of the attitude distribution.

TABLE 4.5.1

*Attitude of a sample of voters toward an issue: computation
of the median.*

Attitude	Number in the sample f_i	Cum f_i
Very much in favor	60	60
In favor	30	90
No preference	10	100
Opposed	15	115
Very much opposed	25	140
Total	140	

4.6 Optional: Algebraic Notes and Shortcut Method

Note 4A: To prove

$$\bar{X} = \frac{\sum \bar{X}_i n_i}{\sum n_i} \tag{4.2.3}$$

From Equation 4.2.2 we obtain $\sum X_i = n\bar{X}$, indicating that the total for a sample of quantities is equal to the product of its mean and the sample size. For a sample of n_i quantities with mean $\bar{X}_i$, we compute the sample total as $\sum X_i = n_i \bar{X}_i$. For several samples, each of size n_i and mean $\bar{X}_i$, we obtain the total for the samples combined as $\sum X_i = \sum n_i \bar{X}_i$ and the size of the combined sample as $\sum n_i$. Then applying Equation 4.2.2, we compute the *general mean* for the samples combined as $\bar{X} = (\sum n_i \bar{X}_i)/\sum n_i$.

Note 4B: Shortcut method for computing the arithmetic mean for a frequency distribution.

Table 4.6.1 presents the frequency distribution of 50 prices from Section 4.2, Table 4.2.3. The class midpoints X_i are transformed to u_i values using the equation

$$u_i = \frac{X_i - k}{g} \tag{4.6.1}$$

where: k = midpoint for the class with the largest frequency or for a class centrally located in the frequency table
g = length of class interval

In Table 4.6.1, $g = 5$, and we will let $k = 22$, the midpoint of the class with $f_i = 15$. (Actually, k may be put equal to *any* class midpoint, however, if the k value is selected as noted, the computational labor will be least.) The first u_i value u_1 is computed according to Equation 4.6.1 as $(12 - 22)/5 = -2$, as shown in the table. The other u_i values are similarly computed. In practice the u_i values may be determined without any computations as follows: Place a 0 in

TABLE 4.6.1

Frequency distribution from Table 4.2.3: computation of the arithmetic mean using the shortcut method.

Midpoint X_i	Frequency f_i	u_i	$u_i f_i$
$12	9	-2	-18
17	10	-1	-10
22	15	0	0
27	9	1	9
32	7	2	14
Total	50		-5

the u_i column as the u_i value associated with the midpoint which equals k. Assign u_i values -1, -2, etc., for succeeding lower midpoints and u_i values 1, 2, etc., for succeeding higher midpoints, as shown in the table.

Then, compute $\bar{u}$, the arithmetic mean of the u_i values, as follows:

$$\bar{u} = \frac{\Sigma u_i f_i}{\Sigma f_i} \qquad (4.6.2)$$

In our illustration, $\bar{u} = -5/50 = -.10$. Finally, the arithmetic mean is obtained as follows:

$$\bar{X} = g\bar{u} + k \qquad (4.6.3)$$

Applying this equation to our problem, we obtain:

$$\bar{X} = (5)(-.10) + 22 = \$21.50$$

This is the same result obtained previously. This shortcut method may be used for population data as well.

Study Problems

1. The arithmetic mean was computed for the number of daily hospital admissions for a sample of 15 days selected from last year's records giving $\bar{X} = 49$. How would you interpret this result?

2. Compute the mean for each of the following populations:
    a.   3, 6, 5, 6
    b.   4, 9, 12, 3, 2
    c.   11, 14, 20, 10, 8, 7, 4

3. Compute the mean for each of the following samples:
    a.   5, 6, 4, 6, 2, 1
    b.   6, 4, 18, 10, 9, 3
    c.   5, 4, 2, 12, 10, 8, 25, 22, 18, 15

4. If a cart weighs 50 lbs empty and 290 lbs when carrying four boys, what is the mean weight per boy?

5. The mean valuation of five paintings is $63,000 per painting; what is the total valuation of the paintings?

6.   If you have $45 to spend on six plants and discover that the mean price is $8 per plant, how much more money do you need to make the purchase?

7.   Given a sample of test scores: 40.3, 52.6, 38.1, 60.8, and 29.9, estimate the population mean test score.

8.   Five feet of steel cable was sold at 90¢ a foot, eight feet at 70¢ a foot, 15 feet at 50¢ a foot, and 11 feet at $1.25 a foot. Compute the mean price per foot for all the steel cable sold.

9.   A sample of 25 students obtained a mean score of 120 on an achievement test, a sample of 20 students obtained a mean of 230, and a sample of 10 students obtained a mean of 90. Compute the mean for the three samples combined.

10.   In a candy factory, the 32 men in plant A were paid a mean wage of $93; the 35 men in plant B were paid a mean wage of $84; and the 43 men in plant C were paid a mean wage of $86. Compute the general mean wage paid per worker in the three plants combined.

11.   Fifteen hundred dollars was invested at five percent interest, $1,000 at six percent interest, and $200 at ten percent interest. Compute the mean rate of interest (the weighted mean) for the total amount invested.

12.   The mean number of young children in a swimming pool was 14 for a sample of four Mondays, 16 for a sample of six Tuesdays, 17 for a sample of ten Wednesdays, 18 for a sample of seven Thursdays, and three for a sample of five Fridays. Estimate the mean number of young children in the swimming pool per day during Monday through Friday.

13.   The number of coding errors was determined for each day in a sample of 12 days for a clerk from last year's production records, and the following frequency distribution was constructed:

Number of errors	Number of days
2–5	3
6–9	5
10–13	2
14–17	1
18–21	1

Estimate the mean number of coding errors per day for this clerk for last year.

14.   The number of divorces per month for a 15-month period was organized into a frequency distribution as follows:

Number of divorces	Number of months
3–5	1
6–8	2
9–11	3
12–14	4
15–17	5

Compute the mean number of divorces per month for the 15-month period.

15.   Given the following frequency distribution of the birth rate per 1,000 population for 27 cities:

Birth rate	Number of cities
12.0–15.9	6
16.0–19.9	10
20.0–23.9	7
24.0–27.9	4

Compute the mean birth rate for the 27 cities.

16.   The reaction-times for a sample of 223 teen-age girls are as follows:

Seconds	Number of girls
.00–.09	31
.10–.19	42
.20–.29	47
.30–.39	44
.40–.49	39
.50–.59	20

Estimate the population mean reaction-time.

17.   Compute the geometric mean:
     a. 4, 16
     b. 9, 4, 6
     c. 1.6, 1.28, 16

18.   Profit after taxes for a firm for a five-year period was

Year	Profit
1	$127
2	234
3	358
4	470
5	509

Find the average yearly rate of growth in profit after taxes for the five-year period.

19.   If the product of three quantities is 68.921, what is the geometric mean of the quantities?

20.   If 3,000 people visited a national park in March and 5,200 people visited this park in May, what was the average monthly rate of increase in the number of visitors to the park during March–May?

21. Expenditures for recreation in a community for a four-month period:

	Expenditures
June	$21,500
July	30,000
August	50,000
September	80,000

Compute the average monthly rate of increase in expenditures.

22. Define (a) median, (b) median position, (c) mode, (d) percentile, and (e) upper quartile.

23. Does the median always represent the central item value in the ranked sequence of a set of quantities? Why?

24. Which average (or averages) has each of the properties listed:
   a. Determined by the specific value of each quantity in a set.
   b. An average of position.
   c. Sensitive to extreme values.
   d. Value with the greatest frequency.
   e. Sum of the absolute deviations is a minimum.
   f. A computed average.
   g. Sum of the squared deviations is a minimum.
   h. Sum of the deviations is zero.
   i. May be determined for quantitative data.
   j. May be determined for ordinal scale data.
   k. May be determined for nominal scale data.

25. Determine the median position and the median:
   a. 5, 6, 3, 6
   b. 3, 9, 12, 2, 4
   c. 8, 14, 4, 11, 7, 10, 20
   d. 4, 5, 2, 6, 6, 1
   e. 6, 10, 9, 3, 18, 4
   f. 8, 12, 4, 18, 5, 15, 2, 10, 25, 22

26. Find the median and the mode:
   a. 19, 20, 21, 20, 18, 20, 23, 20, 19, 20, 22
   b. 29.6, 30.8, 20.1, 40.1, 40.1, 26.3, 28.4, 40.1, 30.8, 34.9, 40.1, 43.7, 40.1, 30.8, 40.1
   c. .005, .009, .003, .012, .017, .005, .005, .017, .020, .019, .005, .005

27. Determine the median and the mode for the following set of weights (in pounds):

Weights	Frequency
62	1
64	3
67	7
69	10
72	24
76	19
78	15
80	13

28.   Given the following distribution of accident rates per 100,000 population:

Accident rate	Frequency
12.0–15.9	6
16.0–19.9	10
20.0–23.9	7
24.0–27.9	4

Compute the median.

29.   Determine the median for the following distribution of the number of units produced in an industrial plant for a sample of 41 weeks:

Number of units produced	Number of weeks
115–117	3
118–120	7
121–123	12
124–126	10
127–129	9

30.   Determine the median for the following distribution of commissions earned by 162 salesmen:

Commission earned	Number of salesmen
$20–$24	3
25– 29	8
30– 34	12
35– 39	21
40– 44	29
45– 49	32
50– 54	30
55– 59	27

31.   Compute Q_1, Q_3, P_{46} for the distribution of accident rates in Problem 28.

32.   Compute Q_1, Q_3, D_3, D_8, P_{18}, and P_{55} for the following distribution of reaction times for an experimental test for 223 subjects:

Reaction times (minutes)	Number of subjects
.00–.09	31
.10–.19	42
.20–.29	47
.30–.39	44
.40–.49	39
.50–.59	20

33.   Compute Q_1, Q_3, D_1, P_{35}, P_{69} for the distribution of commissions in Problem 30.

34.   Determine the modal color of shoes sold: white, red, beige, white, black, yellow, black, red, beige, white, yellow, white, black, red, green, black, red, beige, red, purple, and red.

35.   Determine the median and modal reaction to a drug based on the following distribution for 82 subjects: very mild (7), mild (9), weak (10), moderate (20), strong (25), and very strong (11).

36.   Determine the median and modal citizen attitude toward a bond issue based on the following distribution:

Attitude	Number of citizens
Strongly opposed	50
Moderately opposed	150
Opposed	100
In favor	90
Moderately in favor	60
Strongly in favor	50

37.   Refer to Problem 10. What are the total wages paid in (a) plant A and (b) all three plants?

Supplementary Study Problems

38.   If $S_i = X_i + Y_i$, then for population data $\mu_S = \mu_X + \mu_Y$ (where μ_S = mean for the S_i quantities, etc.). A similar relationship for the mean of sums holds for sample data as well.

   a.   If a class of seniors obtains a mean of 32.6 on part A of an exam and a mean of 57.1 on part B, determine the mean exam grade for the full exam (parts A and B combined).

   b.   A sample of plants grew to a mean height of 1.23 cms during the first week and showed an additional growth of a mean of 1.08 cms during the second week. Compute the mean plant height after the two week period.

39.   If $D_i = X_i - Y_i$, then for population data $\mu_D = \mu_X - \mu_Y$. A similar relationship for the mean of differences holds for sample data as well.

   a.   The mean price for a group of commodities is $20.15. If these prices are reduced by a mean of $2.05 per commodity, what is the mean of the reduced prices?

   b.   The mean time required to travel to a shopping center is 45.3 minutes for a sample of suburban shoppers. These shoppers were able to reduce the travel time by a mean of 10.6 minutes by using a different route. Compute the new mean travel time for this sample of shoppers.

40.   If a constant k is subtracted from each quantity in a set, then the original mean of the set is also reduced by k. Similarly, if a constant k is added to each quantity in a set, then the original mean of the set is also increased by k.

   a.   If the mean score on a research aptitude test is 110.16 and each score is reduced 15 points, what is the new mean score?

   b.   If the price of each toy in a group is increased by $2.00 and the original mean price was $4.10 per toy, what is the new mean price?

41. If each quantity in a set is multiplied (divided) by a constant k, the original mean for the set is also multiplied (divided) by k.

    a. The mean time required to perform a certain task was 115.23 seconds for a sample of adult males. If the time measurements had been recorded in minutes instead of seconds, what would be the mean number of minutes required?

    b. The mean was computed for a set of measurements after each measure was multiplied by 1,000, resulting in a mean of 261.3 cms. Determine the mean for the original set of measurements.

42. The median for a frequency distribution can also be computed using the following equation:

$$M = U - g\frac{d'}{f_M}$$

where: U = upper continuous limit of the median class

    d' = difference between $N/2$ and the more than cum f_i associated with U

    g and f_M have the same meaning as for Equation 4.4.1

Using the foregoing equation, compute the median for (a) Problem 28, (b) Problem 29, and (c) Problem 30.

43. Given the set of quantities: 12, 9, 2, 4, 3.

    a. Verify Equation 4.5.2 using the deviations from the mean and also from 7.

    b. Verify Equation 4.5.3 using the absolute deviations from the median and from 5.

44. Prove Equation 4.2.3 for population data, showing each step.

45. Compute the mean, using the shortcut method, for

    a. The birth rates in Problem 15

    b. The reaction-times in Problem 16

46. Prove Equation 4.5.1 for sample data, showing each step.

47. Refer to Problems 38 and 39.

    a. Prove: The mean of sums is equal to the sum of the means (Problem 38).

    b. Prove: The mean of differences is equal to the difference between the means (Problem 39).

48. Refer to Problems 40 and 41.

    a. Prove: If each quantity in a set is increased (decreased) by a constant, the original mean of the set is also increased (decreased) by the constant (Problem 40).

    b. Prove: If each quantity in a set is multiplied (divided) by a constant, the original mean of the set is also multiplied (divided) by the constant (Problem 41).

Chapter 5

Variation
and Analysis

5.1 Measurement of Variation

Typically, data exhibit variation. For example, the price of a commodity varies
from seller to seller as well as from period to period; the weight per bolt varies
from bolt to bolt even if produced by the same machine; examination grades
vary from student to student; etc. *Variation is the most important measure of
a distribution, from the point of view of statistics.* If we know that all the items
in a set have exactly the same value, for example 50, then we have complete
information about the set and there is no need whatsoever for statistical
methods. If variation exists, statistical methods are required to describe the
distribution of the data. *The extent of variation among the quantities in a population is of major importance in sampling and statistical inference.* This will be
considered in later chapters.

The simplest measure of variation is the *range* (the difference between the highest and lowest values in a set of data). Given the set 2, 3, 4, 5, we compute the range as $5 - 2 = 3$. The interval 2–5, as well as the distance 3, is referred to as the range.

Whereas an average represents a *point* on a scale of values, a measure of variation represents a *distance* along the scale. The range is the simplest measure of variation to compute; however, it is of limited value since it depends only on the highest and lowest values in a set. The two sets 100, 110, 120, 130 and 100, 100, 130, 130 have the same range (30), but clearly are quite different in distribution.

The more useful measures of variation in statistics represent, in some sense, the average distance of a set of values from a point of reference. Consider the set 2, 4, 6, 8 and an arbitrary point of reference, for example 20. Then, the distance of the quantities in the set from 20 are 18, 16, 14, 12. The mean of these distances, 15, represents a measure of variation of these data from the arbitrary point of reference. In other words, on the average, these quantities vary (deviate) from 20 by a distance of 15 units. Such a measure of variation is an improvement over the range in that all the items in a set are involved in the computation; however, a weakness in this measure is the arbitrary nature of the point of reference used. *Measures of variation in statistics use an appropriate average as the point of reference.*

5.2 Standard Deviation

The most important and most frequently used measure of variation is the *standard deviation. This measure of variation uses the arithmetic mean as the point of reference.* Therefore, the standard deviation is the appropriate measure of variation to use when the mean is used as the average.

Consider the set of quantities 10, 11, 12, 13, 14 with mean 12. Deviations from the mean are $10 - 12 = -2$, $11 - 12 = -1$, $12 - 12 = 0$, $13 - 12 = 1$, $14 - 12 = 2$. Let us use the mean of these deviations (from the mean) as a measure of variation for this set of quantities. The sum of these deviations is zero, so that this measure of variation from the mean is zero. Furthermore, if we carried out such a computation for any other set of data, the sum of the deviations from the mean and the measure of variation would also turn out to be zero. Of course, this should be expected since it was noted in Section 4.5 that the sum of the deviations around the mean is always zero (Equation 4.5.1). Clearly, this property of the mean makes this procedure for measuring the variation in a set of data worthless, since it will always equal zero.

A modification of this procedure provides a highly useful measure of variation. This modification requires *squaring* the deviations from the mean and then obtaining the mean of the *squared deviations*. This modified procedure results in the following measures of variation:

Population $$\sigma^2 = \frac{\sum (X_i - \mu)^2}{N} = \frac{\sum x_i^2}{N} \qquad (5.2.1)$$

Sample $$s^2 = \frac{\sum (X_i - \bar{X})^2}{n - 1} = \frac{\sum x_i^2}{n - 1}$$ (5.2.2)

In these equations, σ^2 (σ is the lower case Greek letter sigma) denotes the *variance computed from population data* and s^2 denotes the *variance computed from sample data*. Table 5.2.1 presents the heights of four plants (in inches). Suppose we consider this a population of plant heights. Then, using the data in the table, we compute $\mu = 24/4 = 6$ in. The $X_i - 6$ column in the table shows the deviation of each plant height from the mean. The squared deviations are presented in the $(X_i - 6)^2$ column. Applying Equation 5.2.1 for population data, we compute the *population variance* $\sigma^2 = 20/4 = 5$. Notice from Equation 5.2.1 that the *variance* σ^2 is computed as the *mean of the squared deviations from the mean*. The variance is an important measure of variation; however, the usual practice is to take the square root and compute σ, which is called the *standard deviation*.

TABLE 5.2.1

Heights of four plants (in inches): computation of the standard deviation.

Height (inches) X_i	$x_i = X_i - 6$	$x_i^2 = (X_i - 6)^2$
5	-1	1
7	1	1
3	-3	9
9	3	9
Total 24		20

In our illustration, $\sigma = \sqrt{5} = 2.24$ inches. This indicates how far, on the average, the heights in the set deviate from their mean (the point of reference). Notice from the deviations in the $X_i - 6$ column in Table 5.2.1 that one height is one inch below the mean, another height is three inches above the mean, etc.; however, on the average, a plant height deviates $\sigma = 2.24$ inches above or below the mean. Consequently, *the standard deviation σ measures how far, on the average, the quantities in a set are scattered or dispersed about the mean.* The larger the value of σ, the greater is the scatter of the items in a set around its mean. If $\sigma = 0$, then there is no variation at all in the set and all items are exactly equal to the mean (and so to each other).

It should be observed that it is the distance of each quantity in a set from its mean and not the direction of a deviation which is taken into account in the computation of the standard deviation. This is a consequence of *squaring* the deviations from the mean in the computation. As indicated in Table 5.2.1, the deviations 3 and -3 are both equal to 9 after squaring, so that the sign ($+$ or $-$) plays no part in the computation.

Let us now assume that the plant heights in Table 5.2.1 represent a sample of size 4 selected from a population of heights. Then the column headed $x_i =$

$X_i - 6$ presents the deviations of the plant heights from the sample mean $\overline{X} = 6$ inches and the squared deviations are shown in the $x_i^2 = (X_i - 6)^2$ column. We compute the variance $s^2 = 20/(4 - 1) = 6.67$, according to Equation 5.2.2. Notice from this equation that the variance s^2 is not quite the mean of the squared deviations from the mean, since we divide by $n - 1$ instead of by n. Why do we divide by N for population data and by $n - 1$ for sample data? Let us consider this.

In the case of the arithmetic mean, we distinguished between a population mean and a sample mean by the use of different symbols (μ and $\overline{X}$) and, as previously explained, these means are interpreted differently. This is true, as well, for the variance computed from population data and from sample data; however, in the case of the variance, the computation is also somewhat different, as just pointed out. *The variance computed for population data σ^2 is interpreted as a measure of variation of the quantities in the population about their mean. The variance s^2 computed from sample data is interpreted as an estimate of the variance in the population from which the sample was selected.* It has been shown by mathematical statisticians that if s^2 is computed using n in the denominator of Equation 5.2.2, instead of $n - 1$, the result will tend to be too small when used as an estimate of the population variance. Hence, to correct for this downward bias, we divide by $n - 1$.

Then, the standard deviation computed from sample data is s. Referring to the sample of plant heights in Table 5.2.1, we compute $s = \sqrt{6.67} = 2.58$ inches.

Equations 5.2.1 for σ^2 and 5.2.2 for s^2 are *definitional equations* and are not convenient computationally, except for the simplest problems. Useful *computational equations* for the *standard deviation* are:

Population $\qquad \sigma = \dfrac{1}{N} \sqrt{N \sum X_i^2 - (\sum X_i)^2}$ $\qquad\qquad$ (5.2.3)

Sample $\qquad s = \sqrt{\dfrac{n \sum X_i^2 - (\sum X_i)^2}{n(n - 1)}}$ $\qquad\qquad$ (5.2.4)

For proofs see Section 5.6, Note 5A.

TABLE 5.2.2

Number of arrests in a sample of four neighborhoods: computation of the standard deviation.

Number of arrests X_i	X_i^2
3	9
6	36
8	64
4	16
Total 21	125

Equations 5.2.3 and 5.2.4 may appear formidable to some readers. However, an illustration will show that they are easy to apply. Table 5.2.2 presents the number of arrests reported in a sample of four neighborhoods. We may compute the standard deviation by applying Equation 5.2.4. We determine $\sum X_i = 21$ and $\sum X_i^2 = 125$ as shown in Table 5.2.2. Then, noting that $n = 4$, we obtain

$$s = \sqrt{\frac{(4)(125) - (21)^2}{(4)(3)}} = 2.2$$

Table 5.2.3 presents a frequency distribution of the weekly amount of rainfall in an area for a sample of 25 weeks. How is the standard deviation computed for a frequency distribution? Since a frequency table does not show the individual item values, we make the assumption that all items in a class are equal to the class midpoint, just as we did in the case of the mean. The computational equations for the standard deviation for a frequency distribution are:

Population $\qquad \sigma = \frac{1}{N} \sqrt{N \sum X_i^2 f_i - (\sum X_i f_i)^2}$ (5.2.5)

Sample $\qquad s = \sqrt{\frac{n \sum X_i^2 f_i - (\sum X_i f_i)^2}{n(n - 1)}}$ (5.2.6)

TABLE 5.2.3

Rainfall in an area (in inches) for a sample of 25 weeks: computation of the standard deviation for a frequency distribution.

Rainfall (inches)	Number of weeks, f_i	Midpoint X_i	$X_i f_i$	$X_i^2 f_i$
3–5	3	4	12	48
6–8	5	7	35	245
9–11	10	10	100	1,000
12–14	7	13	91	1,183
Total	25		238	2,476

Notice that Equations 5.2.5 and 5.2.6 are similar to Equations 5.2.3 and 5.2.4. In Equations 5.2.5 and 5.2.6, X_i denotes the class midpoint and f_i the class frequency. Since Table 5.2.3 presents sample data, use Equation 5.2.6 for computation of the standard deviation. We compute $X_i f_i$ for the first class interval by multiplying the midpoint 4 by the frequency 3, to obtain 12, as shown in the table. Compute $X_i^2 f_i$ for this class interval most easily by multiplying $X_i f_i = 12$ by $X_i = 4$ to obtain 48, as shown in the table. Then substituting $\sum X_i f_i = 238$, $\sum X_i^2 f_i = 2,476$, and $n = 25$ into Equation 5.2.6, we obtain

$$s = \sqrt{\frac{(25)(2,476) - (238)^2}{(25)(24)}} = 3.0$$

We may estimate that the population standard deviation for weekly rainfall in the area is 3.0 inches.

Tables 5.2.2 and 5.2.3 present convenient ways to organize the data for computation of the standard deviation for sample data or population data. Furthermore, if it is also desired to compute the arithmetic mean, this may easily be accomplished from these tables. A shortcut method for computing the standard deviation for a frequency distribution is presented in Section 5.6, Note 5B.

5.3 Other Measures of Variation

The *quartile deviation* Q is often the measure of variation used when the median is used as the average. It is half the range between Q_3 (upper quartile) and Q_1 (lower quartile) and is computed as

$$Q = \frac{Q_3 - Q_1}{2} \qquad (5.3.1)$$

Suppose a social adjustment scale for the teen-agers in a city has $Q_1 = 73$ and $Q_3 = 159$. We compute the quartile deviation as $Q = (159 - 73)/2 = 43$. The quartile deviation is also called the *semi-interquartile range*. Sometimes other ranges between position measures are used to measure the variation in a set of data. The *interquartile range* $(Q_3 - Q_1)$ or the *interdecile range* $(D_9 - D_1)$ is sometimes used.

5.4 Relative Variation and Standard Units

The measures of variation we have studied in the previous sections are extremely useful statistical measures, however, taken by themselves, they have serious limitations. For example, $\sigma = 5$ lbs for the weights of a population of steel cylinders and $\sigma = 10$ cms for the diameters of these cylinders. Does this indicate little variation for the population of weights or considerable variation? Should we conclude that the population of cylinders is more variable in diameter than in weight?

Variation measures may be evaluated on a *relative basis, relative to the mean.* If the mean weight of the population of cylinders is 10 lbs per cylinder, then $\sigma = 5$ lbs indicates much more variation than if the mean is 100 lbs. If the mean weight is, for example, 10,000 lbs, then $\sigma = 5$ lbs indicates almost no variation in weight among the cylinders. Relative variation is measured by the *coefficient of variation, V,* as follows:

$$V = \frac{\text{standard deviation}}{\text{mean}} \cdot 100 \qquad (5.4.1)$$

Suppose the mean weight of the cylinders is 50 lbs, then $V = (5/50)100 = 10$ percent indicates that the variation in cylinder weights is 10 percent of the population mean weight. On the other hand, if the mean diameter is 25 cms, then $\sigma = 10$ cms means that the coefficient of variation $V = (10/25)100 = 40$

percent. Clearly, there is considerably more relative variation in diameter (40 percent) than in weight (10 percent) for these cylinders.

If the quartile deviation Q is used as the measure of variation and the median is used as the average, then relative variation may be measured by computing V_Q, the *coefficient of quartile deviation*, as follows:

$$V_Q = \frac{Q_3 - Q_1}{Q_3 + Q_1} \cdot 100 \qquad (5.4.2)$$

As in the case of measures of variation, individual quantities standing alone cannot be evaluated. Suppose a clerk using an entirely new type of card-punching machine produces 50 punched cards per hour. Is this a good rate of production or is it a poor rate? Suppose this same clerk uses a new kind of typewriter and turns out ten letters per hour. Is she a better card-puncher or a better typist?

A quantity X_i may be evaluated more easily after it is transformed to *standard units* z_i as follows:

Population $$z_i = \frac{X_i - \mu}{\sigma} \qquad (5.4.3)$$

Sample $$z_i = \frac{X_i - \bar{X}}{s} \qquad (5.4.4)$$

Quantities expressed in standard units using these equations have a mean of zero and a standard deviation of one. (See Section 5.6, Note 5C for a proof of this statement.)

Suppose a group of clerks punched a mean of 60 cards per hour (μ) using the new card-punching machine in the previous illustration, with a standard deviation (σ) of 5 cards per hour. Then the 50 cards per hour production rate (for the clerk in our illustration) may be transformed to standard units using Equation 5.4.3 as $z = (50 - 60)/5 = -2.00$. Suppose the same group of clerks typed a mean of 7 letters per hour (μ) using the new typewriter, with a standard deviation (σ) of 2 letters per hour. Then, for the clerk in our illustration, the 10 letters per hour production rate may be transformed to standard z_i units as $(10 - 7)/2 = 1.50$.

How are we to interpret and evaluate quantities which have been expressed in standard units? First, notice that z_i may be negative as well as positive, as just shown. Quantities (X_i) below the mean result in negative z_i values; whereas, quantities above the mean result in positive z_i values. An X_i quantity equal to the mean results in $z_i = 0$ (since the numerator in Equation 5.4.3 or Equation 5.4.4 will then be equal to zero).

Quantities transformed to standard z_i values are in standard deviation units. For example, the 50 cards per hour punched by the clerk in our illustration is 10 cards below the mean rate of 60 cards per hour for the group of clerks, and 10 cards amounts to 2.00 standard deviations ($\sigma = 5$ cards per hour). Hence, the corresponding z_i of -2.00 indicates that a card-punch production rate of 50 cards per hour is 2.00 standard deviations below the mean.

We are now able to evaluate the clerk's card-punching and letter production rates and compare them. Her card-punching rate in terms of standard units is

−2.00, indicating that it is 2 standard deviations below the average. On the other hand, her letter-typing rate amounts to 1.50 in terms of z, indicating that it is 1.50 standard deviations above average. We may conclude that this clerk is a better typist than a card puncher.

5.5 Optional: Chebyshev's Inequality

A useful and interesting application of the standard deviation is provided by the well-known inequality developed by the Russian mathematician Chebyshev. Let us use the following notation:

k = a constant equal to one or more

$p = \dfrac{1}{k^2}$, a proportion

$p' = 1 - \dfrac{1}{k^2}$, the complementary proportion

Chebyshev's inequality, which applies to any set of quantities (requiring only that the standard deviation is finite), states that:

1. The proportion of the quantities which are more than k standard deviations from the mean is equal to p or less.
2. The proportion of quantities which are within k standard deviations from the mean is equal to p' or more.

Suppose the mean age of single-family homes in a large city is 18 years, with a standard deviation of 3 years. If we use $k = 2$, then $p = 1/2^2 = .25$ and $p' = 1 - .25 = .75$. Then we may state that at least 75 percent of the single-family homes in the city are within 2σ's or $(2)(3) = 6$ years of the mean age (18 years). In other words, 75 percent or more of the homes are between $18 - 6 = 12$ years of age and $18 + 6 = 24$ years of age. Alternatively, we may state that 25 percent of the homes or less are either less than 12 years of age or more than 24 years of age.

Table B.2 in Appendix B presents the proportions p and p' for various values of k. This is a handy table. For example, for $k = 2$ we can immediately read from this table $p = .25$ and $p' = .75$ (as just computed). Suppose we wish to say something about the ages of the central 50 percent of the houses in our previous illustration. Referring to Table B.2, we find that for $p' = .50, k = 1.41$. Then, we can state that at least 50 percent of the single-family homes are between $\mu - 1.41\sigma$ and $\mu + 1.41\sigma$ in age. More specifically, 50 percent or more of the homes are between $18 - (1.41)(3) = 13.77$ years of age and $18 + (1.41)(3) = 22.23$ years of age.

5.6 Optional: Algebraic Notes and Shortcut Method

Note 5A: To prove

$$\sigma = \frac{1}{N} \sqrt{N \sum X_i^2 - (\sum X_i)^2} \qquad (5.2.3)$$

The definitional equation for σ^2 is

$$\sigma^2 = \frac{\sum x_i^2}{N} \qquad (5.2.1)$$

Expressing the numerator in terms of the original quantities X_i and expanding, we have

$$\sum x_i^2 = \sum (X_i - \mu)^2 = \sum (X_i^2 - 2X_i\mu + \mu^2) \qquad (5.6.1)$$

Applying summation Rule 1 to the last form of this equation, we have

$$\sum x_i^2 = \sum X_i^2 - \sum 2X_i\mu + \sum \mu^2 \qquad (5.6.2)$$

Then, noting that 2 and μ are constants, apply summation Rule 2 and Rule 3 to obtain

$$\sum x_i^2 = \sum X_i^2 - 2\mu \sum X_i + N\mu^2 \qquad (5.6.3)$$

Substituting $\mu = \sum X_i/N$, we obtain

$$\begin{aligned} \sum x_i^2 &= \sum X_i^2 - 2\frac{\sum X_i}{N} \sum X_i + N \left(\frac{\sum X_i}{N}\right)^2 \\ &= \sum X_i^2 - \frac{2(\sum X_i)^2}{N} + \frac{(\sum X_i)^2}{N} \qquad (5.6.4) \\ &= \frac{N \sum X_i^2 - (\sum X_i)^2}{N} \end{aligned}$$

Now, substitute for $\sum x_i^2$ in Equation 5.2.1 according to Equation 5.6.4 to obtain

$$\begin{aligned} \sigma^2 &= \frac{1}{N} \left[\frac{N \sum X_i^2 - (\sum X_i)^2}{N} \right] \\ &= \frac{N \sum X_i^2 - (\sum X_i)^2}{N^2} \qquad (5.6.5) \end{aligned}$$

Finally, the standard deviation σ is obtained by taking the square root of the variance σ^2 to obtain

$$\begin{aligned} \sigma &= \sqrt{\frac{N \sum X_i^2 - (\sum X_i)^2}{N^2}} \\ &= \frac{1}{N} \sqrt{N \sum X_i^2 - (\sum X_i)^2} \qquad (5.6.6) \end{aligned}$$

The proof for Equation 5.2.4 for s is carried out in a similar manner.

Note 5B: A shortcut method for computing the standard deviation for a frequency distribution uses the following equations:

Population $\qquad \sigma = \frac{g}{N} \sqrt{N \sum u_i^2 f_i - (\sum u_i f_i)^2} \qquad (5.6.7)$

$$Sample \qquad s = g \sqrt{\frac{n \sum u_i^2 f_i - (\sum u_i f_i)^2}{n(n-1)}} \qquad (5.6.8)$$

where the u_i values are computed as in Note 4B (Section 4.6) and g is the length of class interval.

Table 5.6.1 presents the frequency distribution of the inches of rainfall in an area for a sample of 25 weeks copied from Table 5.2.3. Using the data from Table 5.6.1 and applying Equation 5.6.8, we compute

$$s = 3 \sqrt{\frac{(25)(24) - (-4)^2}{(25)(24)}} = 2.96$$

This result, $s = 2.96$, is the same as previously obtained for these data. Note that Table 5.6.1 presents a convenient way to organize the computations for computing both the standard deviation and the mean by the shortcut method.

TABLE 5.6.1

Frequency distribution copied from Table 5.2.3: compu-
tation of the standard deviation by the shortcut method.

Midpoint X_i	f_i	u_i	$u_i f_i$	$u_i^2 f_i$
4	3	-2	-6	12
7	5	-1	-5	5
10	10	0	0	0
13	7	1	7	7
Total	25		-4	24

Note 5C: To prove that for a set of data expressed in standard z_i units, the mean is zero and the standard deviation is one.

Let us consider z_i for population data:

$$z_i = \frac{X_i - \mu}{\sigma} \qquad (5.4.3)$$

We may express the mean of the z_i values, μ_z, according to Equation 4.2.1 as

$$\mu_z = \frac{\sum z_i}{N} \qquad (5.6.9)$$

In this equation, substitute for z_i in accordance with Equation 5.4.3 to obtain

$$\mu_z = \frac{\sum \left(\frac{X_i - \mu}{\sigma} \right)}{N} \qquad (5.6.10)$$

Noting that μ and σ are constants, apply summation Rules 2, 1, 3, in that order, to obtain:

$$\mu_z = \frac{\frac{1}{\sigma} \sum (X_i - \mu)}{N}$$

$$= \frac{\frac{1}{\sigma}(\sum X_i - \sum \mu)}{N} \qquad (5.6.11)$$

$$= \frac{\frac{1}{\sigma}(\sum X_i - N\mu)}{N}$$

Substituting $\sum X_i/N$ for μ we have

$$\mu_z = \frac{\frac{1}{\sigma}\left(\sum X_i - N\frac{\sum X_i}{N}\right)}{N}$$

$$= \frac{\frac{1}{\sigma}(\sum X_i - \sum X_i)}{N} \qquad (5.6.12)$$

$$= \frac{\frac{1}{\sigma}(0)}{N}$$

$$= 0$$

Now, write the standard deviation of the z_i values, σ_z, according to Equation 5.2.1 as follows (taking into account that $\mu_z = 0$ as just proven):

$$\sigma_z = \sqrt{\frac{\sum (z_i - 0)^2}{N}} = \sqrt{\frac{\sum z_i^2}{N}} \qquad (5.6.13)$$

Substitute for z_i in accordance with Equation 5.4.3 to obtain

$$\sigma_z = \sqrt{\frac{\sum \left(\frac{X_i - \mu}{\sigma}\right)^2}{N}} \qquad (5.6.14)$$

Noting that σ is a constant, apply summation Rule 2 to obtain:

$$\sigma_z = \sqrt{\frac{\frac{1}{\sigma^2}\sum (X_i - \mu)^2}{N}}$$

$$= \frac{1}{\sigma}\sqrt{\frac{\sum (X_i - \mu)^2}{N}} \qquad (5.6.15)$$

Note that the term inside the square root sign is σ^2 (Equation 5.2.1). Hence, we may write

$$\sigma_z = \frac{1}{\sigma}\sqrt{\sigma^2} = 1 \qquad (5.6.16)$$

The same approach may be used to prove $\overline{X}_z = 0$ and $s_z = 1$ for sample data (Equation 5.4.4).

Study Problems

1.   What is a basic difference between an average and a measure of variation (such as the range or standard deviation)?

2.   Determine the range (interval and distance):
     a.   7, 9, 10, 5, 18, 10, 16
     b.   103, 120, 110, 118, 108, 115, 100
     c.   34, 26, 24, 28, 34, 30, 23, 25

3.   The variance for a population of 50 quantities is 196, find the standard deviation.

4.   The standard deviation for a sample of 25 quantities is 20, find the variance.

5.   The sum of the squared deviations from the mean for a population of 20 measurements is 720. Compute the (a) variance and (b) standard deviation.

6.   The sum of the squared deviations from the mean for a sample of eight quantities is 448. Compute the (a) variance and (b) standard deviation.

7.   If the variance is 38 for a population of 100 quantities, find the sum of the squared deviations from the mean for the population.

8.   If the standard deviation for a sample of 12 quantities is 20, what is the sum of the squared deviations from the mean for the sample?

9.   If a population of six quantities has a sum of squared deviations from the mean equal to 20, find the (a) mean of the squared deviations and (b) standard deviation.

10.   The variance for a population of scores is 142. On the average, how much does a score deviate (above or below) from the population mean score?

11.   A sample of light bulbs showed a mean life of 1,094 hours and a standard deviation of 8.20 hours. What would you estimate is the population mean life and standard deviation?

12.   Given the following populations, compute the standard deviation using (a) the definitional equation and (b) the computational equation:
     (1)   6, 6, 3, 5
     (2)   3, 2, 9, 12, 4
     (3)   8, 4, 7, 20, 10, 14, 11

13.   Given the following samples, compute the standard deviation using (a) the definitional equation and (b) the computational equation:
     (1)   4, 1, 2, 6, 5, 6
     (2)   18, 10, 3, 4, 9, 6
     (3)   8, 5, 4, 2, 15, 12, 10, 25, 22, 18

14.   The number of vacancies for a population of five multi-family structures are 7, 2, 3, 10, 9. Compute the mean, variance, and standard deviation.

15.   The frequency of household accidents recorded for a sample of four days are 4, 2, 6, 1. Compute the mean, variance, and standard deviation.

16.   The number of sales made for a sample of eight Tuesdays are 15, 9, 10, 12, 4, 6, 3, and 8. Determine:
     a.   Range (interval, distance)
     b.   Variance (computational equation)
     c.   Standard deviation

17.   The number of births in a hospital during a nine-day period (population) are 25, 20, 22, 15, 12, 18, 10, 8, and 6. Determine:
     a.   Range (interval, distance)
     b.   Standard deviation (computational equation)
     c.   Variance

18.   Number of trial-cases heard (population) in seven courts:

Number of trial-cases heard	Number of courts
4–5	2
6–7	3
8–9	1
10–11	1

Compute (a) mean, (b) standard deviation, and (c) variance.

19.   Given the following frequency distribution for a sample of eight scores on an attitude scale:

Score	Frequency
9–11	1
12–14	3
15–17	2
18–20	1
21–23	1

Estimate the population (a) mean, (b) variance, and (c) standard deviation.

20.   Weight gains (pounds) for a population of pigs:

Pounds	Number of pigs
2.0–2.9	1
3.0–3.9	2
4.0–4.9	5
5.0–5.9	3
6.0–6.9	2
7.0–7.9	1

Compute the (a) mean, (b) standard deviation, and (c) variance.

21.   Diameter measurements (cm) for a sample of nuts:

Cm	Frequency
.03–.05	1
.06–.08	2
.09–.11	3
.12–.14	5
.15–.17	2

Estimate the population (a) mean, (b) variance, and (c) standard deviation.

22.   If $Q_1 = 20$ and $Q_3 = 80$ for a set of examination grades, compute the (a) quartile deviation and (b) interquartile range.

23.   A racial attitude test administered to the children in an elementary school showed $D_1 = 12$, $Q_3 = 86$, $D_9 = 98$, and $Q_1 = 22$. Compute (a) semi-interquartile range, (b) interquartile range, and (c) interdecile range.

24.   If $Q_3 = 70$ and the quartile deviation is 15, find Q_1.

25.   In a set of stock market prices, the upper quartile is \$132 and the lower quartile is \$86. Compute the quartile deviation.

26.   Compute the coefficient of variation for
   a.   Problem 12 (1)
   b.   Problem 12 (2)
   c.   Problem 13 (1)
   d.   Problem 13 (2)

27.   A population of copper fittings has a mean weight of 30 lbs with a standard deviation of 1.2 lbs. These fittings have a mean diameter of 16 cms with a standard deviation of 0.8 cms.
   a.   Compute the relative variation in copper fitting weights and diameters.
   b.   Does weight have less relative variation than diameter?

28.   Compute the coefficient of quartile deviation for
   a.   Problem 22
   b.   Problem 23
   c.   Problem 25

29.   In Problem 14, transform 3 vacancies to standard units and interpret your result.

30.   In Problem 15, transform 6 household accidents to standard units and interpret the answer.

31.   In Problem 20, transform weight gains of 4.0 lbs and 7.9 lbs to standard units.

32.   In Problem 21, transform diameters of .07 cms and .16 cms to standard units.

33.   The following information was compiled for Jerry (a college student):

	Math	History	Biology
Jerry's grade	60	80	75
Class mean	40	75	70
Class standard deviation	10	5	6

In which subject did Jerry do best and why?

34.   A population of vials contains a mean quantity of 5.02 fluid drams of a drug per vial, with a standard deviation of .04 fluid dram.
   a.   What proportion of the population of vials contains within three standard deviations of 5.02 fluid drams per vial?
   b.   What proportion of the vials contains a quantity of the drug which differs from the mean amount by more than .10 fluid dram?
   c.   What proportion of the vials contain between 4.96 and 5.08 fluid drams per vial?

35.   Certain bioligical measurements have a mean of 22.5 cc (cubic centimeters) and a standard deviation of 2.6 cc and the distribution of measurements is perfectly symmetrical with a central peak.
   a.   Determine upper and lower limits (cubic centimeters) to mark off an interval around the mean which includes at least 25 percent of the measurements.
   b.   Determine the upper and lower quartiles.
   c.   Determine the measurement below which in value is 5 percent of the measurements or less.

36. A large population of grades has a mean of 74 and a standard deviation of 8. What percentage of the grades are less than 65 or more than 83?

Supplementary Study Problems

37. If each quantity in a set (sample or population) is increased (or decreased) by a constant, the standard deviation does not change.
    a. $\sigma^2 = 25$ for a population of prices (in dollars). What is the standard deviation for the population after each price is raised $10?
    b. $s = 4$ feet for a sample of lengths. What is the sample variance after each length is reduced one foot?

38. If each quantity in a set (sample or population) is multiplied (or divided) by k (a constant), the standard deviation of the new quantities is equal to the original standard deviation multiplied (divided) by k.
    a. $s^2 = 400$ for a sample of semester averages. If each semester average is divided by 10, what is the standard deviation of the new semester averages?
    b. $\sigma = 12$ minutes for a population of time measurements. What is the population variance if the measurements are stated in seconds?

39. Prove: Equation 5.2.4 for s.

40. Compute the mean, variance, and standard deviation using the shortcut method:
    a. Problem 19
    b. Problem 20
    c. Problem 21

41. Prove that for sample data, when a set of quantities are transformed to standard units (Equation 5.4.4), the mean is zero and the standard deviation is unity.

42. If σ_i represents the standard deviation for a population, then the general standard deviation for two or more populations combined may be computed as follows:

$$\sigma = \sqrt{\frac{1}{N} \sum_{i=1}^{c} N_i(\sigma_i^2 + \mu_i^2) - \mu^2}$$

where c = number of populations which are combined, N_i = each population size, μ_i = each population mean, $N = \sum N_i$ = size of the combined population, μ = general mean (Equation 4.2.3).

  Given the following data relating to quiz grades for three populations of students, compute the general standard deviation for the three populations combined:

Population	Size	Mean	Standard deviation
A	10	12	2
B	20	15	5
C	10	15	3

43. If s_i represents the standard deviation for a sample, then the general standard deviation for two or more samples combined may be computed as follows:

$$s = \sqrt{\frac{\sum_{i=1}^{c} (n_i - 1)s_i^2 + \sum_{i=1}^{c} \bar{X}_i^2 n_i - \bar{X}^2 n}{n - 1}}$$

where c = number of samples which are combined, n_i = each sample size, $\overline{X}_i$ = each sample mean, $n = \sum n_i$ = size of the combined sample, $\overline{X}$ = general mean (Equation 4.2.3).

Given the following information relating to three samples of weights (pounds), compute the general standard deviation for the three samples combined:

Sample	n_i	$\overline{X}_i$	s_i
I	5	10	3
II	10	8	2
III	5	20	5

Part III
Statistical Basis for Decision Making

Chapter 6

Probability

6.1 Probability and Events

In the previous discussion we touched upon inductive statistics at several points. It was noted that sample data, as well as population data, may be organized into a frequency distribution. We learned that the sample mean $\overline{X}$ may be used as an estimate of the population mean μ and that the sample standard deviation s may be used as an estimate of the population standard deviation σ. It is useful to learn about some of the elementary notions of probability before continuing with the study of inductive statistics. In this chapter, we will examine some of the basic concepts of elementary probability.

We are all familiar with the general notion of probability. We use expressions such as "I may get a promotion" or "there is some chance I will rent a cottage next summer." The common element underlying such statements is a feeling of uncertainty that an event will materialize, since we recognize that there are *alternative possible outcomes*, any one of which could occur. For example, in the statement "I may go to Florida next winter," the *set of alternative possibilities* may be "I will go to Florida next winter" or "I will not go to Florida next winter." On the other hand, the *possibilities set* may be "I will go to Florida this winter," "I will go to Florida next winter," or "I will not go to Florida at all." Clearly, *the possibilities set appropriate depends on the circumstances involved.*

Another important aspect of statements like those just presented is the expression or implication of a *degree of assurance* that an event will occur. "I may miss the train" expresses a small degree of assurance that I will miss the train. "I am sure to be late" expresses a strong degree of assurance that I will be late. Generally, *probability expresses the degree of assurance that one or more of a set of alternative possible outcomes will occur.*

A basic concept in probability is the *trial*. We may say that *a trial represents a circumstance which leads to the occurrence of one of the alternatives in a possibilities set*. The circumstance "considering the rental of a cottage next summer" leads to the occurrence of one of the alternative possibilities "rent a cottage next summer" or "do not rent a cottage next summer." More often, in statistics, *a trial means an experiment which can have a number of possible outcomes.* For example, a chemical experiment may have the possibilities set "explosion," "formation of a new substance," or "no reaction." The experiment "tossing a coin" has the possibilities set "head" or "tail."

The possibilities set is said to make up a *sample space*. In the trial "a die (one of a pair of dice) is tossed," the sample space is the set of possible outcomes 1, 2, 3, 4, 5, 6. Each of the possible outcomes in a sample space is called an *element* of the sample space, a *sample point*, or an *event*. *When we speak of a probability, we always mean the probability that a trial will result in the occurrence of a stated event.* For example, when a die is tossed (the trial), what is the probability of obtaining a 5 (the event)? Or if a card is selected from a deck of playing cards (the trial), what is the probability of obtaining a king (the event)? The sample space contains six elements or sample points when the trial is "tossing a die" and 52 elements when the trial is "selecting a card from a deck of 52 cards."

The sample space (or the possibilities set) relating to a trial depends on the problem at hand (or on the probability of interest). If we are interested in the probability of obtaining a 5 on the roll of a die, then the sample space may consist of the six possibilities 1, 2, 3, 4, 5, 6. On the other hand, if we are interested in the probability of obtaining an odd number on the roll of a die, then the possibilities from this point of view are odd or even. These two sample spaces are shown in Figure 6.1.1. Notice that sample space S_1 contains six elements (events) represented by six dots; whereas, sample space S_2 contains only two elements. Furthermore, each element of S_2 consists of three elements in S_1. The element odd in S_2 consists of the elements 1, 3, 5 in S_1 and the element even in S_2 contains the elements 2, 4, 6 in S_1.

Suppose a box contains three marbles, a light blue marble, a dark blue marble, and an orange marble, and we wish to select two marbles *with replacement*.

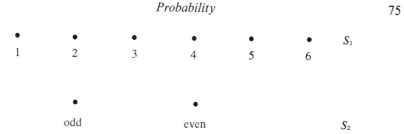

FIGURE 6.1.1

Sample spaces relating to the roll of a die.

This means select a marble, note its color, *replace* the marble to the box, and then select a second marble and note its color. Suppose we are interested in the probability that marbles of specified colors are selected and that these colors are obtained in a specified order (sequence). A specified sequence in statistics is usually called a *permutation*. Figure 6.1.2 presents the sample space S related

	Second marble		
First marble	Light blue	Dark blue	Orange
Light blue	•	•	•
Dark blue	•	•	•
Orange	•	•	•

S

	Second marble		
First marble	Light blue	Dark blue	Orange
Light blue		•	•
Dark blue	•		•
Orange	•	•	

S_1

• • •
0 1 2 S_2

Number of blue marbles selected

FIGURE 6.1.2

Sample spaces relating to the trial "selection of two marbles from a box."

to this problem. Notice that there are three possibilities for the first marble selected (light blue, dark blue, orange) and the same three possibilities for the second marble selected, so that there are 3 · 3 or 9 possible ways to select two marbles with replacement. In other words, the trial "select two marbles" has 9 possible outcomes.

These nine elements (or sample points) of the sample space S are indicated in the figure by the nine dots. The first dot in the first column represents the event (pair of marble selections) "light blue, light blue." The second dot in that column represents the event "dark blue (first), light blue (second)," etc. Notice that, for any color combination, both color permutations or orders are in the sample space. For example, the event "orange (first), light blue (second)," as well as the event "light blue (first), orange (second)," is included, since we stated that we are interested in the color *sequence* as well as the colors selected. Also, since the marbles are selected with replacement, the marbles selected could have the same color (a marble could be selected twice). Of course, when the two marbles have the same color, *permutation* (order) is not involved.

On the other hand, we may specify that selection is to be *without replacement*. That is, a marble is selected and its color noted. Then, without replacing the selected marble to the box, a second marble is selected and its color noted. Sample space S_1 in Figure 6.1.2 presents the possibilities set for this case. Notice that S_1 differs from S only in that the three sample points (events) relating to pairs of marbles with the same color are in S but not in S_1.

We may be interested in the probability of obtaining a specified number of blue marbles when selecting two marbles from the box with replacement, with no interest in the *order* of selection or which of the blue marbles is selected. In this case, the possibilities set contains the events 0, 1, and 2 blue marbles (sample space S_2 in Figure 6.1.2). How would this sample space be affected if selection were without replacement?

It should be emphasized that a sample space is made up of sample points each of which represents a possible outcome of a single trial. For example, in Figure 6.1.1, sample space S_1 represents the six possible outcomes of the trial "rolling a die" and S_2 represents the two possible outcomes for such a trial. These sample spaces differ because our interest in the outcome of the trial differs, as previously explained. In Figure 6.1.2, the trial for each sample space is "selecting two marbles." In this figure, sample spaces S and S_2 relate to the trial when selection is with replacement and sample space S_1 relates to the trial when selection is without replacement. *A sample space always relates to the possible outcomes of a single trial.*

Generally speaking, we will be concerned with the probability of events which are represented by one or more sample points in a sample space. We will denote the sample space, usually, as S (or S_1, S_2, etc.) and $P(S)$ will stand for "the probability that one of the events contained in the sample space will occur." We will often use A, B, C, etc., to denote events made up of one or more sample points in S and $P(A)$ will denote "the probability that event A will occur." Figure 6.1.3 presents the sample space for the possibilities set resulting from the toss of a pair of dice (die 1 and die 2). Event A in this sample space is made

up of two sample points and denotes the event "2 on die 1 and 1 or 2 on die 2."
Event D denotes the event "6 on die 1 and 6 on die 2." Event B is made up of
six sample points and is the event "4, 5, or 6 on die 1 and 2 or 3 on die 2," etc.
We will use N or N_S to denote the number of sample points in S, N_A to denote
the number of sample points in A, etc. Then, in Figure 6.1.3, $N = 36$, $N_A = 2$,
$N_B = 6$, etc.

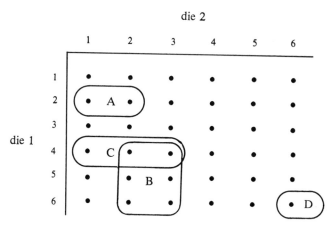

FIGURE 6.1.3

Sample space relating to the toss of a pair of dice.

A *complementary event* A' is the event made up of all sample points in S which
are not in A. The event A' means "outcomes other than A" or more briefly,
"not A." Therefore, in Figure 6.1.3, $N_A = 2$, $N_{A'} = N_S - N_A = 36 - 2 = 34$.
Notice that in the figure, events C and B *overlap* since there are two sample
points which appear in both C and B. These two sample points may be called
the *overlap event* and represent the occurrence of 4 on die 1 and 2 or 3 on die 2.
If this overlap event occurs, then both events C and B occur.

   We will write $P(C \text{ or } B)$ to denote "the probability that C or B or both will
occur." The event $(C \text{ or } B)$ contains all sample points in C and B. In the figure,
the number of sample points in the event $(C \text{ or } B)$, denoted as $N_{(C \text{ or } B)}$, is 3
(in C) + 6 (in B) − 2 (in the overlap) = 7. We subtract the sample points in
the overlap to avoid *double counting*.

   Events which do not overlap (such as events A and D in the figure) have no
sample points in common. Such events are called *mutually exclusive events*.
Since mutually exclusive events have no sample points in common, one and
only one of these events could occur in a single trial. In our example, if A occurs
then D cannot occur in a single toss of a pair of dice. Therefore, we may state
generally that *two or more events are mutually exclusive if the occurrence of one
precludes the occurrence of any of the others*. The event $(A \text{ or } D)$ contains three
sample points (two from A and one from D).

   We will write $P(C \text{ and } B)$ or $P(C, B)$ to denote "the probability that both C
and B will occur." Referring to Figure 6.1.3, notice that both of these events

can occur only if one of the events in the overlap occurs. Hence, the event (C and B) consists only of the sample points in the overlap. An event containing two or more specified events, such as (C and B), is called a *joint event*. The event (A and B) contains no sample points, since A and B do not overlap. *An event which contains no sample points cannot occur and is called a null event.* Using our new symbols, we may write $N_{(C \text{ or } B)} = N_C + N_B - N_{(C \text{ and } B)}$. Then, $N_{(C \text{ or } B)} = 3 + 6 - 2 = 7$.

An event represented by a single sample point is called a *simple event*. Sometimes it is useful to look upon an event as made up of two or more events in different sample spaces. For example, in Figure 6.1.3 event D is the simple event "obtaining 6 on die 1 and 6 on die 2"; however, we may look upon the trial "tossing a pair of dice" as made up of two trials, "tossing die 1" and "tossing die 2." Sample space S_1 in Figure 6.1.1 is the appropriate sample space for each of these trials. Then, from this point of view, event D consists of the two simple events "obtaining a 6 on die 1" from one sample space and "obtaining a 6 on die 2" from the other sample space. Looking at it from this point of view, event D is a joint event made up of two simple events from two sample spaces.

Summary: In this section, we introduced the concept of a trial and indicated that, generally, it means an experiment. The alternative possible outcomes of a trial we called a possibilities set which makes up a sample space (S). A sample point in S or a combination of sample points represents an event. We considered simple and joint events, overlapping and mutually exclusive events, and complementary events. We showed that more than one sample space may be constructed for a trial. The problem at hand and the probability of interest will indicate the appropriate sample space to consider. We introduced the notions of selection with replacement and without replacement. We defined probability in a general way by stating that probability expresses the degree of assurance that a trial will result in the occurrence of a specified event. In the balance of this chapter we will consider how probabilities are determined.

6.2   Definitions

How do we determine the probability that an event will occur? *Mathematically, probability represents a proportion or relative frequency which indicates the degree of assurance that an event will occur.* Suppose a man misses the morning bus 3 percent of the time while going to work, based on records for several years. Then, on the average, we may expect that this man will miss the morning bus 3 times out of 100, if the past experience continues to hold in the future. We estimate that the probability is .03 that he will miss the morning bus on a working day (or the chances are 3 in 100 that he will miss the bus).

Suppose an employment agency finds that 35 percent of the job applicants are college graduates, based on records for a number of years. We may expect, then, that 35 out of 100 job applicants who contact this agency, on the average, will be college graduates, if the relationship of the past continues to hold in the future. Hence, we estimate that the probability is .35 (or the chances are 35 in 100) that a job applicant who applies at this agency is a college graduate. We may express this definition of probability more formally as follows:

Probability (Definition 1): If an event A occurs X times in n trials, the probability that A will occur in a given trial is

$$P(A) = \text{the mathematical limit of } X/n \text{ as } n$$
$$\text{becomes infinitely large}$$

Probability Definition 1 may be explained by reference to Table 6.2.1 and Figure 6.2.1. Notice from the table that the proportion of 5's obtained varies less and less as the number of tosses of a die (the number of trials) increases and that this proportion tends to fluctuate around or approach some particular value. This is more noticeable in the figure. This value is called the *mathematical limit*. Generally speaking, the relative frequency X/n may be expected to approach some limit as the number of trials increases indefinitely. This limit is the probability that event A will occur. More simply, Definition 1 says that $P(A)$ is the proportion of times that event A occurs in the long run. The relative frequency for an infinitely large number of trials (the mathematical limit) is a theoretical concept and can be determined only theoretically.

TABLE 6.2.1

Hypothetical experiment: relative frequencies of 5's obtained when a die is tossed a number of times.

Number of trials, n	Number of 5's obtained, X	$\dfrac{X}{n}$
10	4	.40
20	5	.25
30	7	.23
40	7	.18
50	7	.14
60	11	.18
70	11	.16
80	14	.18
90	14	.16
100	17	.17

On the other hand, the theoretical relative frequency may be *estimated* experimentally. If 50 tosses of a coin (trials) result in 23 heads, we estimate $P(\text{head})$ as $23/50 = .46$. A better estimate is obtained by basing the estimate on a larger number of trials. If an additional 150 tosses results in 79 heads, then using the $50 + 150 = 200$ tosses and the $23 + 79 = 102$ heads obtained, we estimate $P(\text{head}) = 102/200 = .51$. Alternatively, the probability may be hypothesized. For example, if it is assumed that the coin is a fair coin, so that

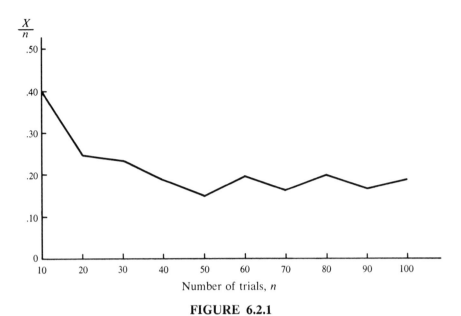

FIGURE 6.2.1

*Hypothetical experiment: relative frequencies of 5's ob-
tained when a die is tossed a number of times (Table 6.2.1).*

heads and tails are equally likely to occur, then it could be theorized that
$P(\text{head}) = .50$.

If it is impossible for event A to occur, then $P(A) = 0$. On the other hand,
if it is certain that event A will occur, then $P(A) = 1$. It should be noted that
we are concerned with the probability of occurrence of an event before the
trial has taken place or, if the trial has already taken place, then before we
know the outcome. If a die is tossed and we know that a 3 was obtained then,
for that trial, $P(3)$ is 1 (certainty) and $P(\text{any other outcome})$ is 0 (impossible).

Actually, determination of probabilities is a matter of controversy among
statisticians. The majority of statisticians today favor *objective probabilities*.
These are probabilities determined in some objective way, such as by the use
of Definition 1. Some statisticians prefer to determine probabilities on a *subjective basis*, since they believe that personal judgments should form the basis
for determining probabilities. Such probabilities are called *personalistic* or
subjective probabilities.

The *objectivists* take the view that trials are repeatable and estimate probabilities based on the notion of *repeated trials*, which is the basis of Definition 1.
The *subjectivists* are not wedded to the concept of repeated trials since, as they
claim, this is not always a possibility. For example, what is the probability that
a horse will come in first in a race? The objectivist would think in terms of
repeated situations (trials) of an identical nature; however, he would find
similar situations from the past and compute the proportion of times the horse

came in first. Using "similar" situations is a compromise with reality, since identical situations are not to be expected.

The subjectivists would point to the uniqueness of the situation and the unreasonableness of conceiving identical situations. For example, horses differ in a great variety of ways, so that identical or even "similar" horses are not to be found. Various aspects of a race such as the track, weather conditions, and the condition and age of the horses are clearly different from race to race. Subjectivists prefer to depend on personal evaluation to develop the required probability.

Nevertheless, there is no controversy as to the *general conditions which must be satisfied* by the probabilities assigned to the sample points (events) in a sample space. These conditions are specified in a set of three *postulates*.

Postulate 1: Let $P(a_i)$ denote the probability assigned to sample point a_i in a sample space. Then

$$0 \leq P(a_i) \leq 1$$

Postulate 2: For a sample space S

$$P(S) = 1$$

Postulate 3: Let A, B, C, etc., represent mutually exclusive events in a sample space. Then

$$P(A \text{ or } B \text{ or } C \text{ or} \ldots) = P(A) + P(B) + P(C) + \ldots$$

Postulate 1 requires that the probability assigned to a sample point is not less than zero and not greater than one. As previously noted, $P(a_i) = 0$ means that it is impossible for the event a_i to occur and $P(a_i) = 1$ means that it is a certainty that a_i will occur. Postulate 2 states that it is a certainty that a trial will lead to the occurrence of one of the events in the possibilities set. Postulate 3 states that the probability that one or the other of a set of mutually exclusive events will occur is equal to the sum of the probabilities of the mutually exclusive events. For example, say on the toss of a die $P(1) = 1/6$, $P(2) = 1/6$, $P(3) = 1/6$. Then since the events 1, 2, 3 are mutually exclusive when a die is tossed once, $P(1 \text{ or } 2 \text{ or } 3) = 1/6 + 1/6 + 1/6 = 3/6$ or .50, according to Postulate 3.

The set of probabilities associated with the sample points in a sample space is called a *probability distribution*.

If A is an event in a sample space S and A' is the complementary event, then A and A' are mutually exclusive and the event $(A \text{ or } A') = S$. Based on Postulate 2, we may write

$$P(S) = P(A \text{ or } A') = 1$$

and based on Postulate 3 we have

$$P(A \text{ or } A') = P(A) + P(A') = 1 \qquad (6.2.1)$$

Let A = obtaining a 4 on the toss of a die. Then, A' = obtaining an outcome other than 4. If $P(A) = P(4) = 1/6$, we have

$$P(A) + P(A') = 1$$

$$\frac{1}{6} + P(A') = 1$$

and
$$P(A') = 1 - \frac{1}{6} = \frac{5}{6}$$

Suppose an event A is made up of the X sample points $a_1, a_2, \ldots, a_X$, we may write

$$P(A) = \sum_{i=1}^{X} P(a_i) \tag{6.2.2}$$

This equation holds based on Postulate 3 and recognizing that the sample points (a_i) in a sample space are mutually exclusive. If the N sample points in a sample space represent *equi-probable events* (events with equal probability) and recognizing that $P(S) = \sum_{i=1}^{N} P(a_i) = 1$ (according to Postulates 2 and 3), we may write

$$P(a_i) = \frac{1}{N} \tag{6.2.3}$$

If $P(a_i)$ is the same for each of the N events, each $P(a_i)$ must equal $1/N$ so that the sum of the $P(a_i)$ is equal to $P(S) = 1$, in accordance with Postulate 2.

Substituting for $P(a_i)$ from Equation 6.2.3 into Equation 6.2.2 and applying summation Rule 3 (Section 2.4), we have

$$P(A) = \sum_{i=1}^{X} \frac{1}{N} = \frac{X}{N} \tag{6.2.4}$$

This is a very useful result. We will call it Definition 2 for the probability that an event A will occur. Suppose a bag contains three red marbles and two white marbles. The bag is shaken and, with your eyes closed, you select a marble. What is the probability that you select a red marble? The related sample space consists of $N = 5$ sample points (one for each marble in the bag). These five events are equi-probable and the event A "selecting a red marble" consists of three of these sample points (three marbles are red), so that we may write $X = 3$. Then, applying Definition 2 (Equation 6.2.4) we have

$$P(A) = P(\text{red}) = \frac{3}{5} \quad \text{or} \quad .60$$

We may write Definition 2 in a more general way as follows:

Probability (Definition 2): Given a trial with a set of N alternative possible outcomes which are mutually exclusive and equi-probable. If X of these possible outcomes fall in the category A, then

$$P(A) = \frac{X}{N} = \frac{\text{Number of possible outcomes in category } A}{\text{Total number of alternative possible outcomes}}$$

What is the probability of selecting a queen from a well-shuffled deck of 52 playing cards, with your eyes closed? Since there are 4 queens, we apply Definition 2 with $X = 4$ and $N = 52$ and obtain $P(\text{queen}) = 4/52 = 1/13$.

Notice that Definition 1 and Definition 2 both express probability as a proportion, but in different ways. Definition 2 expresses $P(A)$ as the proportion of a set of mutually exclusive events in a possibility set which falls in the category A. Definition 1 expresses $P(A)$ as the proportion of times the event A occurs in a series of trials. It is not always possible to determine probabilities by the use of Definition 2. Then, Definition 1 must be used. Suppose an urn contains orange marbles and yellow marbles, what is P (orange marble) when a marble is selected with your eyes closed and after shaking the urn? If we know the proportion of orange marbles in the urn, we could use Definition 2. If not, then Definition 1 must be used. In order to apply Definition 1 we must experiment by selecting marbles from the urn. Suppose 25 selections of a marble from the urn, with replacement, results in an orange marble 10 times. Using Definition 1 we estimate P (orange marble) $= 10/25 = .40$.

Definition 1 uses n to denote the number of trials, since any finite number of trials represents a sample of the unlimited number of trials when n becomes infinite. On the other hand, Definition 2 uses N to denote the number of possible outcomes in the possibilities set, since this represents the population of all possible outcomes of a trial.

Another way of expressing the probability for an event A is to speak in terms of the *odds that the event will occur*. Odds are determined as follows: determine $P(A)$ and $P(A')$ for event A. $P(A') = 1 - P(A)$ according to Equation 6.2.1. Say $P(A) = .25$ and $P(A') = .75$. Then, $P(A)/P(A') = .25/.75 = 1/3$, so that the odds are 1 to 3 that event A will occur; or, 3 to 1 that event A will not occur.

6.3 Probability Rules

Let us consider some useful rules for computing probabilities.

Rule A: General addition rule. Let A and B represent two events. Then,

$$P(A \text{ or } B) = P(A) + P(B) - P(A \text{ and } B)$$

We may derive this probability rule quite easily. We know from our discussion in Section 6.1 that for a sample space made up of N sample points

$$N_{(A \text{ or } B)} = N_A + N_B - N_{(A \text{ and } B)} \tag{6.3.1}$$

Then, according to probability Definition 2

$$P(A \text{ or } B) = \frac{N_{(A \text{ or } B)}}{N} = \frac{N_A}{N} + \frac{N_B}{N} - \frac{N_{(A \text{ and } B)}}{N}$$
$$= P(A) + P(B) - P(A \text{ and } B) \tag{6.3.2}$$

In a certain city, suppose the records show that 30 percent of new home purchases are made by newly married couples, 60 percent are made by professional men, and 10 percent are made by professional men who are newly-married. What is the probability that a house in that city will be purchased by a newly-married couple or a professional man?

Let: A = house to be purchased by a newly-married couple
$\quad\;\;$ B = house to be purchased by a professional man

Then, from the information given and applying probability Definition 1, we have $P(A) = .30$, $P(B) = .60$, and $P(A \text{ and } B) = .10$. Applying Rule A, we obtain

$$P(A \text{ or } B) = .30 + .60 - .10 = .80$$

If events A and B are mutually exclusive (do not overlap), then $(A \text{ and } B)$ is the *null event* (contains no sample points) and $P(A \text{ and } B) = 0$. This leads to the following probability rule for mutually exclusive events:

Rule B: Special addition rule. Let A and B represent mutually exclusive events. Then,

$$P(A \text{ or } B) = P(A) + P(B)$$

Rule B may be extended to any number of mutually exclusive events, as follows:

Rule B extended: Let A_1, A_2, etc., represent mutually exclusive events. Then,

$$P(A_1 \text{ or } A_2 \text{ or } \ldots) = P(A_1) + P(A_2) + \ldots$$

Clearly, Rule B and Rule B extended are restatements of Postulate 3.

Suppose it is known that when a certain type of seed is planted, 40 percent of the flowers grown are tall, 10 percent are tall and pink, and 30 percent are pink. (Assume that each seed planted results in the growth of a flower.) Based on the information given and using Definition 1 we have: $P(\text{tall}) = .40$, $P(\text{tall and pink}) = .10$, and $P(\text{pink}) = .30$. Table 6.3.1 presents the appropriate sample space and probability distribution. (Note: The sample space in Table 6.3.1 is similar to the sample spaces shown in Figure 6.1.2. However, the probabilities associated with the sample points are shown instead of dots.) This sample space involves *two variables*, height and color.

TABLE 6.3.1

	Color		
Height	Pink	Other	Total
Tall	.10	.30	.40
Not tall	.20	.40	.60
Total	.30	.70	1.00

The three known probabilities are shown in the table: $P(\text{tall and pink}) = .10$ is shown for the event (sample point) "tall and pink," $P(\text{tall}) = .40$ is shown in the total column for "tall" since it is the probability of obtaining a tall flower (pink or other color), $P(\text{pink}) = .30$ is shown in the total row for "pink" since it is the probability of obtaining a pink flower (tall or not tall). The probability 1.00 is shown as the sum of the column probabilities as well as the sum of the row probabilities, since it represents $P(S)$ which equals 1.00 according to Postulate 2. The other probabilities are computed by subtracting as appropriate based on Postulate 3 or Equation 6.2.1. For example:

$$P(\text{tall and pink}) + P(\text{tall and other color}) = P(\text{tall})$$
$$.10 + P(\text{tall and other color}) = .40$$

and so $\quad P(\text{tall and other color}) = .40 - .10 = .30$

Also $\quad\quad P(\text{tall}) + P(\text{not tall}) = P(S) = 1.00$
$$.40 + P(\text{not tall}) = 1.00$$

and so $\quad P(\text{not tall}) = 1.00 - .40 = .60$

Notice that the probabilities associated with the four sample points (.10, .30, .20, .40) which define the probability distribution of "height and color of flower grown" add to 1.00, as required by Postulate 2. The probabilities in the total column and in the total row are called *marginal probabilities*. (The probabilities in the total column, .40 and .60, represent the probability distribution for the column variable, "height." The probabilities in the total row, .30 and .70, represent the probability distribution for the row variable, "color.") Based on the probability distribution in Table 6.3.1 we may readily determine various probabilities relating to the flowers grown from the seeds. For example, $P(\text{tall}$ and other color) may be read from the probability distribution as .30. Applying probability Rule A and using the appropriate probabilities from the probability distribution, we may compute

$$P(\text{not tall or pink}) = P(\text{not tall}) + P(\text{pink}) - P(\text{not tall and pink})$$
$$= .60 + .30 - .20 = .70$$

Suppose you are told that a flower grown from a seed is pink, what is the probability that it is tall? Referring to the probability distribution in Table 6.3.1, is this probability $P(\text{tall}) = .40$? Absolutely not! $P(\text{tall})$ is the probability that a flower is tall, *without any regard as to its color*. It is the *unconditional probability* that a flower is tall. The probability that we want is the *conditional probability that a flower is tall, the condition being that it is pink*. The events relating to "other color" cannot occur. The part of the sample space in Table 6.3.1 which relates to our question contains only two sample points: "tall and pink" and "not tall and pink," so we are concerned only with the *reduced sample space* and the associated probabilities.

The probabilities in the *reduced sample space* are $P(\text{tall and pink}) = .10$ and $P(\text{not tall and pink}) = .20$ and the sum of these is .30 ($= P(\text{pink})$). The probability distribution of a sample space must, however, satisfy the three postulates and Postulate 2 requires that the probabilities of a sample space, even for a reduced sample space, must add to 1.00. If each probability in the reduced sample space is divided by the sum of the probabilities (.30), the resulting probabilities will add to 1.00, as shown in Table 6.3.2. Then, the probability distribution (in Table 6.3.2) for the reduced sample space shows that the probability that a pink flower is tall is .33. This is the probability we want. It represents "the *conditional probability* that a flower is tall, *given that it is pink*." We denote this as $P(\text{tall}|\text{pink})$. We may express the computation of this conditional probability as follows:

$$P(\text{tall}|\text{pink}) = \frac{P(\text{tall and pink})}{P(\text{pink})} = \frac{.10}{.30} = .33$$

TABLE 6.3.2

Reduced sample space and probability distribution.

Height	Pink	
Tall	.10/.30 =	.33
Not tall	.20/.30 =	.67
Total		1.00

In general, we may express $P(B|A)$, "the conditional probability that event B will occur, given that event A has occurred," as

$$P(B|A) = \frac{P(A \text{ and } B)}{P(A)} \qquad (6.3.3)$$

Solving this equation for $P(A \text{ and } B)$ we obtain

$$P(A \text{ and } B) = P(A) \cdot P(B|A) \qquad (6.3.4)$$

This is an important result which we will call probability Rule C. We state this rule more generally, as follows:

Rule C: General multiplication rule. Let A and B represent two events. Then,

$$P(A \text{ and } B) = P(A) \cdot P(B|A)$$

The *joint probability*, $P(A \text{ and } B)$, as expressed in Rule C is the probability that A occurs first and B second. The joint probability $P(B \text{ and } A)$ relating to the occurrence of the *reverse permutation* (reverse order) of A and B is as follows:

$$P(B \text{ and } A) = P(B) \cdot P(A|B) \qquad (6.3.5)$$

What is the probability that first a king and then a queen are obtained when two cards are selected without replacement (with your eyes closed) from a well-shuffled deck of 52 playing cards? Applying probability Rule C, we obtain

$$P(\text{king and queen}) = P(\text{king}) \cdot P(\text{queen}|\text{king})$$
$$= \frac{4}{52} \cdot \frac{4}{51} = .006$$

$P(\text{king}) = 4/52$ is determined according to Definition 2, taking into account that the deck of 52 cards contains 4 kings. $P(\text{queen}|\text{king}) = 4/51$ is also determined according to Definition 2, but taking into account the condition that a king has already been selected. Hence, there are only 51 cards in the deck including 4 queens. The probability that a king and queen are selected in the reverse permutation is computed in a similar way and results in the same probability, $P(\text{queen and king}) = .006$.

How do we compute the probability that the two cards selected one at a time without replacement will be a king and a queen, with no stipulation as to the order of selection? Or, if the cards are scattered in a box and two cards are selected simultaneously with your eyes closed, what is the probability of obtaining a king and a queen? In either case, probability Rule C must be applied to each permutation separately and the probabilities added according to probability Rule B since they relate to mutually exclusive events. That is, the joint events (king and queen) and (queen and king) are mutually exclusive. Then the probability that a king and a queen are selected with no regard to order of selection is

$$P(\text{king and queen}) = P(\text{king first and queen second})$$
$$+ P(\text{queen first and king second})$$
$$= .006 + .006 = .012$$

It is instructive to construct the sample space and probability distribution pertinent to the foregoing problem. This is presented in Table 6.3.3. Notice that each selection has three possibilities so that the sample space contains $3 \cdot 3$ or nine sample points. The probability associated with each sample point is determined according to probability Rule C. For example,

$$P(\text{king and other card}) = P(\text{king}) \cdot P(\text{other card}|\text{king})$$
$$= \frac{4}{52} \cdot \frac{44}{51} = .0664$$

Notice that the probability distribution presents the probability for each permutation. (Note that, even though the event corresponding to a sample point is a simple event, we treat it as a joint event for the purpose of computing the probability.)

TABLE 6.3.3

Sample space and probability distribution for the trial "selecting two cards without replacement from a deck of 52 playing cards" with an interest in the selection of a king and a queen.

First card	Second card			Total
	King	Queen	Other card	
King	.0045	.0060	.0664	.0769
Queen	.0060	.0045	.0664	.0769
Other card	.0664	.0664	.7134	.8462
Total	.0769	.0769	.8462	1.000

Rule C may be extended to any number of events. For example, for a joint event made up of three events we may write:

Rule C extended: Let A_1, A_2, A_3 represent three events. Then,

$$P(A_1, A_2, \text{ and } A_3) = P(A_1) \cdot P(A_2|A_1) \cdot P(A_3|A_1 \text{ and } A_2)$$

The conditional probability in Rule C extended $P(A_3|A_1 \text{ and } A_2)$ denotes "the probability that event A_3 will occur, given that events A_1 and A_2 have occurred."

Rule C extended is to be used to compute the probability that a joint event $(A_1, A_2,$ and $A_3)$ occurs in a *specified permutation*. Suppose an urn contains five white marbles, six blue marbles, seven green marbles, and two orange marbles, making a total of 20 marbles. If the urn is well shaken, what is the probability that three marbles selected (with your eyes closed) will be orange, green, and blue, in that order, if selection is without replacement? Applying Rule C extended, we compute

$$P(\text{orange, green, blue}) = P(\text{orange}) \cdot P(\text{green}|\text{orange})$$
$$\cdot P(\text{blue}|\text{orange and green})$$
$$= \frac{2}{20} \cdot \frac{7}{19} \cdot \frac{6}{18} = \frac{84}{6{,}840} = .012$$

If the three marbles are selected simultaneously or if they are selected one at a time and there is no interest in the sequence of the selection, then the probability that the joint event (orange, green, blue) occurs must be computed for each possible permutation of the three colors and the probabilities added.

The number of possible permutations may sometimes be determined most easily by listing them. For example, for three colors we have six permutations:

orange, green, blue	green, blue, orange
orange, blue, green	blue, orange, green
green, orange, blue	blue, green, orange

Or we may use the equation

$$_NP_X = \frac{N!}{(N - X)!} \qquad\qquad (6.3.6)$$

where $_NP_X$ denotes "the number of permutations possible for N different objects taken X at a time." In the foregoing example, we want the number of permutations of $N = 3$ colors taken $X = 3$ at a time. The symbol $N!$ is read "N factorial" and denotes the product $N(N - 1)(N - 2) \dots (1)$. For example, $3! = 3 \cdot 2 \cdot 1 = 6$, $4! = 4 \cdot 3 \cdot 2 \cdot 1 = 24$, etc. By convention, $0! = 1$ and $1! = 1$. Then, for $N = 3$ and $X = 3$, we compute $N! = 3! = 6$ and $(N - X)! = (3 - 3)! = 0! = 1$. Applying Equation 6.3.6, we compute $_NP_X = {}_3P_3 = 3!/0! = 6$. This is the same number of permutations shown in the list.

Another way to determine the number of possible permutations is to construct a *tree diagram* as illustrated in Figure 6.3.1 for the three marbles selected. Starting from an origin point, arrows (branches) are drawn to show the possible choices for each selection. The arrows for the first marble to be selected point to each of the three colors. Each of these arrows is followed by two arrows, since one of the two remaining colors must be chosen for the second marble. Then, each of these "second marble arrows" is followed by a single arrow, since there is only one color left for the third selection. Following the route indicated by a set of connected arrows identifies a possible permutation of the three colors.

If Rule C extended is used to compute the probability of the joint event (orange, green, blue), it will be found to be the same (.012) for each permutation. Hence, the probability that this event occurs, when there is no interest in

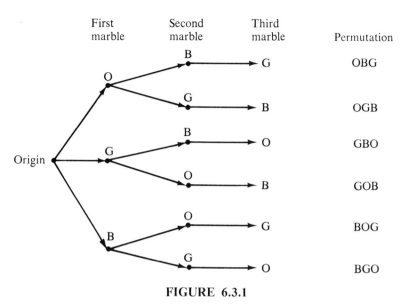

First marble   Second marble   Third marble   Permutation

B → G   OBG

O

G → B   OGB

B → O   GBO

G

Origin

O → B   GOB

O → G   BOG

B

G → O   BGO

FIGURE 6.3.1

Tree diagram for selection of three marbles (O = orange,
B = blue, G = green).

order of occurrence, is most easily determined as $(.012) \cdot {}_3P_3 = (.012)(6) = .072$.

Two events are independent if the occurrence of one has no effect on the probability of occurrence of the other. Given a bag with red marbles and green marbles, if two marbles are selected with replacement, the probability of occurrence of the event "second is red" is not affected by the first selection. Hence, events such as "first marble green" and "second marble red" are independent if selection is with replacement. If two coins are tossed, the events "head on the first coin" and "head on the second" are independent. Events which are not independent are called *dependent events.*

Rule D: Let A and B represent independent events. Then,

$$P(A|B) = P(A)$$

and
$$P(B|A) = P(B)$$

Rule D indicates that, for independent events, the conditional probability is equal to the unconditional probability. Clearly, this must be so since *the condition is irrelevant if the events are independent.* For example, if a coin is tossed twice, $P(\text{head second}|\text{head first}) = P(\text{head second})$. The condition that the first toss was a head is irrelevant since the outcomes of the two tosses are independent of each other.

Taking Rule D into consideration, we may restate Rule C and Rule C extended as follows for independent events:

Rule E: Special multiplication rule. Let A and B represent independent events. Then,

$$P(A \text{ and } B) = P(A) \cdot P(B)$$

Rule E extended: Let $A_1, A_2, \ldots$ represent independent events. Then,

$$P(A_1, A_2, \ldots) = P(A_1) \cdot P(A_2) \cdot \ldots$$

If Tom hits the bull's eye 70 times in 100; Ed, 80 times in 100; and Joe, 40 times in 100, what is the probability that only Joe hits the bull's eye if they each shoot once at a target? Based on the given information and applying Definition 1 and Equation 6.2.1 for complimentary events, we have

P(Tom hits) = .70	P(Tom no hit) = .30
P(Ed hits) = .80	P(Ed no hit) = .20
P(Joe hits) = .40	

Then, applying Rule E extended, since we are dealing with independent events, we compute

$$P(\text{Tom no hit, Ed no hit, Joe hits}) = P(\text{Tom no hit}) \cdot P(\text{Ed no hit}) \cdot P(\text{Joe hits})$$
$$= (.30)(.20)(.40) = .024$$

Note that where permutation is involved, Rule E and Rule E extended are to be used to compute the probability related to a *specified permutation*, as in the case of Rule C and Rule C extended. In the foregoing illustration (hitting the bull's eye), of course, permutation was not involved.

6.4   How to Solve Probability Problems

In this section an attempt is made to synthesize our discussion of probability, probability concepts, and events by solving selected problems.

Example 1: Suppose the dean of graduate students in a university classifies applicants for study grants into five categories: A (social science majors), B (physical science majors), C (engineering majors), D (fine arts majors), and E (other majors). We are interested in determining the probability that an applicant is a major in one of these categories. Table 6.4.1 presents the probability distribution defined on the appropriate sample space. Notice that, instead of showing dots associated with each sample point (A, B, . . .), we show the associated probabilities. The probabilities assigned to each sample point could have been computed according to Definition 1 and must satisfy the requirements of the three postulates of probability. For example, the proportion of applicants for the last year or two falling in each category could be used as estimates of the probabilities. Based on the probability distribution, various probabilities can be computed using the probability rules. For example, event (C or D) represents an applicant who is an engineering major or a fine arts major and the probability for this event may be computed by the use of Rule B (for mutually exclusive events) as follows:

$$P(C \text{ or } D) = P(C) + P(D)$$
$$= .10 + .20 = .30$$

TABLE 6.4.1

Major category (sample space)	P(category)
A	.20
B	.40
C	.10
D	.20
E	.10
Total	1.00

Example 2: In Example 1, suppose two applicants who do not know each other apply for study grants six months apart. What is the probability that one is a social science major (we do not care which applicant) and the other is a fine arts major? That is, determine $P(A$ and $D)$, with no interest in any particular permutation. Notice that in the situation as described A and D are independent events and that two permutations are possible. Applying probability Rule E and multiplying the result by two (two permutations), we obtain

$$P(A \text{ and } D) = 2 \cdot P(A) \cdot P(D)$$
$$= (2)(.20)(.20) = .08$$

Observe that it is not necessary to construct the sample space and probability distribution to determine this probability. True, we obtained $P(A)$ and $P(D)$ from the probability distribution in Table 6.4.1; however, often these probabilities are available without constructing the probability distribution. This is illustrated in Examples 3, 4, and 5.

As an instructive exercise, however, let us construct the sample space and probability distribution relating to the major areas of the two applicants. This is presented in Table 6.4.2. Since there are five possibilities for each applicant, the sample space contains 5 · 5 or 25 sample points. Each sample point may be considered to represent a *joint event*, such as (A for the first applicant and B for the second applicant). As previously noted, the two simple events making up each joint event are independent and Rule E may be used to compute the associated probability, as illustrated for the event (A and D). Each sample point represents a *particular permutation*, so that there is no need to multiply by two.

TABLE 6.4.2

Major area of first applicant	Major area of second applicant					Total
	A	B	C	D	E	
A	.04	.08	.02	.04	.02	.20
B	.08	.16	.04	.08	.04	.40
C	.02	.04	.01	.02	.01	.10
D	.04	.08	.02	.04	.02	.20
E	.02	.04	.01	.02	.01	.10
Total	.20	.40	.10	.20	.10	1.00

Since each joint event is represented by a *single* sample point in the Table 6.4.2 sample space, it is actually a simple event in that sample space; however, this is a technicality which we may ignore. It is mentioned merely to alert the reader to a seeming inconsistency and avoid confusion. Attention is invited to the *marginal probabilities*. The marginal probabilities in the total column make up the *probability distribution* for the "major area of first applicant." The marginal probabilities in the total row make up the *probability distribution* for the "major area of second applicant."

Example 3: It is known that 20 percent of the married male voters expect to vote for Jones and 40 percent of the married female voters expect to vote as their husbands do. What is the probability that George, a randomly selected voter, and his wife expect to vote for Jones?

Let A = George expects to vote for Jones and B = wife expects to vote for Jones. Then, based on the given information, we may write: $P(A) = .20$ and $P(B|A) = .40$. Applying Rule C,

$$P(A \text{ and } B) = (.20)(.40) = .08$$

Notice that there is no need to construct the sample space or probability distribution.

Example 4: Five small radios are packed in identical, unmarked, individual, sealed boxes. Three boxes are on table I and contain two radios made by firm A and one by firm B. Two boxes are on table II and contain one radio made by firm A and one by firm B. If someone moves a box from table I to table II and you arbitrarily select a box from table II, what is the probability that you select a radio made by firm B?

Most problems in probability, such as this one, require careful consideration of the possibilities involved and some ingenuity in computing the required probability. The probability that a firm B radio is selected depends on which radio was moved from table I to table II.

Let A = firm A radio was moved from table I to table II
$\quad B$ = firm B radio was moved from table I to table II
$\quad C$ = firm B radio was selected from table II

Then, if event A occurred, we may compute the joint probability

$$P(A \text{ and } C) = P(A) \cdot P(C|A)$$
$$= \frac{2}{3} \cdot \frac{1}{3} = \frac{2}{9}$$

If event B occurred, then

$$P(B \text{ and } C) = P(B) \cdot P(C|B)$$
$$= \frac{1}{3} \cdot \frac{2}{3} = \frac{2}{9}$$

Finally, recognizing that the joint events (A and C) and (B and C) are *mutually exclusive*, we compute the required probability $P(C)$ using Rule B as

$$P(C) = P(A \text{ and } C) + P(B \text{ and } C)$$
$$= \frac{2}{9} + \frac{2}{9} = \frac{4}{9} \quad \text{or} \quad .44$$

Again, notice that it was not necessary to define the probability distribution or construct the sample space.

Example 5: An artist owns a townhouse in midtown and a villa in a foreign country. The probability that a fire will break out in the townhouse is .10 and for the villa it is .05. What is the probability that a fire will break out in:

(a) At least one of the two homes
(b) Only one of the two homes
(c) Both homes
(d) Neither one

Let T = fire in the townhouse
V = fire in the villa

This problem indicates the need for careful consideration of the events involved. The following probabilities are given: $P(T) = .10$ and $P(V) = .05$.

Part (a): We need to determine $P(T \text{ or } V)$, the probability that T, V, or both will occur. Since T and V are not mutually exclusive, apply Rule A as follows:

$$P(T \text{ or } V) = P(T) + P(V) - P(T \text{ and } V)$$
$$= .10 + .05 - P(T \text{ and } V)$$

How do we determine $P(T \text{ and } V)$? Since T and V are independent events, apply Rule E:

$$P(T \text{ and } V) = P(T) \cdot P(V)$$
$$= (.10)(.05) = .005$$

Substituting this result, we obtain

$$P(T \text{ or } V) = .10 + .05 - .005 = .145$$

Part (b): We need to determine, first, $P(T \text{ and } V')$ and $P(T' \text{ and } V)$, since these are the probabilities of mutually exclusive ways for *only T* or *only V* to occur. We obtain the required probability as the sum of these two probabilities. First determine the following probabilities (Equation 6.2.1):

$$P(T') = 1 - P(T) = 1 - .10 = .90$$
$$P(V') = 1 - P(V) = 1 - .05 = .95$$

Then, using Rule E,

$$P(T \text{ and } V') = (.10)(.95) = .095$$
$$P(T' \text{ and } V) = (.90)(.05) = .045$$

Finally, using Rule B,

$$P(\text{only } T \text{ or only } V) = .095 + .045 = .14$$

Part (c): Applying Rule E,

$$P(T \text{ and } V) = (.10)(.05) = .005$$

Part (d): Applying Rule E,

$$P(T' \text{ and } V') = (.90)(.95) = .855$$

The foregoing probabilities were determined without defining the probability distribution on the appropriate sample space. However, it is worthwhile to observe how much easier these probabilities may be determined if the probability distribution is first defined. Table 6.4.3 presents this probability distribution. Details of construction will be left to the reader, as well as determining the required probabilities.

TABLE 6.4.3

	V	V'	Total
T	.005	.095	.10
T'	.045	.855	.90
Total	.05	.95	1.00

Example 6: A large retail outlet maintains extensive records on customer characteristics, including the following type of data:

Code	Characteristic
I	Income level
L	Low income
M	Middle income
H	High income
R	Location of residence
C	In city
O	Outside of city
F	Credit arrangement
A	Charge account
B	Cash

Suppose it is desired to determine the probability that a customer will have specified characteristics on the three variables: income level (I), location of residence (R), and credit arrangement (F). How can we construct the appropriate sample space? This can be accomplished by constructing a tree diagram as was done for permutations (Figure 6.3.1) and is illustrated in Figure 6.4.1. Following the route indicated by a set of three connected arrows identifies a sample point. The sample space contains 12 sample points (3 income levels · 2 residence locations · 2 credit arrangements), as shown in Figure 6.4.1.

The firm's records provide information on the proportion of customers falling in the 12 three-variable categories corresponding to the sample points. For example, it was found that 21 percent of the customers were low income city residents with charge accounts (sample point LCA, listed first in the "sample

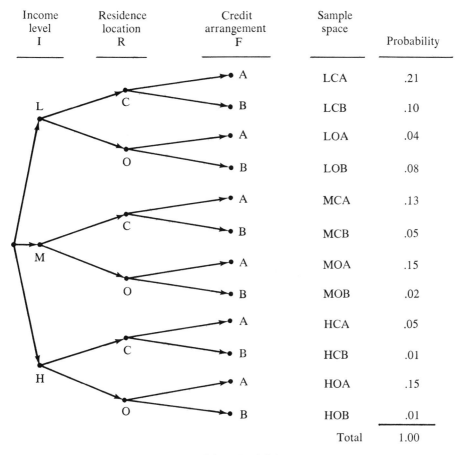

Income level I	Residence location R	Credit arrangement F	Sample space	Probability
		A	LCA	.21
L	C	B	LCB	.10
		A	LOA	.04
	O	B	LOB	.08
		A	MCA	.13
	C	B	MCB	.05
M		A	MOA	.15
	O	B	MOB	.02
		A	HCA	.05
	C	B	HCB	.01
H		A	HOA	.15
	O	B	HOB	.01
			Total	1.00

FIGURE 6.4.1

space" column in the figure). These percentages may be interpreted as probabilities and are shown in the "probability" column in the figure.

(a) Determine the marginal probability distributions (marginal probabilities).

This is a little more difficult when three variables are involved. The approach to be used is to compute the marginal probabilities separately for each variable in the sample space. For example, for the variable "income level," we need to compute the probability for each income level: $P(L)$, $P(M)$, $P(H)$. This may be accomplished for $P(L)$ by adding the probabilities for the sample points which contain L as one of the customer characteristics. This is an application of Rule B extended for mutually exclusive events. Looking down the "sample space" column in Figure 6.4.1, add $P(LCA) = .21 + P(LCB) = .10 + P(LOA) = .04 + P(LOB) = .08$ to obtain $P(L) = .43$. We obtain $P(M)$ by adding the probabilities for the sample points which contain M as one of the customer characteristics. We compute $P(H)$ by adding the probabilities for the sample points which contain H as one of the customer characteristics.

The same approach is followed to obtain the marginal probabilities (*marginal probability distributions*) for the variables "location of residence" and "credit arrangement." These marginal probability distributions are shown in Table 6.4.4. Notice that the sum of each distribution is 1.00.

TABLE 6.4.4

Marginal probability distributions for I, R, and F.

Income level I	Residence location R	Credit arrangement F
$P(L)$ = .43	$P(C)$ = .55	$P(A)$ = .73
$P(M)$ = .35	$P(O)$ = .45	$P(B)$ = .27
$P(H)$ = .22		
Total 1.00	1.00	1.00

(b) What is the probability that a customer will fall in one of the categories: (MOA), (MCA), or (HOB)?

Since these are mutually exclusive events, we use the probabilities in Figure 6.4.1 and compute (using Rule B extended)

$$P(MOA, MCA, \text{ or } HOB) = .15 + .13 + .01 = .29$$

(c) What is the probability that a customer lives outside of the city (O) and pays cash (B)?

The probability we want is $P(O \text{ and } B)$. The probability distribution (Figure 6.4.1) shows this probability for each of the three income levels separately and these events are mutually exclusive. Hence, we compute

$$P(O \text{ and } B) = P(LOB) + P(MOB) + P(HOB)$$
$$= .08 + .02 + .01 = .11$$

(d) We determine that a customer is in the medium income level (M). What is the probability that he lives outside the city (O) and has a charge account (A)?

The required probability is $P(O, A|M)$, the conditional probability that O and A occur given that M has occurred. We are concerned with the *reduced sample space* made up of the four events in Figure 6.4.1 which include the characteristic M (the given condition). These are listed in Table 6.4.5. Since the probabilities of these events add to .35, we divide the four probabilities by .35 to obtain the probability distribution for the reduced sample space. These probabilities now add to 1.00, as shown in Table 6.4.5. Reading from this probability distribution, we determine $P(O, A|M) = .43$.

Summary: Solving problems in probability requires careful consideration of the events involved and of the probability to be determined. Often, there is no need to construct the sample space and define the probability distribution. Still, this is sometimes necessary and other times facilitates solution of a problem.

TABLE 6.4.5

Taken from the sample space in Figure 6.4.1		Reduced sample space	
Events which include M	Probability	Sample point	Probability
MCA	.13	$C, A\|M$	.37
MCB	.05	$C, B\|M$	.14
MOA	.15	$O, A\|M$	.43
MOB	.02	$O, B\|M$	.06
Total	.35		1.00

Probabilities of simple events may be determined by the application of Definition 1 or Definition 2.

In the case of joint events, the probability rules are usually employed. In such cases, determine the events to be taken into consideration, whether they are mutually exclusive, overlapping, independent, or complementary. Often, a probability problem requires first computation of the individual probabilities of mutually exclusive events and then the addition of these probabilities according to Rule B or Rule B extended.

Computation of probabilities such as $P(A$ or $B)$ or $P(A, B,$ or $C)$ require the use of the addition rules (Rules A, B, B extended). Computation of probabilities such as $P(A$ and $B)$ or $P(A, B,$ and $C)$ require the use of the multiplication rules (Rules C, E, E extended). Conditional probabilities are always related to an appropriate reduced sample space. Above all, solution of probability problems requires a degree of ingenuity in analyzing the problem and developing the solution.

6.5 Optional: Comments on Notation

The notation used in this chapter has been kept as simple as possible. In this section, some additional notation is introduced which is widely used in discussions of sets and probability.

The event or *subset* $(A$ or $B)$ may be written as $(A \cup B)$ and read as "the union of A and B" or "A union B." It has the same meaning as $(A$ or $B)$ and is the event consisting of all sample points in A and B together.

The subset $(A$ and $B)$ may be written as $(A \cap B)$ and read as "the intersection of A and B" or "A intersect B." It has the same meaning as $(A$ and $B)$ and is the event consisting of the sample points in *both* A and B.

Given the sample space S in Figure 6.5.1 and events (subsets) A, B, C. The number of sample points in $(A \cup B)$ is $N_A + N_B - N_{(A \cap B)}$ or $4 + 9 - 1 = 12$. The sample points in $(B \cap C)$ are a_{34} and a_{44}. The event $(A \cup B \cup C)'$ contains the six sample points $a_{31}, a_{41}, a_{13}, a_{14}, a_{15},$ and a_{25}. The event $(A \cap C)$ is the null set. The event $(A \cap C')$ is the event A. The event $(A \cap B) \cup (B \cap C)$ consists of the sample points $a_{22}, a_{34},$ and a_{44}.

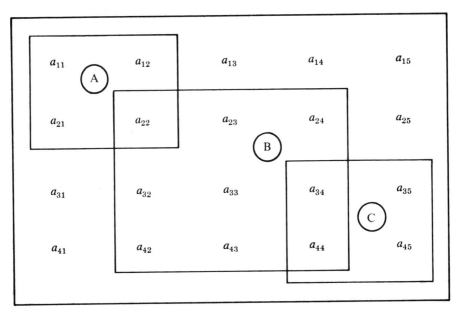

FIGURE 6.5.1

Sample space S.

As an exercise in the use of the new symbols, let us rewrite some of the probability rules:

Rule A: $P(A \cup B) = P(A) + P(B) - P(A \cap B)$

Rule B extended: $P(A_1 \cup A_2 \cup \ldots) = P(A_1) + P(A_2) + \cdots$

Rule C extended: $P(A_1 \cap A_2 \cap A_3) = P(A_1) \cdot P(A_2|A_1) \cdot P(A_3|A_1 \cap A_3)$

Study Problems

1.   Larry is thinking about accepting one of two jobs offered him. Should he take the job offered by a New York firm or the one offered by a Houston firm or should he remain at his present job in Los Angeles?
      a.   Identify the trial.
      b.   State the alternative possible outcomes.
      c.   Show the sample space by a set of dots properly identified.

2.   Define and illustrate: sample space and alternative possible outcomes.

3.   What is another name for sample space?

4.   List three names which are used for the alternative outcomes in a sample space.

5.   A set of three cards contains a red card, a white card, and a blue card. The cards are well shuffled and one is to be selected by a blindfolded person.
      a.   Construct the sample space assuming that the interest is in which card is selected (use properly identified dots).

    b.   What is the trial?

    c.   Construct the possibilities set assuming that the interest is in which cards are not selected.

6.   Define and illustrate: (a) selection with replacement and (b) selection without replacement.

7.   Construct the sample space (using properly identified dots):

    a.   Four coins are tossed and you are interested in the number of heads obtained.

    b.   A coin is tossed and then a die is thrown. (Show the sample space as in Figure 6.1.3.)

    c.   Two coins are tossed.

8.   A deck of cards contains a white card, a red card, and a blue card. A card is to be selected and then a die is to be tossed.

    a.   How many sample points are in this sample space?

    b.   Construct the sample space (as in Figure 6.1.3).

    c.   Identify (by drawing a circle in the sample space) event A = (Blue, 5), obtaining a blue card and 5 on the die.

    d.   Identify event B = (Blue, T), where T = 4, 5, or 6.

    e.   Identify event C = (S, 4), where S = white, red, or blue.

    f.   Are A and B mutually exclusive? Why?

    g.   Are B and C mutually exclusive? Why?

    h.   Are A and C mutually exclusive? Why?

    i.   How many sample points in (A or B)?

    j.   How many sample points in (B or C)?

    k.   How many sample points in (A or C)?

    l.   How many sample points in (A and B)?

    m. How many sample points in (A and C)?

    n.   How many sample points in C'?

    o.   How many sample points in (B or C)'?

    p.   How many sample points in (A and B)'?

9.   In Problem 8, if the interest is in whether the card selected is red or not red and whether the number on the die is 1, 2, 3, or more than 3, construct the sample space (as in Figure 6.1.3).

10.   Three marbles in a bag differ only in color. One is silver and two are gold. If two marbles are selected, construct the sample space (as in Figure 6.1.3) if selection is (a) with replacement and (b) without replacement.

11.   Four cards in a set are differentiated only by the number on the card (1, 2, 3, 4). Two cards are to be selected and we are interested in the permutation of the cards selected.

    a.   Show the sample space (as in Figure 6.1.3) if selection is with replacement.

    b.   Identify (by marking off the event in the sample space) event A = (2, 4), 2 on the first card, 4 on the second card.

    c.   Identify event B = [(2, 3, or 4), (1 or 2)].

    d.   Identify event C = [(1, 2, or 3), (1, 2, or 3)].

    e.   Are A and B mutually exclusive and why?

    f.   Are B and C mutually exclusive and why?

  g. Are B' and A mutually exclusive and why?
  h. How many sample points in $(B$ and $C)$?
  i. How many sample points in $(A, B,$ or $C)$?
  j. How many sample points in $(A$ and $B)'$?
  k. How many sample points in $(A, B,$ or $C)'$?

12. In Problem 11, if selection is without replacement:
  a. Construct the sample space.
  b. How many sample points in $(B$ or $C)$?
  c. How many sample points in $(A, B,$ or $C)'$?

13. In Problem 11, if selection is without replacement and we are not interested in permutation, construct the sample space.

14. What is the null event?

15. Generally speaking, how is probability conceptualized mathematically? Illustrate.

16. How do you interpret the probability statement P(Sue will type a letter without errors) $= .87$, based on the notion of repeated trials?

17. Do the postulates of probability relate to objective or subjective probabilities? Explain.

18. How do you interpret: P(Harold will fail a math exam) $= 0$ and P(a contract will be signed) $= 1$?

19. If a card selected from a deck of 52 playing cards is known to be a king, for that trial, what is P(king)?

20. In Problem 19, what is P(queen) for that trial?

21. Distinguish between objective probabilities and personalistic (or subjective) probabilities.

22. Does the concept of repeated trials underlie objective or subjective probabilities? Explain.

23. When faced with unique events, which is more suitable, objective or subjective probabilities? Why?

24. State the three postulates of probabilities and explain what they mean.

25. Consider the sample space relating to selection of a marble from an urn and the three probability distributions (P_1, P_2, P_3):

Color of selected marble

	Green	Blue	Yellow	Other color
P_1	.20	.30	.10	.40
P_2	.20	1.05	.60	.10
P_3	.20	.30	$-.40$	.90

Are each of these probability distributions acceptable? Why?

26. Consider the relationship $P(T') = 1 - P(T)$. Why does this relationship hold?

27. Why are the events represented by the sample points in a sample space mutually exclusive?

28. Define and illustrate equi-probable events.

29.   Do both Definition 1 and Definition 2 define probability as a proportion? How do they differ?

30.   If *P*(win a game) = .60, what are the odds that the game will be won?

31.   If *P*(rain tomorrow) = .20, what are the odds that it will not rain tomorrow?

32.   If a medication is known to be ineffective 3 percent of the time:
      a.   What is the probability it will be ineffective for a patient?
      b.   Which probability definition did you use?
      c.   What is the probability it will be effective for a patient?
      d.   Show the sample space.

33.   Ten out of fifty telephone calls received by an answering service are for firm A. What is the probability that the next telephone call is for firm A?

34.   How would you determine (estimate) the probability that an automobile passing a certain intersection will have an accident? Which probability definition would you use?

35.   If an urn contains a specified number of red marbles and a specified number of green marbles, how would you determine the probability of selecting a red marble (after shaking the urn and closing your eyes)? Which definition of probability would you use?

36.   In Problem 35, if the number of marbles of each color were not specified, how would you determine the probability and which definition would you use?

37.   A bag contains ten blue chips, five red chips, and five green chips. If you shake the bag vigorously, and with your eyes closed select a chip, what is the probability that the chip selected will be (a) red, (b) blue or green, (c) blue, red, or green, (d) yellow, and (e) not blue.

38.   In Problem 37, what are the odds the chip selected is (a) blue or green, (b) not blue?

39.   In your pocket you have dimes with the following dates of issue: five in 1967, three in 1950, four in 1965, two in 1947, one in 1949, two in 1963, one in 1957, and two in 1956. If you shake your pocket and select a coin, what is the probability the date of issue is (a) 1965, (b) in the 1950's, (c) after 1956, (d) in the 1940's or 1960's, and (e) between 1955 and 1964?

40.   In Problem 39, what are the odds against the coin having an issue date in the 1940's or 1960's?

41.   A department store has 15 clerks as follows: ten men (five new, two recently employed, three old-timers) and five women (one new, three recently employed, one old-timer). If the 15 time cards for these clerks are well shuffled and one card selected (with your eyes closed), what is the probability that the selected card is for:
      a.   A new male clerk
      b.   A recently employed clerk
      c.   A female clerk
      d.   A recently employed clerk or a female clerk
      e.   An old-timer or a male clerk

42.   A study of loan repayments made by college graduates after graduation showed that 40 percent complete their payments in less than 5 months, 30 percent in 5 to less

than 15 months, 20 percent in 15 to less than 30 months, and the balance in 30 months or more. What is the probability that a college graduate will complete repayments in:

    a.   5 to less than 15 months

    b.   Less than 30 months

    c.   5 months or more

    d.   30 months or more

43.   In Problem 42, what are the odds that a college graduate will complete his payments in 30 months or more?

44.   In Problem 37, determine the sample space and probability distribution. (Set up answer as in Figure 6.1.1 but show probabilities instead of dots.)

45.   In Problem 39, determine the probability distribution. (Set up answer as in Figure 6.1.1 but show probabilities instead of dots.)

46.   In Problem 41, determine the probability distribution. (Set up answer as in Figure 6.1.2 but show probabilities instead of dots.)

47.   A box contains red, white, and blue chips. Your eyes are closed and you select three chips one at a time (after the box is well shaken). Determine how many permutations of the three colors taken three at a time are possible by:

    a.   Using Equation 6.3.6

    b.   Constructing a tree diagram

    c.   Listing the possible permutations

48.   A class of four students contains two freshmen (F) and two sophomores (S). If two students are to be selected so that the first selected is class president and the other is class treasurer, how many ways can academic level (F or S or both) be represented in the pair selected? Determine your answer by:

    a.   Listing the possibilities

    b.   Constructing a tree diagram

49.   A large set of bolts contain 5 percent defectives. If D denotes defective bolts and G denotes those not defective and you select three bolts one at a time, in how many different ways can bolt quality (D or G or both) be represented in your selection? Determine your answer by:

    a.   Listing the possibilities

    b.   Constructing a tree diagram

50.   If you have five different books to arrange on a shelf, how many permutations are possible? (Use Equation 6.3.6.)

51.   If three books are to be arranged on a shelf and you can choose from five different books, then Equation 6.3.6 is to be applied with $N = 5$ and $X = 3$. Hence, the number of possible permutations is $5!/2! = 60$. How many permutations of four exam questions are possible, if you can choose from a pool of seven questions?

52.   A bag contains blue, yellow, red, and white marbles (one of each color). You shake the bag and select two marbles, one at a time (eyes closed). How many color permutations are possible? Determine your answer by:

    a.   Listing the possibilities

    b.   Constructing a tree diagram

    c.   Using Equation 6.3.6 as explained in Problem 51.

53.   Are events A and B (as defined below) dependent or independent?

    a. A and B represent heads on two coins.

    b. A and B are mutually exclusive events.

c. A = king and B = queen for two cards selected from a deck with replacement.

d. A = head and B = tail when tossing a coin once.

e. A = Ken votes for Smith and B = his wife votes for Smith, if the wife tends to vote like her husband.

f. A = head on a tossed coin and B = obtaining a two on a tossed die.

54. If $P(A)$ = .70, $P(A \text{ and } B)$ = .50, and $P(B)$ = .20, are A and B independent events? Why?

55. $P(K)$ = .20, $P(M)$ = .40, and $P(K|M)$ = .50:

    a. Are M and K dependent? Why?

    b. $P(M \text{ and } K)$ = ?

56. If a well-balanced die is tossed, what is the probability of obtaining (a) 6, (b) 2 or 4, (c) odd number?

57. What is the probability that a card selected (while blindfolded) from a well-shuffled deck of 52 playing cards is:

    a. Red

    b. A club or spade

    c. A 5, queen, or ace

    d. A king or spade

    e. A club or a 2 card

58. A box contains 20 marbles (four are yellow, six are green, two are blue, and eight are yellow and blue). If you shake the box vigorously and (with your eyes closed) select a marble, what is the probability it is:

    a. Yellow or green

    b. At least partly blue

    c. At least partly yellow or at least partly blue

    d. Yellow or yellow and blue

59. If 30 percent of the used cars sold by a dealer are all cash sales, 80 percent are sold to teen-agers, and 24 percent are sold to teen-agers on an all cash basis, what is the probability that the next used car is sold to a teen-ager or on an all cash basis?

60. In Problem 58, construct the probability distribution (as in Table 6.4.1).

61. Problem 59 involves the two variables T = type of sale (cash or credit) and C = type of purchaser (teen-ager or other). Construct the probability distribution for T and C (as in Tables 6.4.2 and 6.4.3) and show the marginal probability distributions.

62. In Problem 61, let A = next car is sold to a teen-ager, B = next car is sold on credit. Are A and B independent? Why?

63. A collection of 300 applications for admission to a business school includes 150 for typing, 100 for stenography, 25 for accounting, and 25 for management. If these applications are well shuffled and one is selected (with your eyes closed), what is the probability it is for an applicant who desires training in:

    a. Accounting, management, or typing

    b. A field other than typing or stenography

64. A file in a social psychology laboratory is made up of 80 index cards. Each card shows the serial number for an item of equipment and the date purchased. Thirty of these cards are for typewriters (20 purchased in 1966 or earlier and the balance after 1966), 40 are for calculators (10 purchased in 1966 or earlier and the balance after 1966), and 10 are for adding machines (all purchased after 1966). If the cards are well

shuffled and one is selected (while you are blindfolded), what is the probability that the selected card is for:

    a.   An adding machine or an item purchased in 1966 or earlier?

    b.   A calculator or an item purchased after 1966?

    c.   An item other than a calculator and not purchased after 1966?

65.   Problem 64 involves the two variables E = equipment type (typewriter, etc.) and Y = purchase year.

    a.   Construct the probability distribution for E and Y (as in Tables 6.4.2 and 6.4.3) and show the marginal probabilities.

    b.   Let X = typewriter, Y = purchased after 1966, Z = calculator. Are X and Y, also Z and Y, independent? Why?

66.   The probability is .45 that during a 24-hour period in a city no more than one crime will occur, .25 that two crimes will occur, and .10 that three crimes will occur. What is the probability that in a 24-hour period:

    a.   Two or three crimes will occur?

    b.   Less than three crimes will occur?

    c.   More than three crimes will occur?

67.   In Problem 66, construct the probability distribution for the variable "number of crimes in a 24-hour period in the city" (use Table 6.4.1 as a model).

68.   A voter in a municipal election may vote for candidate A (probability .25), candidate B (probability .30), candidate C (probability .20), candidate D (probability .15), or not vote at all. What is the probability a voter will vote for:

    a.   Candidate B, C, or D

    b.   Candidate A or not vote at all

69.   In Problem 68, what is the probability that two voters K and L who do not know each other:

    a.   Will both vote for candidate C?

    b.   K will vote for A and L will not vote?

70.   In Problem 69, construct the probability distribution for the voting by K and L and show the marginal probability distributions.

71.   The probability is .20 that a marble selected from a bag will be red and .40 that a number obtained when spinning a roulette wheel will be less than 100. What is the probability that:

    a.   The selected marble is red and the roulette wheel number is less than 100?

    b.   The selected marble is not red and the roulette wheel number is 100 or more?

    c.   Construct the appropriate probability distribution for this problem.

72.   The probability that the stock of a giant company will increase in price next week is .35 and the probability that you will purchase a few shares of this stock tomorrow is .60. What is the probability that:

    a.   You will purchase the stock and it will go up in price?

    b.   You will purchase the stock and it will not go up in price?

    c.   Either you will purchase the stock or it will go up in price (but not both of these events)?

73.   In Problem 72, (a) construct the probability distribution. (b) What is the probability that the price goes up, given that you purchase a few shares? Show the proba-

bility distribution for the reduced sample space. (c) Does the answer to part (b) follow from Rule D? Why?

74. The probability is .15 that a student will pass a certain exam if he makes no effort and .40 that he will pass if he studies for it. The probability is .70 that he will study for the exam.

    a. What is the probability that he will study for the exam and pass it?

    b. Construct the probability distribution. Hint: Use Rule C to compute *P*(Study and Pass) and *P*(No Study and Pass).

    c. What is the probability he will not study and not pass? (Use the probability distribution.)

    d. Show whether "studying" and "passing" are independent variables. (Use Rule E.)

75. A box of 25 miscellaneous seeds contain five for white flowers, eight for orange flowers, ten for yellow flowers, and the balance for purple flowers. If you have no idea at all as to which seed will produce which flower, and you select two seeds arbitrarily (one at a time), what is the probability that:

    a. Both are for orange flowers, if the seeds are selected without replacement?

    b. Both are for white flowers, if the seeds are selected with replacement?

    c. The first is for an orange flower, the second for a yellow flower (no replacement)?

    d. The first is for a purple flower (replace seed) and the second for a white flower?

    e. One is for an orange flower and one is for a white flower (no replacement)?

    f. One is for a yellow flower and one is for a white flower (with replacement)?

    g. The first is not for a white flower and the second is for a white flower (without replacement)?

76. In Problem 75, if the seeds are selected without replacement, what is the probability the first is not for a yellow flower and the second is not for a white flower?

77. In Problem 75, construct the probability distribution, show the marginal probability distributions, and encircle the probabilities which when added equals *P*(first seed is not for a yellow flower and second seed is not for a white flower).

78. A slot machine in a gambling house has three independently rotating wheels. The wheels are numbered 1, 2, 3 and each contains the sequence 1, 2, 3, 4, lemon, cherry. If a coin is put into the machine, and the wheels are well balanced and honest, what is the probability of obtaining on the three wheels (in the order wheel 1, wheel 2, wheel 3, unless noted otherwise):

    a. 3, cherry, 2

    b. Odd number, even number, odd number

    c. A number less than 3, not 4 or a cherry, not a number

    d. A number less than 4 on all wheels

    e. The second wheel a lemon, the other two 2 or a lemon

    f. 3, 4, lemon (in any order)

79. The probability is .35 that the crime rate will increase next month, .60 that the police commissioner will be replaced if the crime rate increases, and .25 that the police commissioner will be replaced. Construct the pertinent probability distribution. Based on this distribution, what is the probability that:

    a. Crime rate will increase and police commissioner will be replaced?

    b. Either one or both of these events will occur?

    c. Crime rate will increase and the commissioner will not be replaced?

    d. Crime rate will not increase, given that the police commissioner is replaced? Show the probability distribution for the pertinent reduced sample space.

    e. Neither event (crime rate increase, commissioner replaced) will occur?

    f. Crime rate will not increase, if the police commissioner is not replaced? Show the probability distribution for the pertinent reduced sample space.

    g. Are crime rate change and police commissioner replacement independent? Why?

80.   A file of 12 cards was prepared for a committee on minority rights. The cards are identified by serial numbers 1 through 12. Numbers 1 through 3 are for committee members from the North, 4 through 6 for members from the East, 7 through 10 for members from the South, and the balance for members from the West. If the cards are well shuffled and a sample of 3 cards are selected (with your eyes closed), what is the probability you select:

    a. First a Southerner, then two odd numbers (without replacement)

    b. A Southerner, not a Westerner, an Easterner (in the stated permutation, no replacement)

    c. Westerner or Easterner, Northerner, Westerner (in the stated permutation, no replacement)

    d. Northerner, odd number, number 10 or number 12 card (in stated permutation, with replacement)

    e. Number 1 or number 2 card, Northerner, not a Westerner (in stated permutation, without replacement)

81.   The probability is .60 that a subject will score high on a spatial aptitude test, .70 that he will score high on a finger-dexterity test if he scores high on the spatial aptitude test, and .80 that he will score high on a mechanical aptitude test if he scores high on the other two tests. What is the probability that a subject will score high on (a) the first two tests and (b) all three tests?

82.   The probability is .30 that a man entering a department store will buy a shirt, .40 he will buy a tie, and .20 he will buy a suit. If three men (strangers to each other) enter the store, what is the probability:

    a. The first will buy a tie; the second, a shirt; the third, a suit?

    b. One buys a suit, one a tie, one a shirt?

    c. None buy a tie?

83.   The probability is .25 that a door-to-door magazine salesman will make a sale on the first call, .30 he will make a sale on the second call if he makes a sale on the first call, and .40 that he will make a sale on the third call if he succeeds on the first two calls. What is the probability that he will succeed (a) on the first two calls and (b) on all three calls?

84.   If a fair die is tossed twice, what is the probability that the outcome is the same on the two tosses or that the outcomes add to five?

85.   If two fair dice are tossed, what is the probability that you obtain two different odd numbers or that they add to six?

86.   The probability is .40 that a small investor A will buy a certain stock and .30 that a small investor B (a stranger to A) will buy the stock (there is sufficient quantity for all who wish to buy). What is the probability that:

    a.   Only one of the two investors actually buy the stock?

    b.   Both buy the stock?

    c.   Neither one buys the stock?

    d.   How do you interpret the probability obtained as 1 minus the answer to part (b) of this problem? Should this probability be the same as the answer to part (c)? Defend your answer.

87.   A pair of fair dice is tossed three times. What is the probability that the outcomes add to 9 only on the last toss?

88.   A large bag contains ten small darts, three are good darts and seven are defective. The probability that a man will hit a target is .60 for a good dart and .10 for a defective dart. If he arbitrarily selects a dart by withdrawing one from the bag without looking, P(hit the target) = ?

89.   A dress has a fresh stain of unknown origin. The probability is .20 that the stain will be removed by allowing it to dry (with or without the application of a stain remover). If a stain remover is needed, then four in a set of ten unmarked bottles contain a stain remover which will be effective in removing the stain if it is applied immediately and the stain is then allowed to dry. If a bottle is selected arbitrarily and the contents of the bottle applied to the stain, what is the probability the stain will disappear after it dries?

90. A, B, and C are residents of the same city but do not know each other. A municipal issue is to be voted on (for or against) and these three voters each cast a ballot (for or against). We are interested in how A, B, and C voted on this issue.

    a.   How many sample points make the sample space?

    b.   Determine the appropriate sample space, using a tree diagram.

    c.   If the sample points represent equi-probable events, what is the probability that A, B and C do not all vote the same way?

91.   The probability is .40 that you will be offered jobs by firms A and B, .80 that you will be offered a job by firm A or B, and .90 that you will be offered a job by firm B. What is the probability that you will be offered a job by firm A? (Hint: Use Rule A.)

92.   The probability is .20 that a salesman will sell to customer A, .10 that he could get customer A a price reduction if he (customer A) places an order, and .30 that the price will be reduced anyway. What is the probability that:

    a.   Customer A places an order or the price is reduced?

    b.   Customer A places an order, given that the price is reduced?

93.   The probability is .30 that a man will vote for candidate A, .20 his wife will vote for A, and .40 that the man or his wife will vote for A. What is the probability that:

    a.   The man will vote for A, if his wife does?

    b.   The wife will vote for A, if her husband does?

    c.   Both vote for A?

94.   If H, K, L represent three events in a sample space, express the following in set notation:

    a.   Union of H and L

    b.   Intersection of K and H

     c. Complement of L
     d. Complement of the union of H and K
     e. Union of H and the complement of the union of K and L
     f. Intersection of H, K, and the complement of L
     g. Intersection of the union of L and H and the intersection of K and H
     h. Complement of the union of L and the intersection of H and K

95.   Express the following in words:

     a. $L \cup M$
     b. $A \cap (K \cup G)$
     c. $(A \cap M) \cup (K \cap G)$
     d. $(A \cup G)' \cap (K \cap L)'$

Chapter 7

Random Sampling

7.1 General Comments

Collection and study of population data is practical only for reasonably small populations where the data are easily obtainable and at moderate cost. In the usual situation, it is more efficient and less expensive to collect and study sample data.

Suppose a soap manufacturer wants to know how many of the eight housewives in his employ prefer to presoak their clothes before washing, using a certain compound. Clearly, the required information may be obtained simply by asking each housewife. On the other hand, if he wishes to obtain this information for all the housewives in a large city, the problem is very much different.

The population of housewives is much larger and widely scattered throughout the city.

A large library maintains a file card for each book in its collection. Suppose it is desired to determine the proportion of its collection published before a specified date, as noted on the cards. If the number of cards in the file is not too large, such information may be obtained by reviewing each card, otherwise, a review of each card is not practical. Moreover, if the data to be collected from each card were at all extensive, it would be practical to review each card only if the number of cards were quite small.

Finally, suppose a tire manufacturer wishes to determine the average life of a new line of tires. Determining the life of a tire means using it until it wears out and is no longer useful. Clearly, it would not be sensible to test all the tires produced, since then there would be none to sell.

In many situations, such as those just described, the practical thing to do is to collect sample data. Samples, if properly selected, may be expected to provide useful and adequate information about the population. For example, if the soap manufacturer found that 25 percent of the housewives in his employ prefer to use the given compound to presoak their clothes, would he be justified in drawing the inference that 25 percent of all the housewives in the city have such a preference? Suppose an employee in the library grabbed a handful of cards from the file and determined from these cards the proportion of books which were published before a certain date and the proportions in other categories of interest. Would it be reasonable to infer that these proportions hold for the entire collection of books? More specifically, are the samples described *representative* of the population from which they were selected? This is the important consideration in sample selection. *A representative sample is one which reflects the characteristics of the population from which it was selected in true population proportions.*

Sometimes an attempt is made to select a representative sample on a judgment basis. Suppose it is desired to determine the average amount of money spent per family for food during a given week in New York City. Suppose an "expert" selects a sample of families to represent the population of families in the city with respect to such expenditures. How can he know that this *judgment sample* is representative? If he knew enough about the spending habits of the population of families and of the sample selected to make such a judgment, then he would have sufficient information to make a study of such expenditures unnecessary. If he has any less information, his judgment on this matter cannot be accepted.

Actually, without perfectly accurate and complete knowledge about a population, a representative sample cannot be guaranteed. However, if a sample is selected by a *random selection procedure*, it is more likely that a representative sample will be obtained than if any other method of selection is used. Such samples are called *random samples*. When we speak of samples in this text, we will always mean random samples. In the next section, we will consider some random selection procedures.

7.2   Random Selection Procedures

It is the method used in selecting a sample which makes it a random sample, not the items included. It is possible that the identical sample items could be selected on a judgment basis as on a random selection basis; however, *only with a random sample may we apply the methods of statistical inference to generalize from the sample to the population on a probability basis*. Random sampling without replacement from a finite population is the most frequently used sampling procedure. We will also mention briefly selection of random samples with replacement from finite populations and sampling from an infinite population.

When sampling without replacement from a finite population, a random sample is obtained if the items in the population have equal probabilities of being selected on the first draw and, as items are selected, the remaining items in the population have equal probabilities of being selected in successive draws. Let us illustrate such a *random selection procedure*. Suppose it is desired to select a random sample of $n = 3$ TV sets from a population of $N = 10$ sets. Suppose the sets are numbered from 1 to 10 and we number ten identical marbles also from 1 to 10. Then, place the ten marbles in a box, shake the box vigorously and, while blindfolded, select three marbles one at a time without replacement. The numbers on the three marbles selected identify a random sample of three TV sets. This selection procedure provides equal probability of selection for each item (TV set) in the population (or remaining in the population) in a given draw.

The probability that a particular group of three TV sets will be selected may be determined according to Rule C extended and Rule B extended as $(1/10)(1/9)(1/8)_3P_3 = 1/120$. Notice that we multiplied by $_3P_3$ to take into account all possible permutations, since a particular sequence of selection is usually of no interest in sample selection. This random selection procedure gives each possible sample of size n from the population an equal probability of being selected.

When sampling from a finite population of size N with replacement, a random sample of size n is obtained if, in each trial (draw), each item in the population has a probability of $1/N$ of being selected and the trials are independent. In our previous illustration, if three marbles are selected one at a time with replacement, after vigorously shaking the box of ten marbles before each draw and while you are blindfolded, a random sample of three TV sets is identified by the numbers on the selected marbles. This *random selection procedure* implies independent trials (the selected marble is replaced) and the probability that a particular TV set is selected in any trial is $1/N = 1/10$. When sampling with replacement, of course, a particular TV set may be selected more than once. If set 1 is selected twice and set 2 once, the random sample of three TV sets is made up of set 1 considered twice and set 2 considered once.

When sampling with replacement, a finite population takes on some of the characteristics of an infinite population. That is, it cannot become exhausted and the population distribution never changes, no matter how many items are selected. Therefore, the concept of random sampling from an infinite population is similar to random sampling with replacement from a finite population. An important difference is that we cannot specify that the probability of selection of an item is $1/N$ in a trial, since N is infinitely large.

When sampling from an infinite population, a random sample is obtained if the probabilities of item selection are constant from trial to trial and the trials are independent. For example, the outcomes of five tosses of a die represent a random sample of $n = 5$ from the population of outcomes which would be obtained if the die were tossed an infinite number of times. Clearly, the probabilities of obtaining each of the possible alternative outcomes of a toss are constant from trial to trial and the trials are independent. The question of sampling with or without replacement is not pertinent when the population is infinite.

7.3 Sample Design and Selection

A plan for collecting a sample and preparing estimates of population characteristics based on the sample is called a *sample design*. There are a variety of ways to collect a sample. The simplest and basic type of sample design is a *simple random sample* or, more briefly, a *random sample*. This involves selecting a specified number of items from the population using one of the random selection procedures presented in Section 7.2. Suppose a library has 800 cards in its card file and it is desired to select a random sample of $n = 50$ without replacement. A simple random sample may be obtained by shuffling the 800 cards thoroughly and selecting 50 cards, one at a time, with your eyes closed. Clearly, this is not a desirable way to select the sample. A more practical way is to use a table of random digits such as Table B.3 in Appendix B.

The digits in a table of random digits were determined by some type of randomizing device which gives each of the ten digits $(0, 1, 2, \ldots, 9)$ an equal chance of being selected. For example, we could print each of the digits on a marble, place the ten marbles in a bag, shake the bag vigorously and select a marble while blindfolded. The number printed on the selected marble becomes the first digit in such a table. Then, replace the marble in the bag and repeat the procedure. The number on the second marble selected becomes the second digit in the table, etc. There are more sophisticated procedures to produce such tables, however, in any procedure, the ten digits $0, 1, 2, \ldots, 9$ have equal probability of selection in each trial.

How is a table of random numbers used? First, each item in the population must be identified by a number. In our illustration of 800 file cards in a library, each card must be identified by a number from 001 to 800. (Notice that we use 001, not merely 1. Since the highest number (800) contains three digits, each number must be expressed in terms of three digits.) This may be accomplished in various ways. One way is to actually write the identifying number on each card, or another way is to agree that the cards are identified by the numbers 001 through 800 in the same sequence in which they appear in the files, without actually writing the numbers on the cards. No matter which method is used, *it is absolutely essential that it be firmly decided, in advance, which number identifies which card.*

The next step is to decide *where to begin* in the table of random numbers. The digits in Table B.3 are grouped into blocks of five rows and five columns for user convenience. The numbers (1), (2), etc., in the heading of the table identify groups of five columns each. The numbers 1, 2, etc., in the stub identify the row or line number. We may enter the table at any point. We may start with the digit 1 in the first row, first column or we may start with the digit 9 in the sixth row, third column. Suppose we use the latter starting point. We must then decide how to form three-digit numbers: move across the table row by row to form the numbers 921, 069, etc., or move down the table column by column to form the numbers 953, 549, etc. Any defined method is acceptable. Suppose we use the following method: the digits in the sixth row and in the third, fourth and fifth columns will form the first number (921). Then, moving down these three columns read off additional numbers: 562, 301, 579, etc. *It is absolutely essential that the starting point and the method of forming the required numbers be fully and precisely determined before looking into the table of random digits.*

Following the specified procedure, numbers are copied out of Table B.3 until 50 *acceptable numbers* are obtained. *Acceptable numbers are those which are consistent with the population and the sample design.* For example, the population has been identified with the numbers 001 through 800. Therefore, any number read out of Table B.3 which is larger than 800 is not acceptable and should not be used. The sample design specified sampling without replacement. Therefore, a number is acceptable only the first time it is selected. Finally, the 50 acceptable numbers selected from Table B.3 identify the sample of 50 cards to be selected from the card file.

Other types of sample designs will be mentioned very briefly, as general background for the reader. Notice that in each type of sample, simple random sampling is included in one way or another.

Suppose it is desired to estimate the average salary of management employees in a large firm based on sample data. We could divide the population of management employees into relatively homogeneous subpopulations with respect to salary, so that salary will vary considerably less within a subpopulation than for the population as a whole. This is an often used procedure since a smaller sample is needed when variability is reduced, thus, permitting the survey to be carried out at lower cost and in less time. Subpopulations are called *strata*. Appropriate strata for estimating average salary are senior management employees, middle management employees, etc. Then a simple random sample is selected from each *stratum* separately, as indicated previously for a simple random sample. Each stratum is treated as a separate and independent population. Such a sampling procedure is called *stratified random sampling*. The average salary for the total population is estimated from the samples collected in all the strata.

Cluster sampling is frequently used for sample selection. In the foregoing illustration, suppose the personnel office has a card for each management employee showing salary and other data, and these cards are filed alphabetically in ten drawers. The cards in each drawer may be considered a *cluster* (or group of cards). *Cluster sampling* involves selecting a random sample of clusters (drawers) and basing the estimate of average salary on the salaries shown in the clusters of cards selected.

We may select a *systematic random sample* of the management employees. This could be accomplished by using the card file in the personnel office. First specify an interval, for example, 10. Then select a number within this interval (from 01 to 10) on a random basis. Suppose 06 is selected. Starting with the first card in the file, count until you reach the sixth card. This is the first card in the sample. Every tenth card is also in the sample (since the interval is 10): the 16th card, 26th card, etc. The selected cards identify a systematic random sample of management employees (and their salaries).

The systematic sampling method may be used with a stratified sampling design by selecting a systematic random sample from each stratum. This is called a *stratified systematic random sample.*

A vital part of any sample design is to carefully and precisely define the population to be sampled and to make provision for the sample to be selected from the defined population. This may seem to be a simple matter to those not experienced in statistical sampling, however, this aspect of a sample design is full of pitfalls.

Suppose it is desired to select a sample of medical doctors to estimate average income and that a random sample of doctors is to be selected from a directory published by a medical association. Would this procedure result in a random sample of all doctors? The answer is "no," unless it is determined that all doctors are members of the association and that they are all listed in the directory. Doctors not listed may include nonmembers, members who submitted required information too late to be included in the directory, those who became doctors after the directory was issued, etc. Furthermore, the definition of "medical doctor" must be carefully stated. For example, should the population of doctors to be studied include interns, doctors who do not see patients and spend all their time in the laboratory, part-time practicing doctors, retired doctors, chiropractors?

Sometimes the population to be sampled is restricted for cost and related reasons. For example, a college teacher may wish to sample all college seniors for attitude toward a certain issue, but restricts his sample to seniors from his own university to reduce the cost and effort involved. He will then have to restrict his statistical inferences from the sample to the population of seniors in his university. Any generalization of the findings to all college seniors must be made on a judgment basis, not a statistical basis.

7.4 Sampling Distributions

A statistical measure such as the mean, median, or standard deviation is called a *parameter* if computed from population data and a *statistic* if computed from sample data. *A parameter is a constant for a given population.* For example, there is only one mean and only one standard deviation for a given population. On the other hand, *a statistic is a variable*, since it may be expected to vary from sample to sample.

Let us consider the population of five quantities: 2, 4, 6, 8, 10, with mean $\mu = 6$ and standard deviation $\sigma = 2.83$. Suppose we do not know μ and wish to estimate it based on the $\overline{X}$ value computed from a random sample of size $n = 2$ selected without replacement. What are the different $\overline{X}$ values possible?

First, let us consider how to compute "the number of possible *combinations* of N different objects taken X at a time" which we will denote by the symbol $\binom{N}{X}$. A *combination* represents a collection of objects without regard to order. For example, the possible combinations of 3 letters (A, B, C) taken 2 at a time, $\binom{3}{2}$, are AB, AC, BC, so that we have

$$\binom{3}{2} = 3 \text{ combinations}$$

Of course, if we are interested in the number of possible *permutations*, we would have to take into consideration that *each combination* of two letters permits two permutations, since $_2P_2 = 2!$ (see Equation 6.3.6). For example, the combination AB permits the two permutations AB and BA. Hence, the number of possible permutations of 3 different objects taken 2 at a time may be computed as

$$_3P_2 = \binom{3}{2} 2!$$

Solving for $\binom{3}{2}$, we obtain

$$\binom{3}{2} = \frac{_3P_2}{2!}$$

In general, we may express "the number of possible combinations of N different objects taken X at a time" (taking into consideration Equation 6.3.6) as follows:

$$\binom{N}{X} = \frac{_NP_X}{X!} = \frac{N!}{X!(N - X)!} \qquad (7.4.1)$$

Returning to our population of 5 quantities, the number of possible samples of size 2 selected without replacement may be computed as the number of possible combinations of 5 quantities taken 2 at a time:

$$\binom{5}{2} = \frac{5!}{2!3!} = 10 \text{ combinations or samples}$$

Table 7.4.1 presents the 10 possible samples and sample means $\overline{X}$. These results are summarized in Table 7.4.2. The first column shows all the possible $\overline{X}$ values. The second column shows the frequency with which each $\overline{X}$ value occurs. For example, $\overline{X} = 5$ could occur in two ways (if the sample is made up of the quantities 2 and 8 or 4 and 6).

TABLE 7.4.1

All possible samples of n = 2 selected without replacement from the population 2, 4, 6, 8, 10 and $\overline{X}$'s computed from the samples.

Items in sample	$\overline{X}$
2, 4	3
2, 6	4
2, 8	5
2, 10	6
4, 6	5
4, 8	6
4, 10	7
6, 8	7
6, 10	8
8, 10	9

Notice from Table 7.4.2 that the sample mean $\overline{X}$, which may be used as an estimate of μ (= 6), could be as low as 3 and as high as 9. Only two of the possible 10 samples result in $\overline{X}$ = 6. However, six of the possible 10 samples result in $\overline{X}$ between 5 and 7, thus, providing close estimates of μ. The last column in Table 7.4.2 presents the relative frequencies and indicates the probability of obtaining each of the possible $\overline{X}$ values. We may look upon the selection of a random sample as an experiment and the possible $\overline{X}$ values as the alternative possible outcomes. Then, the column of $\overline{X}$ values in Table 7.4.2 represents a sample space and the probabilities in the last column make up a probability distribution (as discussed in Chapter 6). When the variable in a probability distribution is a statistic, such as $\overline{X}$, the distribution is called a *sampling distribution*.

TABLE 7.4.2

Sampling distribution of the mean for n = 2 (based on data in Table 7.4.1).

$\overline{X}$	f	$P(\overline{X}$ = specified value) $= f/10$
3	1	.10
4	1	.10
5	2	.20
6	2	.20
7	2	.20
8	1	.10
9	1	.10
Total	10	1.00

The sampling distribution in Table 7.4.2 is the *sampling distribution of the mean*, since the variable is the sample mean $\overline{X}$. Based on this sampling distribution, we may make probability statements relating to the possible value of $\overline{X}$ computed from a random sample. For example, $P(\overline{X} = 6) = .20$, which indicates that the probability is .20 that a random sample of $n = 2$ selected without replacement from the population of 5 quantities will result in $\overline{X} = 6$. Or $P(5 \leq \overline{X} \leq 7) = .60$, which indicates that the probability is .60 that $\overline{X}$ will have a value between 5 and 7, both limits included. This is obtained by adding $P(\overline{X} = 5) = .20$, $P(\overline{X} = 6) = .20$, $P(\overline{X} = 7) = .20$, according to probability Rule B extended for mutually exclusive events.

The sampling distribution of the mean (or any sampling distribution) is actually a population distribution, since it is the distribution of a population of sample means. Therefore, we will denote the mean and standard deviation of a sampling distribution by the symbols we have reserved for populations. We will use $\mu_{\overline{x}}$ and $\sigma_{\overline{x}}$ to denote the mean and standard deviation, respectively, for the sampling distribution of the mean. (We will continue to use μ and σ for the corresponding population measures.) Referring to the sampling distribution of the mean in Table 7.4.2, we compute the mean and standard deviation and obtain $\mu_{\overline{x}} = 6$ and $\sigma_{\overline{x}} = 1.73$. Notice that the mean of the sampling distribution of $\overline{X}$ is equal to the population mean μ ($= 6$). This is always true. The standard deviation of the sampling distribution is less than the population standard deviation σ ($= 2.83$). This is always true for samples of $n = 2$ or more.

The standard deviation $\sigma_{\overline{x}}$ measures the extent to which sample means vary around the mean of the sampling distribution $\mu_{\overline{x}}$, however, since $\mu_{\overline{x}} = \mu$, $\sigma_{\overline{x}}$ is also a measure of the variation of sample means around the population mean μ. This is of great importance in statistical inference. When a sample mean $\overline{X}$ is used as an estimate of μ, we expect that there will be some error. The extent of error, of course, will vary from sample to sample. The standard deviation $\sigma_{\overline{x}}$ is a measure of the error to be expected, on the average, when the mean of a random sample is used as an estimate of μ. In other words, $\sigma_{\overline{x}}$ is a measure of the estimation error due to random sampling or chance. Consequently, $\sigma_{\overline{x}}$ is usually called the *standard error of the mean*. The standard deviation of a sampling distribution is usually called the *standard error*.

The sampling distribution presented in Table 7.4.2 is based on samples of size $n = 2$. *There is a different sampling distribution for each sample size.* Consider all possible samples of size $n = 3$ from the population of 5 quantities previously discussed. The number of possible samples are $\binom{5}{3} = 10$. Table 7.4.3 presents these samples and the sample means. These data are summarized in Table 7.4.4 where the sampling distribution of the mean for samples of size $n = 3$ is presented. We may determine that $\mu_{\overline{x}} = 6$ (as we found for samples of size $n = 2$) and $\sigma_{\overline{x}} = 1.16$ (compared with $\sigma_{\overline{x}} = 1.73$ for samples of size $n = 2$). The error to be expected due to chance (random sampling) when a sample mean is used as an estimate of the population mean is less for larger samples.

TABLE 7.4.3

All possible samples of n = 3 selected without replacement from the population 2, 4, 6, 8, 10 and $\overline{X}$'s and M's computed from the samples.

Items in sample	$\overline{X}$	M
2, 4, 6	4.00	4
2, 4, 8	4.67	4
2, 4, 10	5.33	4
2, 6, 8	5.33	6
2, 6, 10	6.00	6
2, 8, 10	6.67	8
4, 6, 8	6.00	6
4, 6, 10	6.67	6
4, 8, 10	7.33	8
6, 8, 10	8.00	8

TABLE 7.4.4

Sampling distribution of the mean for n = 3 (based on data in Table 7.4.3).

$\overline{X}$	f	$P(\overline{X} =$ specified value) $= f/10$
4.00	1	.10
4.67	1	.10
5.33	2	.20
6.00	2	.20
6.67	2	.20
7.33	1	.10
8.00	1	.10
Total	10	1.00

So far we considered only the sampling distribution of the mean. *There is a sampling distribution for each statistic.* If we compute the *sample median* for each possible sample of a given size, we may construct the *sampling distribution of the median*. In Table 7.4.3 we present the median (M) computed for each sample of size $n = 3$. These results are summarized in Table 7.4.5 where the sampling distribution of the median is presented. The mean of this sampling distribution is $\mu_M = 6$ and the standard error is $\sigma_M = 1.55$. Notice that $\mu_M = \mu_{\overline{x}} = \mu$. However, the standard error of the median $\sigma_M = 1.55$ is larger than the standard error of the mean $\sigma_{\overline{x}} = 1.16$. In other words, whereas the sample median as well as the sample mean may be used as an estimate of μ, the estimation error due to chance is greater when the sample median is used.

TABLE 7.4.5

Sampling distribution of the median for n = 3 (based on data in Table 7.4.3).

M	f	P(M = specified value) = f/10
4	3	.30
6	4	.40
8	3	.30
Total	10	1.00

7.5 Theoretical Sampling Models

Sampling distributions of the mean and median were developed based on sampling from a very small population, in the previous section. This provided a convenient basis for presenting the concept of a sampling distribution. Typically, we deal with large or very large populations so that it is not realistic to think in terms of listing all possible samples and to construct a sampling distribution as in Tables 7.4.2, 7.4.4, and 7.4.5.

Generally, we deal with *theoretical sampling distributions* developed on the notion of repeated sampling from an infinite population, as follows. Imagine that a sample of size n is selected and $\overline{X}$ is computed, and then a second sample of size n is selected and $\overline{X}$ computed. Imagine that this is repeated an infinite number of times so that we have an infinite number of $\overline{X}$ values. We summarize this information in a table similar to Table 7.4.2 and construct the sampling distribution by computing the relative frequencies. Such theoretical sampling distributions are determined mathematically in texts on mathematical statistics.

Formulation of statistical inferences relating to a population based on sample data requires a knowledge of the population distribution and of the pertinent sampling distribution. Such information is usually not available. The typical approach is to relate the problem at hand to an appropriate *theoretical sampling model*. A *theoretical sampling model* specifies a theoretical population and describes the sampling distribution of a statistic. Of course, no problem in statistical inference encountered in practice can precisely fit any theoretical sampling model. Still, when properly applied, such models provide powerful tools for the solution of statistical problems. Several theoretical sampling models will be presented in the following chapter. We will also present two important theoretical population distributions, the *normal distribution* and the population of 1's and 0's.

Study Problems

1. What is meant by (a) representative sample and (b) judgment sample?
2. Under what circumstances can a representative sample be guaranteed?

3.   Generally speaking, what is a random sample?

4.   List three types of random selection procedures.

5.   Generally, when is it permissible to apply the methods of statistical inference to generalize from a sample to the population on a probability basis?

6.   Illustrate random sampling from a finite population, without replacement.

7.   Illustrate random sampling from a finite population, with replacement.

8.   Illustrate random sampling from an infinite population.

9.   Given a population of eight illiteracy rates (the illiteracy rates for eight cities). If sampling is without replacement, how many different samples are possible, if (a) $n = 2$ and (b) $n = 3$?

10.   In Problem 9, what is the probability that the sample will be made up of the illiteracy rates for (a) two specified cities, if $n = 2$ and (b) three specified cities, if $n = 3$?

11.   Given a population of 12 test scores. If sampling is without replacement, how many different samples are possible, if (a) $n = 5$ and (b) $n = 3$?

12.   In Problem 11, what is the probability that a specified sample of test scores will be obtained if (a) $n = 5$ and (b) $n = 3$?

13.   Assume that a roulette wheel will remain in perfect balance if spun an infinite number of times. Suppose this roulette wheel is spun ten times. Would the ten outcomes represent a random sample? Why? What is the population involved?

14.   What is a sample design?

15.   Give an illustration to show (briefly) how you would select the specified type of sample, if sampling is with replacement:
    a.   Simple random sample
    b.   Stratified random sample
    c.   Cluster sample
    d.   Systematic random sample
    e.   Stratified systematic random sample

16.   In a table of random numbers, such as Table B.3, what is the probability that a three-digit number arbitrarily selected (with your eyes closed) will be 404?

17.   Refer to the table of random numbers, Table B.3. What is the probability that the number in:
    a.   A specified row and column is 0, 3, or 5?
    b.   A specified row and in the third and fourth columns (a two-digit number) is 00, 99, or 12?

18.   A random sample of ten families is to be selected (without replacement) from a population of 1,538 families by the use of Table B.3. It is agreed to select random numbers by starting with the digit in the first column, first row and forming appropriate random numbers by reading down the first column, then down the second column, etc. How would you describe acceptable numbers for this problem?

19.   A large sample of city voters is to be selected by use of a table of random numbers and selecting names from the city telephone directory. The sample of voters will be asked which candidate they favor for mayor and the results obtained will be used to forecast next week's election. Comment on the adequacy of the sample and the appropriateness of using the sample results as noted.

20. A sociologist determined, based on sample data, that 85 percent of a certain minority group in a given city opposed a bill being debated in the Congress. Based on this finding, he announced to the press that the nationwide sentiment of this minority group was against the bill. Comment.

21. Given the population of units produced: 8, 6, 10, 4.
    a. How many samples of $n = 2$ are possible for sampling without replacement?
    b. Determine the sampling distribution of the mean for $n = 2$ and sampling without replacement.
    c. Compute μ, $\mu_{\bar{X}}$, σ, $\sigma_{\bar{X}}$.
    d. $P(\bar{X} < \mu) = ?$

22. Given the population of scores: 2, 1, 3, 5, 4.
    a. Determine the sampling distribution of the mean for $n = 3$ and sampling without replacement.
    b. Compute μ, $\mu_{\bar{X}}$, σ, $\sigma_{\bar{X}}$.
    c. $P(\bar{X} - \mu = 0) = ?$

23. In Problem 22:
    a. Determine the sampling distribution of the median for $n = 3$ and sampling without replacement.
    b. Compute μ_M, σ_M.
    c. $P(M < \mu) = ?$

24. The steps to follow when using a table of random digits are: (1) identify each item in the population by a number; (2) decide on the point of entry into the table; (3) decide on the method to be used to form numbers that will correspond to the indentifying numbers in (1); and (4) copy a sufficient number of acceptable numbers. For the following sets of data (to be considered as populations), select a random sample (as specified) using Table B.3. Estimate the population mean and standard deviation by computing $\bar{X}$ and s. State precisely how Table B.3 was used, in accordance with steps (1) through (4) previously noted.
    a. Sample of $n = 5$ selected without replacement, data from Problem 13, Chapter 3.
    b. Sample of $n = 7$ selected without replacement, data from Problem 12, Chapter 3.
    c. Sample of $n = 10$ selected without replacement, data from Problem 11, Chapter 3.

25. Define and illustrate (a) statistic and (b) parameter.

26. How many combinations are possible of seven different objects taken four at a time?

27. How many committees of five students each can be formed from a class of 11 students?

28. How many samples of $n = 4$ may be selected (without replacement) from a population of $N = 12$?

29. Explain (briefly) the meaning of standard error of the mean.

30. What is a theoretical sampling model and why are such models needed?

Chapter 8

Population and Sampling Models

8.1   Normal Distribution

The most important theoretical distribution in statistics is the *normal distribution*, often called the *normal curve*. This is a continuous distribution of an infinite population. That is, the variable of the distribution X_i, or simply X, is a continuous variable which can take on any value from − infinity to + infinity. Even though we will not go into the mathematics of the normal curve, it is useful to present its equation.

$$Y = \frac{1}{\sqrt{2\pi}\sigma} e^{-\frac{1}{2}\left(\frac{X-\mu}{\sigma}\right)^2} \tag{8.1.1}$$

where: Y = height of the curve at a given value of X
   σ = standard deviation of the population
   μ = mean of the population

$\pi = 22/7 = 3.1416$, a mathematical constant
$e = 2.7183$, a mathematical constant
$X = $ a continuous variable

The normal distribution has two parameters μ and σ. Equation 8.1.1 defines a *family of normal curves* and when values are specified for μ and σ a particular normal curve is identified. Let $\mu = 100$ and $\sigma = 10$. This identifies the normal distribution in Figure 8.1.1, on the left. Notice that the normal curve is perfectly symmetrical with a central peak, so that it resembles a bell. The mean μ is located at the center of the distribution. Owing to the symmetry of the distribution and the central peak, the mean, median, and mode are all equal. The curve is highest in the center and approaches the horizontal axis in each tail of the distribution, stretching out indefinitely in each direction and approaching but never touching the horizontal axis.

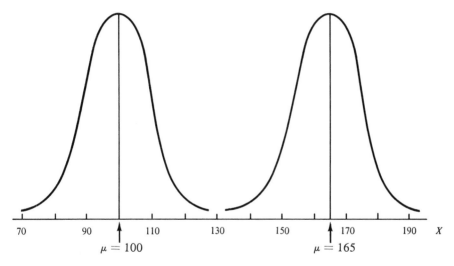

FIGURE 8.1.1

Normal distributions with the same standard deviation and different means.

How is the normal curve affected if μ and σ are changed? In Figure 8.1.1, if μ is changed from 100 to 165 and σ remains the same (10), the curve shifts to a higher position on the X axis. Otherwise the distribution does not change, as indicated in Figure 8.1.1, on the right. Figure 8.1.2 presents two normal curves with the same mean but with different standard deviations. The distribution with the larger σ has a greater spread. Reducing the standard deviation results in a contraction of the central portion of the distribution so that there is a greater concentration of the distribution around the mean. The smaller σ, the higher is the central portion of the curve and the closer is the curve to the X axis in the tails.

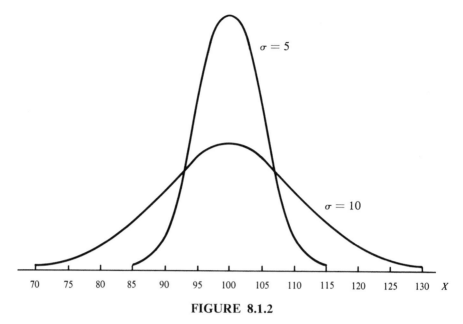

FIGURE 8.1.2

Normal distributions with the same mean and different standard deviations.

The most important use of the normal distribution in statistics is related to the *area under the curve*. It will be recalled from Section 3.2, that the area under a continuous curve represents a relative frequency. The total area under the curve represents the total of the relative frequencies for the distribution, which is 1.00. Karl Friedrich Gauss was associated with the early development of the normal distribution, so that it is sometimes called the *Gaussian distribution*. It is also called the *normal curve of errors* because it has been found to approximate very closely the distribution of errors made when a measurement procedure is repeated a large number of times. For example, if the length of a rod is measured a large number of times using the same yardstick, the distribution of the measurements around the mean will tend to follow the normal curve.

Generally, we deal with the *standard normal distribution*, where the variable X is expressed in standard units (z), as illustrated in Figure 8.1.3. Equations 5.4.3 and 5.4.4 are used to transform a variable X to z units. This results in a distribution with a mean of 0 and a standard deviation of 1. Notice that the horizontal scale in Figure 8.1.3 is labeled z, not X, and that the mean is at $z = 0$. As discussed in Section 5.4, z values are in standard deviation units, so that the point $z = 1$ in the figure is one σ above the mean and $z = -1$ is one σ below the mean. The distribution of the area under the normal curve shown in Figure 8.1.3 is the same for all normal distributions. As indicated in the figure, 34 percent of the area is contained between the mean and one standard deviation above (or below) the mean; 14 percent of the area is contained between one standard deviation and two standard deviations above (or below)

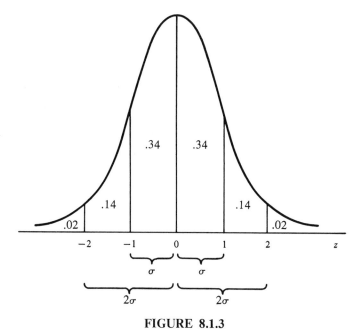

FIGURE 8.1.3

The standard normal distribution ($\mu = 0$, $\sigma = 1$).

the mean; and, 2 percent of the area lies beyond two standard deviations from the mean (above or below).

8.2 Applications of the Normal Distribution

Formulation of statistical inferences is generally based on the proportion of the area under the normal curve for specified intervals along the horizontal axis. Such applications are facilitated by the use of a table of areas under the normal curve. Table B.4 in Appendix B is such a table. It is based on the *standard normal curve*. The shaded area at the head of Table B.4 indicates that the table presents the proportion of the total area under the curve between the mean ($z = 0$) and specified z values. Suppose we wish to determine the area between the mean and $z = 1.00$ (one standard deviation above the mean). Look down the column headed "z" until you reach 1.0, then move across that row to the column headed ".00". The value .3413 obtained, corresponding to $z = 1.00$, is the proportion of the area under the curve between the mean and one standard deviation above the mean. This agrees with .34 in Figure 8.1.3, after rounding to two decimals.

Table B.4 presents areas under the normal curve only for positive values of z. We may determine the area between the mean and $z = -1.96$ as follows: Look down the column headed "z" until you reach 1.9, then move across that row to the column headed ".06". The value .4750 obtained, corresponding to

$z = 1.96$, is the proportion of the area under the curve between the mean and 1.96 standard deviations above the mean. It is also the proportion of the area under the curve between the mean and 1.96 standard deviations below the mean ($z = -1.96$), due to the symmetry of the normal curve.

What is the area under the normal curve between $z = -1.50$ and $z = 1.39$, as shown in Figure 8.2.1? Consider this area as made up of two parts, one part is the area below the mean and the other is the area above the mean. Referring to Table B.4, we determine the area corresponding to $z = -1.50$ as .4332 and the area corresponding to $z = 1.39$ as .4177. Adding these two areas, we obtain .8509 as the area under the curve between $z = -1.50$ and $z = 1.39$. Find the area between $z = .49$ and $z = 1.71$ (Figure 8.2.2). Referring to Table B.4, we determine that the area corresponding to $z = .49$ is .1879 and the area corresponding to $z = 1.71$ is .4564. Subtracting the smaller area from the larger, we obtain .2685 as the proportion of the total area under the curve between $z = .49$ and $z = 1.71$.

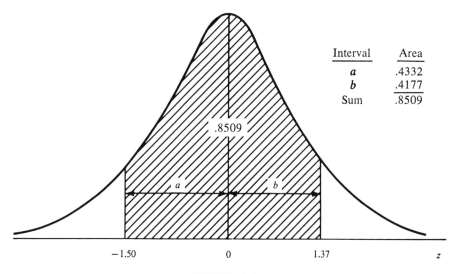

Interval	Area
a	.4332
b	.4177
Sum	.8509

FIGURE 8.2.1

Suppose the time required for a family in a certain region to travel to a newly-constructed recreation center is approximately normal in distribution, with mean travel time 22.483 minutes and standard deviation 3.950 minutes. How can we estimate the proportion of families with travel time between 18.500 minutes and 21.000 minutes? Since travel time is approximately normal in distribution, use the normal curve for this purpose. First, transform the limits 18.500 minutes and 21.000 minutes to standard units to obtain

$$z_1 = \frac{18.500 - 22.483}{3.950} = -1.01$$

$$z_2 = \frac{21.000 - 22.483}{3.950} = -.38$$

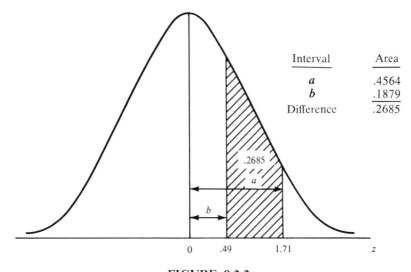

Interval	Area
a	.4564
b	.1879
Difference	.2685

FIGURE 8.2.2

Referring to Table B.4, read out .3438 as the area between the mean and z_1 and .1480 as the area between the mean and z_2. Then, subtracting the smaller area from the larger, we obtain .1958 as the proportion of families with travel times between 18.500 minutes and 21.000 minutes. Notice that the limits (18.500 and 21.000) were transformed to standard units (Equations 5.4.3 and 5.4.4) so that we could enter Table B.4.

Since the area under the normal curve corresponding to an interval along the horizontal axis represents a proportion, it may also be interpreted as a probability. This is the principal use of the normal distribution in statistics. In the foregoing illustration relating to family travel times, we may state that the probability is .1958 (or about .20) that for a family selected at random the travel time is between 18.500 minutes and 21.000 minutes. What is the probability that the travel time for a family selected at random is more than 29.999 minutes? Transforming to standard units, we obtain

$$z = \frac{29.999 - 22.483}{3.950} = 1.90$$

Figure 8.2.3 presents this z value on the standard normal curve. We determine from Table B.4 that the area between the mean and $z = 1.90$ is .4713. Since the area under the normal curve is symmetrical, half the area (.5000) is above the mean. Hence, $.5000 - .4713$ gives .0287 as the area above $z = 1.90$. Then, .0287 or 2.87 percent of the travel times are more than 29.999 minutes. Also, .0287 (or .03) is the probability that the travel time for a family selected at random is more than 29.999 minutes.

Suppose a large population of measurements is approximately normal in distribution with a mean of 118.35 cm and a standard deviation of 22.10 cm. Find the measurement value below which 10 percent of the measurements fall. Figure 8.2.4 presents a normal curve which represents the measurement dis-

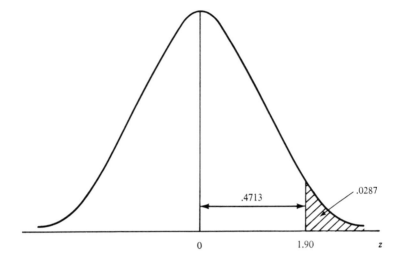

.4713

.0287

0   1.90   z

FIGURE 8.2.3

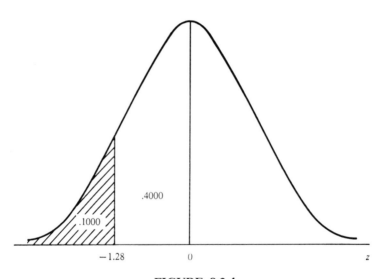

.4000

.1000

−1.28   0   z

FIGURE 8.2.4

tribution in standard units. The shaded area in the lower tail of the distribution represents the lower 10 percent or .1000 of the measurements. Therefore, since .5000 of the total area is to the left of the mean, the area between the mean and the shaded portion is equal to .5000 − .1000 or .4000, as shown in the figure. Referring to Table B.4, find .4000 or a value as close to this as possible in the body of the table. The closest value is .3997 and the corresponding z is 1.28; however, since we are concerned with the lower tail, $z = -1.28$. The area between the mean and $z = -1.28$ makes up 40 percent (actually 39.97 percent) of the area under the curve and the area to the left of $z = -1.28$ makes up 10 percent of the total area. Therefore, $z = -1.28$ is the measurement value we want, but it is in standard units. Substituting in the equation for z (Equation 5.4.3), we have

$$-1.28 = \frac{X - 118.35}{22.10}$$

Solving for X, we obtain $X = 90.06$. Then, 10 percent of the population of measurements are less than 90.06 cms.

A large population of typing students made a mean of 14 errors per student when typing certain material, with a standard deviation of 4 errors. The distribution of "number of errors" is approximately normal. What proportion of the typing students made exactly 20 errors? In this problem we are dealing with a *discrete variable*, $X =$ number of errors. In order to use the normal curve (which is a *continuous* distribution) to determine the required proportion, we must convert the *discrete variable* X to a *continuous variable*. This is called *correction for continuity*. This is accomplished by closing the gaps between the possible values of the discrete variable. For example, X can take on only integral values such as 10, 11, 12, etc. Since fractional values of X do not occur, there are gaps between these values.

When converting to a continuous variable, we close the gaps by replacing each *discrete value* by a *continuous interval*. The discrete value 10 is replaced by the continuous interval 9.5–10.5; 11 is replaced by 10.5–11.5; etc. We make the assumption that X can take on any value within an interval even though the data shows only integral values for X. When converted, the discrete value 20 becomes the continuous interval 19.5–20.5, as indicated by the shaded area in Figure 8.2.5. Then, transforming 19.5 and 20.5 to standard units we obtain $z_1 = 1.38$ and $z_2 = 1.63$, respectively. Referring to Table B.4, we find the area between the mean and z_1 (.4162) and between the mean and z_2 (.4484). Subtracting, we obtain .0322 as the estimated proportion of typing students who made exactly 20 errors.

In the foregoing illustration, what is the probability that a typing student selected at random made 20 errors or more? Remembering that the discrete value 20 must be converted to the continuous interval 19.5–20.5, we find the area in the tail of the normal curve to the right of $X = 19.5$ (after transforming to $z = 1.38$). This is the required probability. On the other hand, what is the probability that a typing student selected at random made more than 20 errors? Since more than 20 errors does not include 20 errors, we obtain the required probability by determining the area in the tail of the normal curve to the right of $X = 20.5$.

The normal curve, which is a distribution of a continuous variable for an infinite population, is a theoretical concept. Hence, probabilities and proportions determined for real world problems on the basis of the normal curve are estimated values. The normal distribution can only approximate a real distribution, where the population, no matter how large, is never infinite.

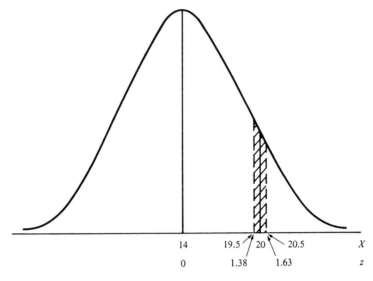

FIGURE 8.2.5

8.3   Population of 1's and 0's

A discrete population of considerable importance in statistics is one where the items included fall into one of two categories. For example, a population of adults classified only as male or female, a population of prices classified only as "above $10" or "$10 and below," a population of employees classified as part-time or full-time, or a population of bills classified as paid or unpaid. Such a population is often called a *dichotomous (two-category) population*.

It is useful to assign the code 0 (zero) to each of the items in one category and the code 1 to each of the items in the other category. The usual practice is to assign code 1 to the category of primary interest. For example, we may be interested in studying the extent to which men use electric shavers. Then, assign code 1 to men who use electric shavers and code 0 to all other men. In other words, we set up a population of 1's and 0's. Statisticians often refer to a dichotomous population as a population of 1's and 0's.

Clearly then, a population of 1's and 0's is made up of only two points. That is, the population variable X_i, or more simply X, is a discrete variable which can take on only the values 0 and 1. The development of statistical theory relating to the *binomial probability distribution* (Section 8.5) and descriptive measures of a dichotomous population are based on the treatment of the codes 0 and 1 as if they are true item values.

Suppose we are interested in the extent to which a population of 100 youngsters include youngsters who belong to minority groups. The youngsters fall into two classes, those who belong to minority groups (code 1) and all other

TABLE 8.3.1

Population of 1's and 0's: youngsters in a group of 100 who belong to minority groups (code 1) and all other youngsters in the group (code 0).

Code X_i	f_i	$X_i f_i$	$X_i^2 f_i$
0	78	0	0
1	22	22	22
Total	100	22	22

youngsters in the population (code 0), as presented in Table 8.3.1. We may compute the mean of the population of 1's and 0's as follows:

$$\mu = \frac{\sum X_i f_i}{N} = \frac{22}{100} = .22$$

Notice that the numerator 22 is actually f_1, the frequency (number) of 1's in the population. Therefore, the mean μ of a population of 1's and 0's represents the proportion of the population in the category coded 1. Usually p is used to denote this proportion, so that we may write

$$p = \mu = \frac{f_1}{N} \tag{8.3.1}$$

Actually, whenever we are interested in a proportion, we imply a population of 1's and 0's. If we are interested in the proportion of voters who favor candidate Smith, we are thinking in terms of the two categories "voters who are for Smith" and "all other voters."

The standard deviation of a population of 1's and 0's may be computed using Equation 5.2.5:

$$\sigma = \frac{1}{N} \sqrt{N \sum X_i^2 f_i - (\sum X_i f_i)^2}$$

$$= \frac{1}{100} \sqrt{(100)(22) - (22)^2} = .41 \tag{8.3.2}$$

A simpler equation may be determined by noting that $\sum X_i^2 f_i$ and $\sum X_i f_i$ are both equal to f_1 (= 22, see Table 8.3.1). Substituting in Equation 8.3.2, we obtain

$$\sigma = \frac{1}{N} \sqrt{N f_1 - (f_1)^2}$$

$$= \sqrt{\frac{N f_1 - (f_1)^2}{N^2}} \tag{8.3.3}$$

$$= \sqrt{\frac{f_1}{N} - \left(\frac{f_1}{N}\right)^2}$$

Substituting p for f_1/N, we obtain

$$\sigma = \sqrt{p(1 - p)} \tag{8.3.4}$$

where p = the proportion in the population coded 1 and $1 - p$ = the proportion coded 0.

8.4 Sampling Distribution of the Mean

Sampling distributions and sampling models were discussed in Chapter 7. *Statistical inferences are essentially probability statements made on the basis of sampling distributions. Theoretical sampling models,* developed mathematically on the basis of probability theory, are presented in this chapter which describe sampling distributions based on random sampling from theoretical populations. We are not concerned with the mathematical development of the theoretical sampling models, but with an understanding of the models and their application.

Model A

If samples of size n are selected from a normal population with mean μ and standard deviation σ, the *sampling distribution of the mean* is normal, with

mean $\qquad\qquad \mu_{\bar{X}} = \mu$

standard error
of the mean $\qquad\quad \sigma_{\bar{X}} = \dfrac{\sigma}{\sqrt{n}}$

A population of completion times for a chemical process is normally distributed with $\mu = 160$ sec and $\sigma = 18$ sec. What is the probability that a sample of size 36 selected from this population will have a mean greater than 165 sec? The sampling distribution of the mean $\bar{X}$ for this problem, according to Model A, is normal with mean $\mu_{\bar{X}} = 160$ sec and standard error $\sigma_{\bar{X}} = 18/\sqrt{36} = 3$ sec. The required probability is represented by the shaded area under the appropriate normal curve as shown in Figure 8.4.1. In this normal curve the variable is $\bar{X}$, the mean is $\mu_{\bar{X}} = 160$ sec, and the standard deviation is the standard error $\sigma_{\bar{X}} = 3$ sec.

First, transform $\bar{X} = 165$ sec to standard units. The equations to transform a variable to z (Equations 5.4.3 and 5.4.4) may be expressed in words as follows:

$$z = \frac{\text{(specified value of a variable)} - \text{(mean of the variable)}}{\text{(standard deviation (or standard error) of the variable)}} \qquad (8.4.1)$$

Applying Equation 8.4.1 and the equations in Model A to the transformation of $\bar{X}$ values to standard units, we obtain

$$z = \frac{\bar{X} - \mu_{\bar{X}}}{\sigma_{\bar{X}}} = \frac{\bar{X} - \mu}{\sigma/\sqrt{n}} \qquad (8.4.2)$$

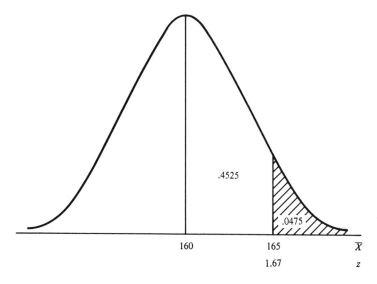

FIGURE 8.4.1

Then, using Equation 8.4.2 transform $\overline{X} = 165$ to z to obtain

$$z = \frac{165 - 160}{3} = 1.67$$

Referring to Table B.4, we find that the area corresponding to $z = 1.67$ is .4525 so that the area in the tail above $\overline{X} = 165$ sec is .0475. Therefore, the probability is .0475, or about 48 chances in 1,000, that the mean of a random sample of 36 completion times will be more than 165 sec.

Let us examine some of the important features of Model A. Notice that the sample size n is not restricted in any way, so that this model is applicable for any sample size. This is not true in all sampling models. The mean of the sampling distribution of the mean $\mu_{\overline{X}}$ is equal to the population mean μ. This is a characteristic common to other sampling models we will consider. Also, we observed this relationship in our discussion of the sampling distribution of the mean in Section 7.4. Finally, examine the equation for the standard error of the mean $\sigma_{\overline{X}}$. Notice that $\sigma_{\overline{X}}$ is directly proportional to σ, the population standard deviation, and inversely proportional to $\sqrt{n}$, the square root of the sample size. In other words, the larger σ, the larger $\sigma_{\overline{X}}$; whereas the larger n, the smaller $\sigma_{\overline{X}}$.

It is to be expected that the greater the population variation (σ large), the greater will be the variation among sample means ($\sigma_{\overline{X}}$ large). Also, the larger the sample size, the more information is available and the greater is our expectation that $\overline{X}$ will be a good estimate of μ. Recall (from our discussion in Section 7.4) that $\sigma_{\overline{X}}$ measures the variation of sample means around μ, so that the smaller $\sigma_{\overline{X}}$ the more may we expect that a particular $\overline{X}$ will be close to μ and provide a good estimate of it. Finally, note that Model A is applicable only if the population sampled is normal (or approximately normal) in distribution.

Model B

If samples of size n, where n is large, are selected from an infinite population with mean μ and standard deviation σ, the *sampling distribution of the mean* is approximately normal, with

mean $\mu_{\bar{X}} = \mu$

standard error
of the mean $\sigma_{\bar{X}} = \dfrac{\sigma}{\sqrt{n}}$

Model B is based on the well-known *central limit theorem* and represents one of the most remarkable models in statistics. In particular, notice that the population could have any type of distribution (within very wide limits) for this model to be applicable. The population must, however, be infinite (or very large) and σ must be finite (which is practically always true in problems encountered in practice).

Model B requires that n be large, which is usually taken to mean that the sample size must be 30 or more. We will speak of samples with $n \geq 30$ as large samples and those with $n < 30$ as small samples. Actually, the size of sample required by Model B depends on how close the population being sampled is to the normal distribution. The closer the population is to the normal distribution the smaller is the required sample size. If it is nearly normal, then Model A is applicable and any sample size satisfies the model. The very broad applicability of Model B to statistical problems, indicating that for a wide range of problems the sampling distribution of the mean is approximately normal, is

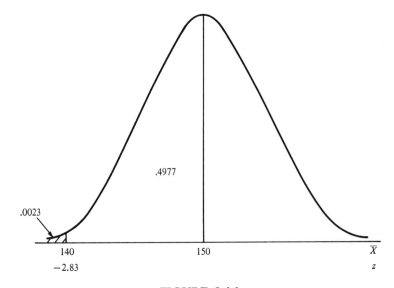

FIGURE 8.4.2

an important reason for the dominant position of the normal distribution in statistics.

The equations for $\mu_{\overline{X}}$ and $\sigma_{\overline{X}}$ are the same in Model B as in Model A so that the transformation of $\overline{X}$ to z is accomplished in the same way for both models (Equation 8.4.2). What is the probability that a sample of 50 social adjustment scores selected from a very large population of such scores ($\mu = 150$ and $\sigma = 25$) will result in a sample mean less than 140? Model A is not applicable, since it is not specified that the sampled population is normal (or approximately normal). Model B may be used, since $n > 30$. Figure 8.4.2 presents the appropriate normal curve and indicates that the required probability is .0023. The details of computation will be left as an exercise for the reader.

8.5   Binomial Probability Distributions

Certain useful theoretical sampling models are based on sampling from an infinite population of 1's and 0's. Consider the bolts produced by a machine in an endless state of production, so that we have an infinite population of bolts. We are interested in the proportion of bolts which are defective. Defective bolts are coded 1 and other bolts are coded 0. Thus, we have an infinite population of 1's and 0's. Let the proportion of defective bolts produced by this machine equal p. Then, p is also the probability that a bolt produced will be defective. Consider the production of a bolt as a "trial" (or experiment). When sampling from such a population the probability of obtaining a defective bolt, p is constant from trial to trial and the trials are independent. A random sample of size n is obtained from this infinite population by selecting the bolts produced in n trials and determining the *frequency* (number) of 1's and 0's in the n trials. (Refer to random sample selection from an infinite population in Section 7.2.)

When sampling from a population of 1's and 0's, interest is typically centered on one of two *equivalent statistics*: the number of 1's obtained in the sample or the proportion of 1's obtained. It is customary to call selection of an item coded 1 a "success" and selection of an item coded 0 a "failure." We may speak of a sample of size n as "n trials." Using these terms, we are interested in the "frequency of successes (or failures) in n trials" or the "proportion of successes (or failures) in n trials." Sampling models based on sampling from an infinite population of 1's and 0's, which we will study, will describe the two equivalent sampling distributions: the sampling distribution of a frequency (X, number of successes) and the sampling distribution of a proportion (X/n, proportion of successes). These sampling distributions are also called *binomial probability distributions*.

We may summarize some of the concepts and the notation as follows:

$p =$ proportion of items coded 1 in the population. Also, $p =$ the probability of a success (obtaining a 1) in a single trial. (In an infinite population, p is constant from trial to trial.)

$1 - p =$ proportion of items coded 0 in the population. Also, $1 - p =$ the probability of a failure (obtaining a 0) in a single trial.

$n =$ sample size or the number of trials.

$X =$ frequency (number) of successes in the sample.

$n - X =$ frequency (number) of failures in the sample.

$X/n =$ proportion of successes in the sample.

$(1 - X)/n =$ proportion of failures in the sample.

Consider a sample of $n = 4$ bolts produced by the machine (4 trials). The possible values of the statistic X (frequency of successes) for such a sample are 0, 1, 2, 3, or 4. The *sampling distribution of a frequency* for $n = 4$ is constructed by computing the probabilities associated with each of the possible values of X: $P(X = 0)$, $P(X = 1)$, $P(X = 2)$, etc. Let us consider how to compute $P(X = 2)$, the probability that 2 defective bolts are obtained in a sample of 4. First, what is the probability that the first two bolts selected are defective and the second two are not defective? Applying probability Rule E extended for independent events and using the foregoing notation, we have

$$
\begin{aligned}
P(1, 1, 0, 0) &= P(1) \cdot P(1) \cdot P(0) \cdot P(0) \\
&= p \cdot p \cdot (1 - p) \cdot (1 - p) \qquad\qquad \textbf{(8.5.1)} \\
&= p^2(1 - p)^2
\end{aligned}
$$

Of course, the two defective bolts could be some other pair, such as the first and third bolts selected, etc. Many combinations of $X = 2$ defective bolts in a sample of $n = 4$ are possible. Applying Equation 7.4.1, we compute the number of possible combinations as

$$
\binom{4}{2} = \frac{4!}{2!2!} = 6 \text{ possible combinations}
$$

Each of these possible combinations represents a sample of 4 bolts containing a specified pair of defective bolts. They represent *mutually exclusive* possible outcomes of the selection of 4 bolts. The probability that a particular one of these possible outcomes will occur may be computed according to Equation 8.5.1. Then applying Rule B extended for mutually exclusive events the probability that a sample of 4 bolts will include 2 defectives (with no interest in which 2 bolts are defective) may be computed as

$$
P(X = 2) = \binom{4}{2} p^2(1 - p)^2
$$

We may write, in general, the probability of obtaining X successes in n trials as

$$
P(X) = \binom{n}{X} p^X(1 - p)^{n-X} \qquad\qquad \textbf{(8.5.2)}
$$

This equation represents the general term of the binomial probability distribution. More specifically, this is the general term of the sampling distribution of a frequency. $\binom{n}{X}$ denotes the number of possible combinations of n trials with

X successes. p^X denotes the probability of obtaining X successes. $(1 - p)^{n-X}$ denotes the probability of obtaining $n - X$ failures.

Suppose, in our illustration, the probability that a bolt is defective is $p = 1/6$. Then, the probability that $X = 2$ defective bolts in a sample of $n = 4$ may be computed using Equation 8.5.2 as

$$P(X = 2) = \binom{4}{2}\left(\frac{1}{6}\right)^2\left(\frac{5}{6}\right)^2 = .116$$

The probability that 3 bolts out of a sample of 4 will be defective is computed as

$$P(X = 3) = \binom{4}{3}\left(\frac{1}{6}\right)^3\left(\frac{5}{6}\right) = .015$$

In the same way, we may compute $P(X = 0)$, $P(X = 1)$, and $P(X = 4)$. These probabilities are presented in Table 8.5.1.

TABLE 8.5.1

Binomial probability distributions: sampling distribution of a frequency (X) and sampling distribution of a proportion (X/n) for a sample of $n = 4$ from an infinite population of 1's and 0's with $p = 1/6$.

X	X/n	$P(X) = P(X/n)$
0	0	.482
1	.25	.386
2	.50	.116
3	.75	.015
4	1.00	.001
	Total	1.000

We may look upon the X values in the first column of Table 8.5.1 as a set of *sample points* making up a *sample space*. The associated probabilities in the last column, $P(X)$, represent the probability distribution of the statistic X. This probability distribution is the *sampling distribution of a frequency*. Clearly, if $X = 1$ and $n = 4$, the associated sample proportion of successes is $X/n = 1/4$ or .25. In other words, the statistics X and X/n are equivalent and represent different ways of saying the same thing. If $P(X = 1) = .482$, as shown in Table 8.5.1, $P(X/n = .25)$ must also equal .482. That is why the last column in the table is headed $P(X) = P(X/n)$. Consequently, the probabilities in this column are also associated with the statistic X/n and represent the *sampling distribution of a proportion*. The two sampling distributions are identical, except that the variable is different (X and X/n).

Model C

If samples of size n are selected from an infinite population of 1's and 0's, where the probability of success in a given trial (p) is constant from trial to trial and the trials are independent:

1. The *sampling distribution of a frequency* (X, the number of successes in the sample) has

mean $\qquad \mu_X = np$

standard error
of a frequency $\qquad \sigma_X = \sqrt{np(1 - p)}$

2. The *sampling distribution of a proportion* (X/n, the proportion of successes in a sample) has

mean $\qquad \mu_{X/n} = p$

standard error
of a proportion $\qquad \sigma_{X/n} = \sqrt{\dfrac{p(1 - p)}{n}}$

Suppose it is known that 40 percent of the very large population of borrowers in the central library of a city are of Oriental ancestry. What is the probability that 60 percent of a sample of 5 borrowers are of Oriental ancestry? Our interest in a proportion (or percentage) permits us to represent the population of borrowers by a population of 1's and 0's, with code 1 assigned to borrowers of Oriental ancestry, code 0 assigned to all other borrowers, and $p = .40$. We want to compute $P(X/n = .60)$. However, Equation 8.5.2 can be used only to compute the probability of a specified frequency of successes $P(X)$, not the probability of a specified proportion of successes $P(X/n)$. Since $X/n = .60$, X equals $.60n = .60(5) = 3$ borrowers with Oriental ancestry in a sample of 5. We may apply Equation 8.5.2 and obtain

$$P\left(\frac{X}{n} = .60\right) = P(X = 3) = \binom{5}{3}(.40)^3(.60)^2 = .23$$

Therefore, the probability is .23, or 23 chances in 100, that a sample of 5 borrowers will include 60 percent (or 3) borrowers with Oriental ancestry. According to Model C we may compute the mean and standard error for the sampling distribution of a frequency for our illustration, as follows:

$$\mu_X = (5)(.4) = 2$$
$$\sigma_X = \sqrt{(5)(.4)(.6)} = 1.1$$

What do these measures mean? Consider a very large number of samples of 5 borrowers each selected from the population of borrowers in our illustration. The value of X (the number of borrowers in a sample with Oriental ancestry) may be expected to vary from sample to sample. In some samples it will equal 0; in some, 1; etc. The mean of all the X values obtained will tend to equal $\mu_X = 2$ and the standard deviation of all the X values will tend to equal $\sigma_X = 1.1$. We may compute such measures for the sampling distribution of a proportion, according to Model C as follows:

$$\mu_{X/n} = p = .40$$
$$\sigma_{X/n} = \sqrt{\frac{(.40)(.60)}{5}} = .22$$

Then, $\mu_{X/n} = .40$ represents the mean of all sample proportions X/n obtained for a very large number of samples of size 5 and $\sigma_{X/n} = .22$ is the standard deviation of these sample proportions.

8.6   Normal Approximation to Binomial Probabilities

What is the probability that a sample of $n = 100$ selected from a very large (infinite) population of 1's and 0's with $p = .5$ will contain $X/n = .7$ or $X = 70$ successes? Applying Equation 8.5.2, we have

$$P\left(\frac{X}{n} = .7\right) = P(X = 70) = \binom{100}{70}(.5)^{70}(.5)^{30}$$

The computation of binomial probabilities for large n, such as the foregoing, according to Equation 8.5.2 is very laborious. Moreover, interest usually lies in the probability that X/n or X will lie in some interval. As an example, in the foregoing illustration, what is the probability that the sample will include at least 70 successes? That is, determine $P(X/n \geq .7)$ or $P(X \geq 70)$. This requires computing $P(X = 70)$, $P(X = 71), \ldots, P(X = 100)$ and then adding in accordance with Rule B extended for mutually exclusive events. Certainly, this involves considerable labor.

It has been shown mathematically that the binomial probability distribution is closely approximated by the normal curve under certain conditions, thus, providing a convenient basis for determining binomial probabilities. The binomial probability distribution is a two parameter distribution, with parameters n and p (Equation 8.5.2). How well the normal distribution approximates the binomial probability distribution depends on the values of these parameters. Examine Figure 8.6.1, parts A, B, C. In these three distributions $p = .5$, and $n = 4$ in part A, $n = 10$ in part B, and $n = 20$ in part C. Each distribution is symmetrical around a central peak (since $p = .5$). Notice that as n increases the distribution resembles the normal curve more closely. For small (or large) p values, the distribution is markedly skewed as indicated in Figure 8.6.2, parts A, B, C, and D. In each of these distributions $n = 20$, whereas p varies: $p = .04$ in part A, $p = .05$ in part B, $p = .1$ in part C, and $p = .2$ in part D. Notice how the distribution approaches symmetry as p approaches .5.

Only when $p = .5$ are p and $1 - p$ equal, resulting in a symmetrical distribution. When $p = .5$ and n is large, the normal curve very closely approximates the binomial probability distribution. As p deviates from .5 (above or below), the binomial probability distribution becomes skewed and the approximation becomes less acceptable, unless n is increased. Generally, then, n must be sufficiently large for the normal curve to provide useful approximations of binomial probabilities. As a rule of thumb, n should be 100 or more, however, if p deviates considerably from .5 (less than .3 or greater than .7), then an n of 500 or more is desirable.

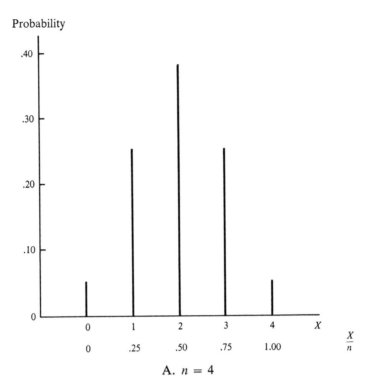

A. $n = 4$

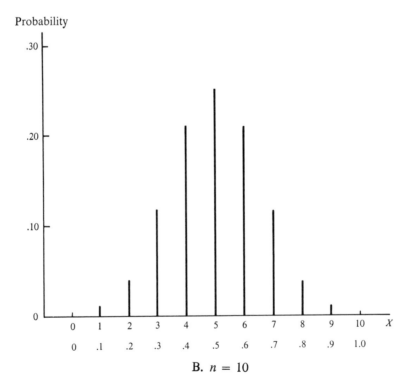

B. $n = 10$

140

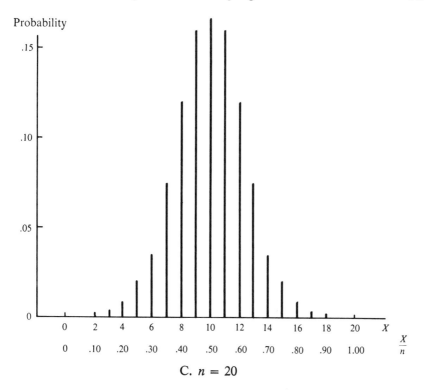

C. *n* = 20

FIGURE 8.6.1(A–C)

Binomial probability distribution: sampling from a popu-lation with p = .5.

Useful, but rough, normal approximations to binomial probabilities may be obtained when *p* is not too close to 0 or 1 for sample sizes (*n*) which satisfy both of the following inequalities:

$$np \geq 5 \qquad \qquad (8.6.1)$$

$$n(1 - p) \geq 5 \qquad \qquad (8.6.2)$$

These inequalities provide a means for determining the *minimum sample size* to use for a given *p* value; however, often considerably larger samples are re-quired. If *p* is less than 1 − *p*, then it is necessary to make only the comparison in Equation 8.6.1. If 1 − *p* is less than *p*, then only Equation 8.6.2 need be used. For example, when sampling from a large population of 1's and 0's with *n* = 60 and *p* = .1, is it appropriate to use the normal approximation? Since *p* = .1 is less than 1 − *p* = .9, compute (60)(.1) = 6 according to Equation 8.6.1. Since this product (6) is larger than 5, we conclude that the normal approximation is useful (in the minimum sense explained). Clearly, there is no need to use Equation 8.6.2, since (60)(.9) must be larger than 6.

Let us compare some binomial probabilities computed according to Equation 8.5.2 with the normal curve approximations. When sampling from a large

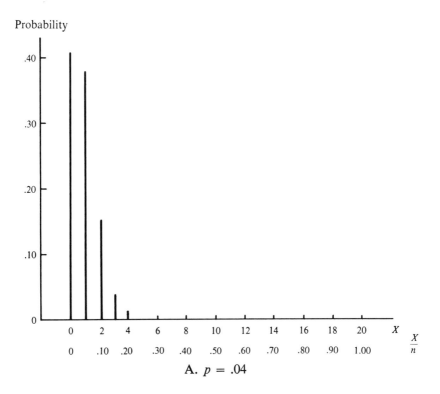

A. $p = .04$

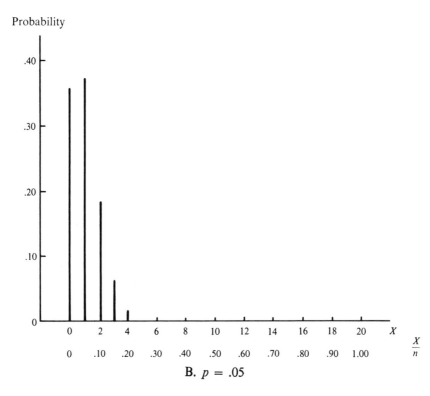

B. $p = .05$

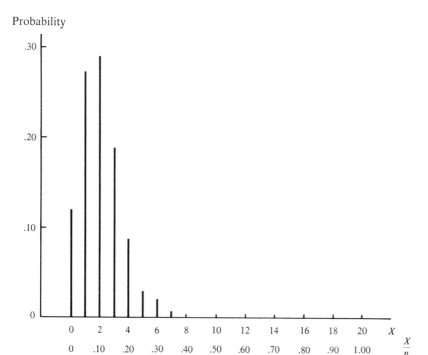

C. $p = .1$

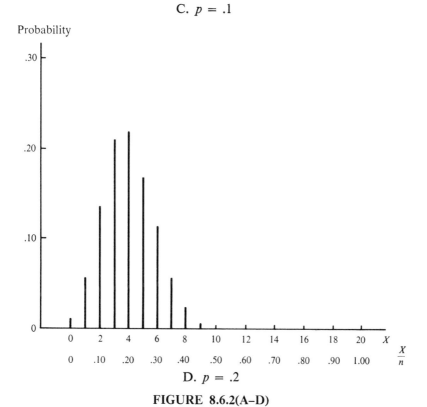

D. $p = .2$

FIGURE 8.6.2(A–D)

Binomial probability distribution: sampling from a population with n = 20.

143

population of 1's and 0's with $p = .5$, what is the probability that a sample of size 10 will include 7 successes? The mean and standard error of the corresponding *sampling distribution of a frequency*, computed according to Model C are

$$\mu_X = (10)(.5) = 5$$
$$\sigma_X = \sqrt{(10)(.5)(.5)} = 1.58$$

Then, $P(X = 7)$ may be estimated based on the area under the normal curve between $X = 6.5$ and $X = 7.5$. (The use of the normal curve to estimate probabilities for *discrete variables* was discussed in Section 8.2. See Figure 8.2.7.) Transformation of $X = 6.5$ and $X = 7.5$ to standard units is accomplished by applying Model C and expressing Equation 8.4.1 as follows:

$$z = \frac{X - \mu_X}{\sigma_X} = \frac{X - np}{\sqrt{np(1 - p)}} \qquad (8.6.3)$$

Transforming $X = 6.5$ and $X = 7.5$ to z we obtain

$$z_1 = \frac{6.5 - 5}{1.58} = .95$$

$$z_2 = \frac{7.5 - 5}{1.58} = 1.58$$

Referring to Table B.4, the area corresponding to z_1 is .3289 and the area corresponding to z_2 is .4429, so that the required probability is the difference, .1140. Using Equation 8.5.2, we compute $P(X = 7) = .1172$. Consequently, the normal curve estimate (.1140) is .0032 less than the exact probability (.1172).

In the foregoing example, what is the probability that a sample of 10 will include at least 7 successes, $P(X \geq 7)$? We may obtain the normal curve approximation by determining the area in the upper tail of the normal curve to the right of $X = 6.5$ or in standard units, $z = .95$ (previously computed). This results in the estimate $P(X \geq 7) = .1711$. We may compute the exact probabilities $P(X = 7)$, $P(X = 8)$, $P(X = 9)$, $P(X = 10)$ using Equation 8.5.2 and add them to obtain the exact value for $P(X \geq 7)$. This gives $P(X \geq 7) = .1719$. The difference between the normal approximation and the exact probability is .0008.

When sampling from a large population of 1's and 0's with $p = .25$ and $n = 20$, determine $P(X/n \geq .50)$, the probability that at least 50 percent of the sample will be successes. The mean and standard error of the corresponding *sampling distribution of a proportion*, computed according to Model C are

$$\mu_{X/n} = .25$$

$$\sigma_{X/n} = \sqrt{\frac{(.25)(.75)}{20}} = .097$$

Then, $P(X/n \geq .50)$ may be estimated based on the appropriate area under the normal distribution with a mean of .25 and standard deviation .097. Of course, X/n is a discrete variable, just as X is discrete, since the possible values of X/n

are obtained by dividing the possible values of X by n. Consequently, we must take into account the area under the normal curve in a continuous interval around .50. When $X/n = .50$ for a sample with $n = 20$, X equals 10 and the continuous interval around X is 9.5 to 10.5. Let us write these limits as

$$X = 10 - .5 = 9.5$$
$$X = 10 + .5 = 10.5$$

Now, divide each of these X values by $n = 20$ to obtain the corresponding limits for the sample proportion $X/n = .50$. This gives

$$\frac{X}{n} = \frac{10}{20} - \frac{.5}{20} = \frac{9.5}{20} = .475$$
$$\frac{X}{n} = \frac{10}{20} + \frac{.5}{20} = \frac{10.5}{20} = .525$$

Hence, to determine the normal approximation for $P(X/n = .50)$, we must find the area under the normal curve for the continuous interval .475 to .525. In general, to estimate the probability that the sample proportion X/n takes on a specified value when a sample of size n is selected from a large population of 1's and 0's, we find the area under the normal curve between

$$\frac{X}{n} = \text{(specified sample proportion)} - \frac{.5}{n} \qquad \textbf{(8.6.4)}$$

and
$$\frac{X}{n} = \text{(specified sample proportion)} + \frac{.5}{n} \qquad \textbf{(8.6.5)}$$

Applying these equations to our illustration, we obtain

$$\frac{X}{n} = .50 - \frac{.5}{20} = .475$$

$$\frac{X}{n} = .50 + \frac{.5}{20} = .525$$

Of course, these are the same limits as previously computed. In our problem where $n = 10$ and $p = .25$, $P(X/n \geq .50)$ is represented by the area under the normal curve to the right of $X/n = .475$. Transformation of X/n to standard units is accomplished by applying Model C and expressing Equation 8.4.1 as follows:

$$z = \frac{\frac{X}{n} - \mu_{X/n}}{\sigma_{X/n}} = \frac{\frac{X}{n} - p}{\sqrt{\frac{p(1-p)}{n}}} \qquad \textbf{(8.6.6)}$$

Transforming $X/n = .475$ to standard units, we obtain

$$z = \frac{.475 - .25}{.097} = 2.32$$

Finally, referring to the table of normal curve areas, we estimate $P(X/n \geq .50) = .0102$ or somewhat over one chance in a hundred.

Model D

If samples of size n (sufficiently large) are selected with equal and independent probabilities from an infinite population of 1's and 0's, where p (the proportion of successes in the population) is not very close to 0 or 1:

1. The *sampling distribution of a frequency* (X, the number of successes in a sample) is approximately normal in distribution with

mean	$\mu_X = np$
standard error of a frequency	$\sigma_X = \sqrt{np(1-p)}$

2. The *sampling distribution of a proportion* (X/n, the proportion of successes in a sample) is approximately normal in distribution with

mean	$\mu_{X/n} = p$
standard error of a proportion	$\sigma_{X/n} = \sqrt{\dfrac{p(1-p)}{n}}$

Notice that Model D requires that p must not be very close to 0 or 1 and n must be sufficiently large (as previously discussed). The requirement that samples are to be selected with *equal and independent probabilities* is another way of stating that p is constant from trial to trial and the trials are independent (as stated in Model C). When the conditions of the model are met, the binomial probability distributions are sufficiently similar to the normal distribution to provide useful estimates of binomial probabilities based on the normal curve.

Suppose a five-ton supply of seeds includes 8 percent of a special type. What is the probability that a sample of 1,000 seeds will include 9 percent to 10 percent of the special type? The statistic involved in this problem is X/n and the probability required is $P(.09 \le X/n \le .10)$. The mean and standard deviation of the appropriate normal curve may be determined according to Model D as follows:

$$\mu_{X/n} = .08$$

$$\sigma_{X/n} = \sqrt{\frac{(.08)(.92)}{1,000}} = .0086$$

We are interested in the area under the normal curve for the continuous interval determined according to Equations 8.6.4 and 8.6.5 as follows:

$$\frac{X}{n} = .09 - \frac{.5}{1,000} = .0895$$

$$\frac{X}{n} = .10 + \frac{.5}{1,000} = .1005$$

Then, transforming these limits to standard units (Equation 8.6.6), we obtain

$$z_1 = \frac{.0895 - .08}{.0086} = 1.10$$

$$z_2 = \frac{.1005 - .08}{.0086} = 2.38$$

Referring to Table B.4, the corresponding areas are .3643 (for z_1) and .4913 (for z_2). Subtracting the smaller area from the larger, we estimate $P(.09 \le X/n \le .10) = .1270$ or 127 chances in 1,000.

8.7 Sampling Models: A Basis for Statistical Inference

Inductive statistics provide methods for making statistical inferences about a population based on sample data and for evaluating the risks involved, as noted in Section 2.2. *The sampling distribution provides the conceptual basis for development of the methods of inductive statistics.* Application of these methods requires a knowledge of the distribution of the population from which a sample was selected and description of the sampling distribution of the statistic involved. Such information is typically not available. The usual procedure is to relate the problem at hand to an appropriate *theoretical sampling model*, thus, making available the full scope of information contained in the model and permitting the application of inductive statistical procedures. Of course, no problem encountered in practice can precisely fit a theoretical model. Therefore, we must consider how such models can be used to solve statistical problems. This will be considered in subsequent chapters.

Four statistical sampling models were presented in this chapter. Additional models will be introduced from time to time as needed.

Study Problems

1. What are the parameters of the normal distribution?

2. In the normal distribution: (a) Is the variable X discrete or continuous? (b) What is the range of values for X? (c) How large is the theoretical normal population?

3. How is the normal curve affected if (a) μ is increased, (b) μ is decreased, (c) σ is increased, and (d) σ is decreased.

4. What is the variable in the standard normal distribution?

5. Using Table B.4, determine the proportion of the area under the normal curve:
    a. Between the mean and $z = 1.20$
    b. Above $z = 1.20$
    c. Below $z = -1.20$
    d. Between $z = -1.40$ and $z = 1.54$
    e. Below $z = -1.90$ and above $z = 1.90$
    f. Between $z = -1.96$ and $z = 1.96$

6. Using Table B.4, determine the proportion of the area under the normal curve:
    a. Between $z = -2.20$ and the mean
    b. Above $z = -2.20$
    c. Below $z = 2.40$
    d. Between $z = 2.28$ and $z = 2.52$
    e. Between $z = -2.40$ and $z = -1.96$

7. A very large population of weights is approximately normal in distribution, with a mean of 200 lbs and a standard deviation of 10 lbs. Estimate the percentage of the weights in the population which are:

    a.   Below 180 lbs
    b.   Above 212 lbs
    c.   Between 210 lbs and 220 lbs
    d.   Between 193 lbs and 220 lbs
    e.   Between 175.50 lbs and 179.77 lbs
    f.   Less than 193 lbs or more than 221.21 lbs
    g.   Within 15.32 lbs of the mean
    h.   Over 200 lbs

8.   A large population of time-reaction measurements is approximately normal in distribution, with a mean of 75 sec and a standard of 5 sec. What is the probability that a time-reaction measurement selected at random from this population will be

    a.   Less than 60 sec
    b.   Between 62 sec and 71 sec
    c.   Between 60.5 sec and 76.5 sec
    d.   More than 79.62 sec
    e.   Between 80.3 sec and 87.4 sec
    f.   More than 7.6 sec from the mean
    g.   Over 75 sec

9.   The time required by a job applicant to complete an application for employment has been found to be normally distributed, with a mean of 25.5 minutes and a standard deviation of 8.0 minutes. What is the probability that a randomly selected job applicant will require

    a.   More than 35 minutes
    b.   20 minutes or less
    c.   Ten minutes or more above the average

10.   The number of fittings per box, for a very large population of boxes, is approximately normal in distribution, with a mean of 50 fittings per box and a standard deviation of 5 fittings per box. Taking into account that the variable "number of fittings per box" is discrete, what is the probability that a randomly selected box will contain

a. Exactly 60 fittings	b. At least 60 fittings
c. At most 60 fittings	d. Over 60 fittings
e. Less than 60 fittings	f. 40 to 45 fittings
g. 40 to 53 fittings	h. 62 to 69 fittings

11.   The number of cases handled per day (excluding Saturdays and Sundays) by a small claims court is considered to be well approximated by a normal curve with a mean of 12 cases per day and a standard deviation of three cases per day. What is the probability that on a randomly selected day

    a.   Ten cases will be handled?
    b.   Over 12 cases will be handled?
    c.   Between 10 and 12 cases will be handled?

12.   The time required to audit an account is known to be normally distributed, with a mean of 35 minutes and a standard deviation of 4 minutes. Determine:

    a.   The most time required for the slowest 30 percent of the account audits.
    b.   The least time required for the fastest 5 percent of the account audits.

13.   In the distribution of 1's and 0's: (a) Is the variable X discrete or continuous? (b) What are the possible values of X?

14.   Why is a dichotomous population called a population of 1's and 0's?

15.   How would you interpret the mean of a dichotomous population?

16.   An agricultural scientist is interested in the proportion of farms in a state which are owner-operated. Explain how this implies a population of 1's and 0's.

17.   A political science student is interested in the proportion of small cities in the United States which have a city manager form of government. Explain how this implies a population of 1's and 0's.

18.   Compute the mean and standard deviation for the population of 1's and 0's implied:

   a.   A city of 5,000 registered voters includes 2,000 voters who are naturalized citizens (category 1).

   b.   350 out of a total of 1,400 IQ scores are in the "below normal" category (coded 1).

   c.   It was found that 10 out of 50 doctors in a city do not make house calls (category 1).

19.   Statistical inferences are essentially probability statements based on what type of distributions?

20.   Describe the sampling distribution of the mean (by specifying the shape of the distribution, the mean, and the standard error) for samples of any size selected from a normal population. Which sampling model is appropriate?

21.   Describe the sampling distribution of the mean (as noted in Problem 20) for large samples selected from a population with unknown distribution. Which sampling model is appropriate?

22.   Distinguish between a large sample and a small sample.

23.   When sample size changes, what is the effect on the mean and standard error of the sampling distribution of the mean?

24.   How is the size of the standard error of the mean affected by changes in the population standard deviation?

25.   It is not always necessary for sample size to be at least 30 for Model B to be applicable. Explain.

26.   A machine is set to produce bolts with a mean diameter of 1.50 cms and a standard deviation of .20 cms. Diameters of bolts produced by the machine are approximately normal in distribution. What is the probability that a sample of 16 bolts produced by the machine will have a mean of

   a.   1.42 cms or less   b.   At least 1.55 cms
   c.   At most 1.55 cms   d.   Between 1.45 cms and 1.47 cms

27.   Which sample model did you use for Problem 26?

28.   A very large population of time measurements is normally distributed, with a mean of 66 seconds and a standard deviation of ten seconds. What is the probability that a random sample of 25 measurements from this population will have a mean between 65 seconds and 67 seconds?

29.   In Problem 28, compute the required probability if
   a. $n = 49$ b. $n = 100$
   c. $\sigma = 5$ sec, $n = 25$ d. $\sigma = 20$ sec, $n = 25$

30.   A large population of test scores is approximately normal in distribution with a mean of 90 and a standard deviation of 10. If a sample of 36 scores is randomly selected (and $\overline{X}$ is considered a continuous variable), what is the probability that

    a. $\overline{X} \geq 92$ b. $\overline{X} < 87$
    c. $88 \leq \overline{X} \leq 92$ d. $\overline{X} > 89$

31.   In Problem 30, suppose $n = 16$. What is the probability that
        a. $\overline{X} \geq 92$ b. $88 \leq \overline{X} \leq 92$

32.   In Problem 30, suppose $n = 100$.
        a. $P(\overline{X} \geq 92) = ?$ b. $P(88 \leq \overline{X} \leq 92) = ?$

33.   In Problem 30, suppose $\sigma = 5$, $n = 16$.
        a. $P(\overline{X} \geq 92) = ?$ b. $P(88 \leq \overline{X} \leq 92) = ?$

34.   In Problem 26, if you have no information as to the type (shape) of the distribution of the diameter of the bolts produced by the machine, could you compute the required probabilities by the application of Model A or Model B? Why?

35.   A large population of weights has a mean of 93.68 lbs and a standard deviation of 17.44 lbs. What is the probability that the mean of a sample ($n = 40$) selected from this population
        a. Will lie between 90 lbs and 95 lbs
        b. Will be 90 lbs or less
        c. Which sampling model did you use

36.   In Problem 35, if $n = 160$, what is the probability that $\overline{X}$ will lie between 90 lbs and 95 lbs?

37.   A large population of production rates has $\mu = 124.6$ and $\sigma = 30.4$. What is the probability that in a sample of $n = 100$ $\overline{X}$ will
        a. Lie between 119.5 and 129.5
        b. Equal 130 or more
        c. Equal no more than 118

38.   In Problem 37, if $n = 200$, find $P(119.5 \leq \overline{X} \leq 129.5)$.

39.   The general term of the binomial probability distribution may be expressed as the product of three factors: the number of possible combinations of n trials with X successes, the probability of obtaining X successes, and the probability of obtaining $n - X$ failures. Identify each of these factors in Equation 8.5.2.

40.   Given an infinite (or very large) population of 1's and 0's, with $p = .30$. For samples of size 9, compute and interpret:
        a. μ_X b. σ_X
        c. $\mu_{X/n}$ d. $\sigma_{X/n}$

41.   Given an infinite (or very large) population of 1's and 0's, with $p = .20$. Is it useful (in a minimum sense) to use the normal approximation to compute binomial probabilities, if (a) $n = 5$, (b) $n = 10$, and (c) $n = 25$? Why?

42.   Describe the sampling distribution of a frequency X (by specifying the shape of distribution, the mean, and the standard error) for samples selected from an infinite population of 1's and 0's. Which sampling model did you use? What restrictions are contained in the model?

43.   Describe the sampling distribution of a proportion (as noted in Problem 42) for samples selected from an infinite dichotomous population. Which sampling model did you use? What restrictions are contained in the model?

44.   A study of 500 medium-sized cities showed that 200 recorded a drop in the crime rate last month (category 1). Compute the mean and standard error for the

   a. Sampling distribution of a frequency for samples of size 10
   b. Sampling distribution of a proportion for samples of size 10

45.   In Problem 44, assume sampling with replacement. What is the probability of
   a.   Two successes, if $n = 3$ b.   No successes, if $n = 4$
   c.   At least one success, if $n = 4$ d.   At most one success, if $n = 3$

46.   In an infinite population of 1's and 0's with $p = .4$, what is the probability of
   a.   25 percent successes, if $n = 4$ b.   40 percent successes, if $n = 5$
   c.   .5 successes, if $n = 4$ d.   1/3 successes, if $n = 6$

47.   It is known that 350 out of a lot of 1,400 TV tubes are defective. If defective
tubes are coded 1 and the others 0, compute the mean and standard error for the sampling distribution of a
   a.   Frequency $(n = 25)$ b.   Proportion $(n = 25)$

48.   In Problem 47, if a sample of four TV tubes is selected (with replacement), what
is the probability that
   a.   They are all defective?   b.   50 percent are defective?
   c.   One-fourth are defective?   d.   One or two are defective?

49.   In a certain region, it was found that 10 out of 50 private clubs practiced some
type of discrimination (category 1), whereas the other 40 had no discrimination
(category 0).
   a.   Compute the mean and standard deviation for the appropriate population of 1's and 0's.
   b.   Compute the mean and standard error for the sampling distribution of
   a frequency $(n = 8$, sampling with replacement).
   c.   Compute the mean and standard error for the sampling distribution
   of a proportion $(n = 8$, sampling with replacement).
   d.   What is the probability of success in a single trial for this population?

50.   A test scoring machine is known to make errors 20 percent of the time. Compute
the probability that
   a.   4 out of $n = 5$ tests scored will be in error.
   b.   Less than 4 out of $n = 4$ tests scored will be in error.
   c.   3/4 of $n = 4$ tests scored will be in error.
   d.   Less than 1/4 of $n = 4$ tests scored will be in error.
   e.   At least one out of $n = 4$ tests scored will be in error.

51.   A machine is known to produce a defective item 30 percent of the time.
   a.   Determine the sampling distribution of X (the number of defective
   items) for $n = 4$.
   b. $P(X > 2) = ?$
   c. $P(X < 3) = ?$
   d. $P(X/n > .25) = ?$
   e. $P(0 < X/n < .75) = ?$

52.   If 60 percent of the felons out on bail are known to commit crimes, determine
the sampling distribution of X/n (the proportion of felons out on bail who commit
crimes) for $n = 5$.

53.   In Problem 52:
   a. $P(X/n \geq .80) = ?$ b. $P(X/n < .40) = ?$
   c. $P(X = 3) = ?$ d. $P(2 \leq X \leq 4) = ?$

54.   In a large metropolitan area, it is known that 75 percent of the families have two or more children. Using the normal curve approximation, compute the probability that in a random sample of 200 families the number having two or more children are

      a.   147

      b.   153 or more

      c.   Between 146 and 150

55.   In Problem 54, compute the probability that the percentage of families in the sample with two or more children are

      a.   73 percent

      b.   Over 74 percent

      c.   Between 74 percent and 76 percent

56.   A study indicated that 64 percent of married women vote as their husbands do. Compute the probability that a random sample of 150 married women includes

      a.   No more than 95 who vote like their husbands.

      b.   Between 97 and 110 who vote like their husbands.

      c.   Between 60 percent and 70 percent who vote like their husbands.

      d.   Over 70 percent who vote like their husbands.

57.   It has been determined that 8 percent of the cars passing through a certain very busy intersection skid. What is the probability that no more than 7 percent of the next 2,500 cars passing through this intersection will skid?

58.   In Problem 57, what is the probability that no more than 5 percent of the 2,500 cars will skid?

59.   In Problem 57, compute and explain the meaning of

      a.   The mean and standard deviation of the appropriate dichotomous population

      b.   The mean and standard error of the sampling distribution of a frequency

      c.   The mean and standard error of the sampling distribution of a proportion

Supplementary Study Problems

60.   If samples of size $n \geq 30$ are selected from a normal population with mean (and median) μ and standard deviation σ, the sampling distribution of the median is approximately normal with mean $\mu_M = \mu$ and standard error of the median $\sigma_M = 1.25\sigma/\sqrt{n} = 1.25\sigma_{\bar{x}}$. In Problem 30:

      a.   What is the probability that the sample median M is less than 87?

      b.   What is the probability that the sample median $M > 89$?

61.   In Problem 30 (and Problem 60), compute the standard error of the median and explain what it means.

Part IV
Statistical
Decision Making

start here

Chapter 9

Confidence Intervals

9.1 Estimation

An important application of inductive statistics is the estimation of population parameters, such as means and proportions, based on random samples. For example, a businessman may wish to estimate the weekly mean number of merchandise returns, a chemist may wish to estimate the mean reaction time of a chemical process, a politician may wish to estimate the proportion of voters who favor his candidacy. An estimate may be one of two types, a *point estimate* or an *interval estimate*. For example, a sociologist may wish to determine the mean score for a population of adult females on a racial acceptance scale. Suppose he selects a random sample of females from the population and determines the sample mean score $\overline{X} = 150$. He then estimates the population mean μ as 150. This is a *point estimate. The sample mean $\overline{X}$ is used as a point estimate of the population mean.*

155

Of course, it is not likely that a sample mean will exactly equal the population mean. As we have seen, there is a full distribution of $\overline{X}$ values possible when a random sample is selected from a population (the sampling distribution of the mean). It is customary to construct an *interval estimate* to indicate how good an estimate of μ is provided by a sample mean. For example, the sociologist may estimate that he is 95 percent confident (the probability is .95) that the interval 150 ± 5 or 145–155 includes the population mean score on the racial acceptance scale. Notice that *an interval estimate specifies a range of values wherein, with a stated degree of confidence, it is claimed that the parameter lies.*

Point and interval estimates are also computed for a population proportion. Suppose an economist determines that .85 of a sample of businessmen expect an increase in the gross national product (GNP) next year. He would use the sample proportion $X/n = .85$ as a point estimate of the population proportion p. He may also state that he is 98 percent confident that the interval $.85 \pm .03$ (or .82–.88) includes the population proportion of businessmen who expect a rise in the GNP next year. An interval estimate as described is called a *confidence interval.*

9.2   Means and Total Amounts: Large Samples

Random sampling from an *infinite population* implies independent trials and constant probability of selection from trial to trial (Section 7.2). These conditions of sampling are closely approximated when sampling without replacement from a large finite population, if the *sampling ratio* n/N is small (for example, less than .05). Suppose a random sample of 100 weights is to be selected from a population of 5,000 weights (so that $n/N = 100/5,000 = .02$). On the first trial, when a weight is to be selected from the 5,000, the probability is $1/5,000 = .0002$ that a given weight will be selected. On the second trial, when a weight is to be selected from the remaining 4,999, the probability is $1/4,999$ that one of the remaining weights will be selected. This probability is only negligibly different from .0002. Even when the last sample weight is to be selected (the 100th weight), the probability is $1/4,901 = .000204$ that a given one of the remaining weights will be selected. Clearly, the probability is virtually constant from trial to trial and the distribution of weights in the population is not appreciably affected during sample selection. Hence, if n/N is less than .05, sampling without replacement from a large finite population may be considered similar to sampling from an infinite population.

Consequently, if we restrict ourselves to selection of large samples ($n \geq 30$), such sampling situations may be represented by Model B (which specifies an infinite population and large samples). We know from this model that the sampling distribution of the mean is approximately normal, with mean $\mu_{\overline{X}} = \mu$ and standard error $\sigma_{\overline{X}} = \sigma/\sqrt{n}$ as shown in Figure 9.2.1. Any $\overline{X}$ value along the horizontal scale in the figure represents a possible sample mean. When a sample is selected to estimate a population mean and the sample mean $\overline{X}$ is used as a point estimate of μ, we do not know how accurate an estimate we have. Fortunately, most of the $\overline{X}$ values possible are concentrated around the population mean μ, as indicated by the central peak in the sampling distribution, so the chances that a particular $\overline{X}$ provides an acceptable estimate of μ are quite good.

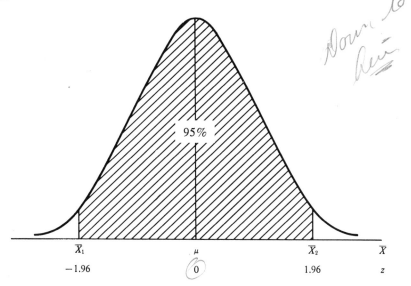

FIGURE 9.2.1

Let $\overline{X}_1$ and $\overline{X}_2$ represent possible sample means, as shown in the figure. These $\overline{X}$ values were selected so that $\overline{X}_1$ corresponds to $z = -1.96$ and $\overline{X}_2$ corresponds to $z = 1.96$. Hence, $\overline{X}_1$ and $\overline{X}_2$ are limits which mark off an interval wherein 95 percent of the possible sample means lie. Alternatively, the probability is .95 that the mean of a random sample $\overline{X}$ will lie in this interval. We may express this probability as follows:

$$P(\overline{X}_1 < \overline{X} < \overline{X}_2) = .95 \qquad (9.2.1)$$

Transforming to standard units, this equation becomes

$$P\left(-1.96 < \frac{\overline{X} - \mu}{\sigma_{\overline{X}}} < 1.96\right) = .95 \qquad (9.2.2)$$

When a random sample is selected in order to estimate a population mean, not only is μ unknown but σ is usually unknown as well. Consequently, we cannot compute the standard error of the mean $\sigma_{\overline{X}} = \sigma/\sqrt{n}$ and must use the *estimated standard error of the mean*

$$s_{\overline{X}} = \frac{s}{\sqrt{n}} \qquad (9.2.3)$$

Substituting in Equation 9.2.2, we obtain

$$P\left(-1.96 < \frac{\overline{X} - \mu}{s_{\overline{x}}} < 1.96\right) = .95 \qquad (9.2.4)$$

The probability expressed in Equation 9.2.2 (.95) continues to hold in Equation 9.2.4 because we are dealing with large samples.

Consider the term inside the parentheses in Equation 9.2.4

$$-1.96 < \frac{\overline{X} - \mu}{s_{\overline{x}}} < 1.96 \qquad (9.2.5)$$

Multiplying through by $s_{\overline{x}}$ and then subtracting $\overline{X}$ from each term, we obtain

$$-\overline{X} - 1.96 s_{\overline{x}} < -\mu < -\overline{X} + 1.96 s_{\overline{x}} \qquad (9.2.6)$$

Finally, multiplying through by -1 (which requires that the inequality signs be reversed) and then reversing the sequence of terms for convenience, we obtain

$$\overline{X} - 1.96 s_{\overline{x}} < \mu < \overline{X} + 1.96 s_{\overline{x}} \qquad (9.2.7)$$

Equation 9.2.7 is a modified version of Equation 9.2.5, which was removed from the parentheses in Equation 9.2.4. Returning Equation 9.2.7 to the parentheses, we obtain

$$P(\overline{X} - 1.96 s_{\overline{x}} < \mu < \overline{X} + 1.96 s_{\overline{x}}) = .95 \qquad (9.2.8)$$

What does Equation 9.2.8 say? This equation states that the probability is .95 that the *variable limits* $\overline{X} - 1.96 s_{\overline{x}}$ and $\overline{X} + 1.96 s_{\overline{x}}$ include the population mean μ. If we think in terms of *repeated sampling* from a population, we may visualize such a pair of limits computed from each sample. According to Equation 9.2.8, 95 percent of such pairs of limits computed for all possible samples of a given size will define intervals which will include μ. Alternatively, the probability is .95 that such an interval computed from a given random sample will include μ. Such an interval is called a *confidence interval* and the related probability (.95) is called the *confidence coefficient*. The limits which define the interval are called *confidence limits*. We may write the equations for these limits as:

Lower Limit $\qquad \text{LL} = \overline{X} - 1.96 s_{\overline{x}} = \overline{X} - 1.96 \dfrac{s}{\sqrt{n}} \qquad (9.2.9)$

Upper Limit $\qquad \text{UL} = \overline{X} + 1.96 s_{\overline{x}} = \overline{X} + 1.96 \dfrac{s}{\sqrt{n}} \qquad (9.2.10)$

A more compact way of expressing these limits is

$$\overline{X} \pm 1.96 \frac{s}{\sqrt{n}} \qquad (9.2.11)$$

Considering the fact that 95 percent of the intervals so constructed may be expected to include μ (based on the notion of repeated sampling), we speak of a particular interval estimate constructed according to Equation 9.2.11 as a .95 confidence interval.

A sample of 144 workers in a region has a mean weekly wage of $148 and a standard deviation of $28. Construct a .95 confidence interval for the mean weekly wage in the population. Applying Equation 9.2.11, we compute:

$$148 \pm 1.96 \frac{28}{\sqrt{144}} = 148 \pm 4.57$$

Then,
$$\text{LL} = 148 - 4.57 = \$143.43$$
$$\text{UL} = 148 + 4.57 = \$152.57$$

Consequently, we are 95 percent confident that the interval $143.43–$152.57 includes the population mean weekly wage. (There is a 5 percent risk that this interval does not include μ.) Note that $\overline{X} = \$148$ represents a *point estimate of* μ and $1.96s/\sqrt{n} = \$4.57$ represents the *maximum amount* by which this point estimate could be off (above or below μ), at the .95 level of confidence. Generally, $1.96s/\sqrt{n}$ represents the maximum amount by which $\overline{X}$ could be in error, at the .95 level of confidence, when it is used as an estimate of μ.

Confidence intervals may be constructed at any desired level of confidence. It will be recalled that we used $z = -1.96$ and $z = 1.96$ in Figure 9.2.1, so that the area between these two limits makes up 95 percent of the area under the curve. If we want to construct a .98 confidence interval, we must find z values which include 98 percent of the area in the central portion of the curve. Such values are $z = \pm 2.33$. Enter these z values into Equations 9.2.9, 9.2.10, and 9.2.11 in place of ± 1.96. To construct a confidence interval at any stated level of confidence, we must use the appropriate z values. We may write the confidence limit equations more generally as

$$\text{LL} = \overline{X} - z \frac{s}{\sqrt{n}} \qquad (9.2.12)$$

$$\text{UL} = \overline{X} + z \frac{s}{\sqrt{n}} \qquad (9.2.13)$$

or
$$\overline{X} \pm z \frac{s}{\sqrt{n}} \qquad (9.2.14)$$

Together, the width of a confidence interval and the confidence coefficient provide a measure of how good an estimate we have. A confidence interval such as 50 ± 10 at the .95 level of confidence is not as good an estimate of μ as the same confidence interval at the .99 level. The interval 50 ± 5 represents a much better estimate of μ than the wider interval 50 ± 10, if both are at the same level of confidence. Of course, confidence intervals at the 1.00 level of confidence would be preferred, since then we would be certain that the interval includes the mean μ. Unfortunately, this is not possible since we can never be perfectly sure of estimates based on sample data.

The higher the level of confidence, the greater is the z value to be used in computing the confidence interval and the wider and less informative is the interval estimate. The z value to use is 1.96 for a .95 confidence interval, 2.33 for a .98 confidence interval, and 2.58 for a .99 confidence interval. The most frequently used confidence level is .95, however, .98 and .99 are also used frequently. Notice that the sample size n appears in the denominator of the expression $z(s/\sqrt{n})$ in Equation 9.2.14. This is the expression for the maximum error expected at a specified confidence level. It is this expression which determines the width of a confidence interval. Hence, the larger the n, the smaller is the expected error and the narrower is the confidence interval computed for any stated level of confidence.

A sample ($n = 50$) is selected from a population of 2,000 weights (in pounds) and the .98 confidence interval for the mean weight in the population is computed as 30 ± 2.5 lbs. What is the point estimate of the total of the 2,000 weights? Construct a .98 confidence interval for the total weight of the population.

The *total amount* for a population (A) is estimated as

$$A = N\overline{X} \tag{9.2.15}$$

The confidence interval for a *total amount* is computed as

$$N\overline{X} \pm Nz \frac{s}{\sqrt{n}} \tag{9.2.16}$$

Applying these equations to our example, the point estimate of the *total population weight* is

$$A = (2,000)(30) = 60,000 \text{ lbs}$$

The .98 confidence interval for the *total population weight* is

$$(2,000)(30) \pm (2,000)(2.5) = 60,000 \pm 5,000$$

Then, the confidence limits for the *total weight* are

$$LL = 60,000 - 5,000 = 55,000 \text{ lbs}$$
$$UL = 60,000 + 5,000 = 65,000 \text{ lbs}$$

Consequently, we are 98 percent confident that the point estimate of 60,000 lbs is off by no more than 5,000 lbs either way and that the interval 55,000 lbs–65,000 lbs includes the true total population weight.

Summary: This section presents a method for constructing confidence intervals for a population mean and total amount where the:

1. Population is large.
2. Sample is large ($n \geq 30$).
3. Sampling ratio $n/N < .05$.

9.3 Means and Total Amounts: Small Samples

When a small sample ($n < 30$) is selected without replacement from a large finite population to estimate the population mean, Model A (which has no sample size restriction) may be used to represent the sampling situation. Model A, as well as Model B, requires an infinite population. This requirement is satisfied because the sampling ratio n/N is always less than .05 when a small sample is selected from a large population. (Refer to the discussion at the beginning of Section 9.2.) To apply Model A the population sampled must be normal or approximately normal in distribution.

We know that for sampling situations, which fit Model A, the sampling distribution of the mean is normal with mean μ and standard error $\sigma_{\bar{x}} = \sigma/\sqrt{n}$. We could set up the same equation as Equation 9.2.2 relating to small samples:

$$P\left(-1.96 < \frac{\overline{X} - \mu}{\sigma_{\bar{x}}} < 1.96\right) = .95 \qquad (9.3.1)$$

If, however, the estimated standard error $s_{\bar{x}} = s/\sqrt{n}$ is substituted for $\sigma_{\bar{x}}$ in this equation (as was done in Equation 9.2.4), the probability .95 no longer holds because we are dealing with small samples. This is due to the fact that when $s_{\bar{x}}$ is used to estimate $\sigma_{\bar{x}}$, the resulting standard unit transformation $(\overline{X} - \mu)/s_{\bar{x}}$ is approximately normal in distribution for large samples but not for small samples. We denote this transformation by the symbol z for large samples and use the symbol t for small samples. When dealing with small samples, if the population standard deviation σ is unknown and must be estimated by s, the following sampling model is appropriate:

Model E

If samples of size n are selected from a normal population with mean μ and estimated standard deviation s, the sampling distribution of the statistic

$$t = \frac{\overline{X} - \mu}{s_{\bar{x}}} = \frac{\overline{X} - \mu}{\dfrac{s}{\sqrt{n}}}$$

is like the t distribution, with $n - 1$ degrees of freedom.

Note that Model E, like Model A, requires that the sampled population have a normal (or approximately normal) distribution and that there is no sample size restriction. Nevertheless, our interest in the use of Model E is for small samples, since Model B is usually applied to sampling problems which involve large samples.

The t statistic is also called *Student's t*, since it was introduced by W. S. Gosset who published his findings under the pen name "Student." Figure 9.3.1 presents a comparison of the t distribution and the normal (z) distribution. Notice that both distributions are symmetrical around the mean ($= 0$) and each has a central peak. The t distribution is not as high as the normal distribution in the central portion, but it is higher in the tails, so that the t distribution is more spread out. Therefore, a smaller proportion of the area under the curve is concentrated around the mean for the t distribution than for the normal distribution and a larger proportion is in the tails. Consequently, if the $\pm z$ limits shown in Figure 9.3.1 mark off a specified proportion of the area under the normal curve, *the corresponding t limits must be further in the tails of the t distribution* to mark off an *equal proportion* of the total area under the t distribution curve. This relationship between the z limits and the t limits is shown in Figure 9.3.1.

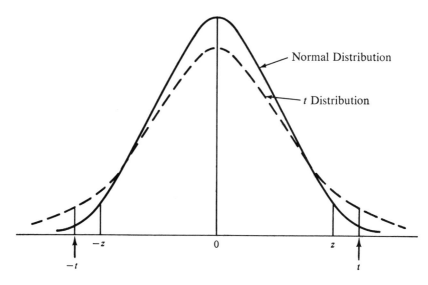

FIGURE 9.3.1

Diagrammatic comparison of the t distribution and the normal distribution.

Actually, there are many *t* distributions. A particular *t* distribution is identified by the number of *degrees of freedom* (df) involved in a problem. In Model E, df = $n - 1$ (one less than the sample size). Generally, df relate to the "freedom to vary" in a set of data. For example, we know that the sum of the deviations around the mean for a set of data is zero (Equation 4.5.1). Suppose three of the deviations from the mean for a set of four prices (in cents) is chosen arbitrarily as 2, -4, 1. The fourth deviation must be 1, so that the four deviations add to zero. If the mean price is 5¢, then the set of four prices, based on the foregoing deviations is (in cents) 7, 1, 6, 6. In this sense, a sample of $n = 4$ has df = 3, since only three of the four quantities (or, their deviations from the mean) may be chosen with complete freedom.

It is useful to compare the *t* values for various df with the corresponding *z* values. If the *z* values in Figure 9.3.1 are ± 1.96, then the area under the normal curve between these limits is .95 and the area in both tails combined is $1.00 - .95$ or .05. If the *t* values in Figure 9.3.1 are ± 1.98 for a *t* distribution with df = 120, the area under the *t* curve between these limits is also .95 and the area in both tails combined is .05. Notice that these *t* limits (± 1.98) are nearly equal to the corresponding *z* limits (± 1.96); however, the *t* limits required to include .95 of the area under the *t* curve increases in magnitude (moves further into the tails) as the number of degrees of freedom is reduced. For example, the corresponding *t* limits are ± 2.00 for df = 60, ± 2.04 for df = 30, and ± 2.23 for df = 10. Clearly, for large samples ($n \geq 30$) where df are large (29 or more), the *t* limits are not too different from the corresponding *z* limits. In other words, for large samples, the *t* distribution is sufficiently similar to the normal distribution so that the differences between the two distributions may be ignored.

Table B.5 in Appendix B presents limits for t distributions corresponding to different numbers of degrees of freedom and for a variety of areas included between the limits. The first column shows df from 1 through 29. The symbol ∞ denotes an *infinite number* of degrees of freedom. The other column headings specify the area in *both tails* of the t distribution combined. These areas represent probabilities, as in the case of the normal distribution. The first probability (or area) column is .5 so that the t values shown in this column cut off (*in both tails combined*) .5 of the area under the t curve (half of .5 or .25 in each tail). Referring to the probability column .05, we note that for 10 df the t limits are ± 2.228. Therefore, $1 - .05$ or .95 of the area under the t curve for 10 df is included between $-t = -2.228$ and $t = 2.228$ and .05 is the area in both tails combined.

It has already been observed that as df (or sample size) become larger the t distribution becomes more and more like the normal distribution. In the limit, with df infinite, the two distributions coincide. Notice that for df $= \infty$, the t values are exactly equal to the corresponding z values. As an example, for .05 of the area in both tails (or .025 in each tail), $t = z = 1.96$.

Returning to Equation 9.3.1, we take into account that substituting $s_{\bar{X}}$ for $\sigma_{\bar{X}}$ results in the variable $(\bar{X} - \mu)/s_{\bar{X}}$ which has the t distribution, by writing:

$$P\left(-t_{.05} < \frac{\bar{X} - \mu}{s_{\bar{X}}} < t_{.05}\right) = .95 \qquad (9.3.2)$$

The symbol $t_{.05}$ denotes a t value such that, for a stated number of degrees of freedom, the limits $-t_{.05}$ and $t_{.05}$ cut off (in both tails combined) .05 of the total area under the curve (or, .025 in each tail). These are the t values in Table B.5, column headed .05. The interpretation of Equation 9.3.2 is analogous to the interpretation of Equation 9.2.4 for large samples. Following the same approach as for large samples, we may write the confidence limits for a population mean based on a small sample as follows:

$$\bar{X} \pm t s_{\bar{X}} \qquad \text{or} \qquad \bar{X} \pm t \frac{s}{\sqrt{n}} \qquad (9.3.3)$$

Notice that in the foregoing equation we use t instead of $t_{.05}$ to indicate a t value in general for any number of degrees of freedom and any desired level of confidence. If confidence limits are to be computed for the .95 level of confidence $(1 - .95$ or .05 in both tails of the t distribution combined), then substitute $t_{.05}$ for t. If confidence limits are to be computed for the .98 level of confidence, then substitute $t_{.02}$ for t, etc.

Suppose a sample of 16 workers selected from a large population shows a mean weekly salary $\bar{X} = \$101.30$ and a standard deviation $s = \$18.80$ and the weekly salary per worker is nearly normal in distribution. We may construct a .98 confidence interval for the population mean by applying Equation 9.3.3, since we are using small sample data. We determine df $= 16 - 1 = 15$. Then, for a .98 confidence interval we refer to Table B.5, column headed .02 (since $1 - .98 = .02$) and we see $t = 2.602$ corresponding to 15 df. Making the proper substitutions into Equation 9.3.3 we obtain

$$101.30 \pm (2.602) \frac{18.80}{\sqrt{16}} = 101.30 \pm 12.23$$

Finally, the .98 confidence limits are

$$LL = 101.30 - 12.23 = \$89.07$$
$$UL = 101.30 + 12.23 = \$113.53$$

In the foregoing illustration, prepare a point estimate of the total amount of wages paid to the population of 3,000 workers for a week. Also, construct a .98 confidence interval for the weekly wage total. Applying Equation 9.2.15, we compute a point estimate of the *total weekly wage (A)* as

$$A = (3,000)(101.30) = \$303,900$$

A confidence interval for a *total amount* based on a small sample is computed as

$$N\bar{X} \pm Nt \frac{s}{\sqrt{n}} \qquad (9.3.4)$$

Applying this equation to our problem, we compute a .98 confidence interval for the total weekly wage as

$$(3,000)(101.30) \pm (3,000)(2.602) \frac{18.80}{\sqrt{16}} = 303,900 \pm 36,688$$

Then, the confidence limits are

$$LL = 303,900 - 36,688 = \$267,212$$
$$UL = 303,900 + 36,688 = \$340,588$$

Hence, we are 98 percent confident that the point estimate of $303,900 for the total weekly wage bill is off by no more than $36,688 either way and that the interval $267,212–$340,588 includes the true population total weekly wage paid.

Summary: This section presents a method for constructing confidence intervals for a population mean and total amount where the:

1. Population is large and approximately normal in distribution.
2. Sample is small ($n < 30$).

9.4 Proportions and Total Number

Model D generally provides a suitable theoretical description of the sampling situation when a sample is selected without replacement from a large finite population in order to estimate the *population proportion p*. This model requires that p is not too close to 0 or 1 and that the sample size is sufficiently large, as discussed in connection with Model D (Section 8.6). Furthermore, the sampling ratio n/N must be less than .05, so that sampling without replacement from a large finite population may be approximated by sampling from an infinite population. Many estimation problems relating to proportions encountered in practice may be adequately represented by Model D.

According to the model, the sampling distribution of a proportion X/n is approximately normal, with mean $\mu_{X/n} = p$ and standard error $\sigma_{X/n} = \sqrt{p(1-p)/n}$. We may set up an equation for the statistic X/n similar to Equation 9.2.2 for $\overline{X}$ as follows:

$$P\left(-1.96 < \frac{\frac{X}{n} - p}{\sigma_{X/n}} < 1.96\right) = .95 \qquad (9.4.1)$$

This equation states that the probability is .95 that a sample proportion expressed in standard units will lie between the limits $z = \pm 1.96$. Of course, when a sample is selected from a population to estimate the population proportion, the parameter p is not known. We cannot compute $\sigma_{X/n}$ and must estimate it by computing

$$s_{X/n} = \sqrt{\frac{\frac{X}{n}\left(1 - \frac{X}{n}\right)}{n}} \qquad (9.4.2)$$

which is the *estimated standard error of a proportion*. When $s_{X/n}$ is substituted into Equation 9.4.1, the probability .95 continues to hold to an acceptable approximation. Therefore, we may write

$$P\left(-1.96 < \frac{\frac{X}{n} - p}{s_{X/n}} < 1.96\right) = .95 \qquad (9.4.3)$$

Using the same approach as for the mean (Equation 9.2.4), we obtain the following equation based on Equation 9.4.3:

$$P\left(\frac{X}{n} - 1.96s_{X/n} < p < \frac{X}{n} + 1.96s_{X/n}\right) = .95 \qquad (9.4.4)$$

We may write the general expression for the confidence interval for a population proportion (using z in place of 1.96) as follows:

$$\frac{X}{n} \pm z s_{X/n} \qquad \text{or} \qquad \frac{X}{n} \pm z \sqrt{\frac{\frac{X}{n}\left(1 - \frac{X}{n}\right)}{n}} \qquad (9.4.5)$$

A sample of 500 voters in a large city showed that 45 voters opposed tax incentives to attract new business firms to the city. Construct a .99 confidence interval for the population proportion opposing such tax incentives. We determine a point estimate for p by computing $X/n = 45/500 = .09$. Using Equation 9.4.2, compute

$$s_{X/n} = \sqrt{\frac{.09(1 - .09)}{500}} = .013$$

Noting that $z = \pm 2.58$ includes 99 percent of the area under the normal curve and applying Equation 9.4.5, compute

$$.09 \pm (2.58)(.013) = .09 \pm .034$$

Finally, the limits defining a .99 confidence interval are

$$LL = .09 - .034 = .056$$
$$UL = .09 + .034 = .124$$

We may conclude based on the sample data that we are 99 percent confident (nearly certain) that the interval .056–.124 contains the population proportion of voters opposing tax incentives. We may also state that, if .09 is used as an estimate of p, .034 represents the maximum error we could expect to make at the .99 level of confidence.

In the foregoing illustration, prepare a point estimate of the total number of voters in the city who oppose tax incentives. Also, construct a .99 confidence interval for the total number who oppose tax incentives. The total number of all voters in the city is 150,000.

The *total number* in a category for a population (N') is estimated as

$$N' = N\frac{X}{n} \tag{9.4.6}$$

The confidence interval for the *total number* in a category (N') is computed as

$$N\frac{X}{n} \pm Nzs_{X/n} \tag{9.4.7}$$

Returning to our illustration, we estimate the total number of voters opposing tax incentives (Equation 9.4.6) as

$$N' = (150,000)(.09) = 13,500 \text{ voters}$$

Applying Equation 9.4.7, we compute a .99 confidence interval for the total number of voters who oppose tax incentives as

$$(150,000)(.09) \pm (150,000)(.034) = 13,500 \pm 5,100$$

Then, the .99 confidence limits are

$$LL = 13,500 - 5,100 = 8,400 \text{ voters}$$
$$UL = 13,500 + 5,100 = 18,600 \text{ voters}$$

Therefore, we are nearly certain (99 percent confident) that the estimate of 13,500 voters opposing tax incentives is off (either way) by no more than 5,100 voters and that the interval 8,400–18,600 includes the total number of voters in the city who oppose tax incentives.

Summary: This section presents a method for constructing confidence intervals for a population proportion and the total number in a category where the:

1. Population is large.
2. p is not too close to 0 or 1.
3. Sample size is sufficiently large ($n \geq 100$; see discussion of Model D in Section 8.6).
4. Sampling ratio $n/N < .05$.

9.5 Optional: Finite Populations

It is more likely that the sampling ratio n/N will be small when the sample is drawn from a large population, than drawn from a small population. *When the sampling ratio is appreciable, for example $n/N \geq .05$, we must take into account the effects of sampling without replacement from a finite population.* We cannot ignore the changes in the population distribution which occur as items are removed during sample selection. Suppose a random sample of 100 students is to be selected without replacement from a population of 400 students. When the first selection is made, the probability is $1/400 = .0025$ that a specified student will be selected; however, the probability is $1/301 = .0033$ that a specified student will be selected after 99 students have been chosen without replacement. Clearly, changes in the probability of selection from trial to trial cannot be ignored and successive selections cannot be considered to represent independent trials. (Review the discussion at the beginning of Section 9.2.)

When sampling without replacement from a finite population, the standard errors $\sigma_{\bar{x}}$ and $\sigma_{X/n}$ should be adjusted by multiplying by a finite population correction factor (fpc) as follows:

$$\text{fpc} = \sqrt{\frac{N - n}{N - 1}} \tag{9.5.1}$$

Then, we may write the adjusted equations for the *standard error of the mean* ($\sigma_{\bar{x}}$) and the *standard error of a proportion* ($\sigma_{X/n}$) as follows:

$$\sigma_{\bar{x}} = \frac{\sigma}{\sqrt{n}} \cdot \sqrt{\frac{N - n}{N - 1}} \tag{9.5.2}$$

$$\sigma_{X/n} = \sqrt{\frac{p(1 - p)}{n}} \cdot \sqrt{\frac{N - n}{N - 1}} \tag{9.5.3}$$

The adjusted equations for the corresponding *estimated standard errors* are:

$$s_{\bar{x}} = \frac{s}{\sqrt{n}} \cdot \sqrt{\frac{N - n}{N - 1}} \tag{9.5.4}$$

$$s_{X/n} = \sqrt{\frac{\frac{X}{n}\left(1 - \frac{X}{n}\right)}{n}} \cdot \sqrt{\frac{N - n}{N - 1}} \tag{9.5.5}$$

Typically, the sample size n is greater than one. Then, $N - n$ in Equation 9.5.1 is smaller than $N - 1$, so that the fpc is less than one. However, the smaller the sample size n relative to the population size N, the larger and closer to one is the fpc. As an example, for $n = 10,000$ and $N = 88,000$

$$\text{fpc} = \sqrt{\frac{88,000 - 10,000}{88,000 - 1}} = .94$$

On the other hand, if $n = 2,500$

$$\text{fpc} = \sqrt{\frac{88,000 - 2,500}{88,000 - 1}} = .99$$

When n is small relative to N (i.e., n/N is less than .05) the fpc is so close to one that we may neglect the fpc when computing the standard error, since it is nearly the same as multiplying by one.

On the other hand, when n/N is not small ($n/N \geq .05$), it is better to use the fpc when computing the standard error. Suppose a sample of 250 accounts out of a population of 1,000 shows a mean balance of $82.30, with a standard deviation of $14.10. Construct a .98 confidence interval for the population mean balance. Since we are using large sample data, Equations 9.2.12 and 9.2.13 are to be used to determine the confidence limits, however, since $n/N = 250/1,000 = .25$, the fpc should not be ignored. Applying Equation 9.5.4 compute

$$s_{\bar{x}} = \frac{14.10}{\sqrt{250}} \cdot \sqrt{\frac{1,000 - 250}{1,000 - 1}} = .67$$

Then, using $z = 2.33$, we obtain

$$\text{LL} = 82.30 - 2.33(.67) = 82.30 - 1.56 = \$80.74$$
$$\text{UL} = 82.30 + 2.33(.67) = 82.30 + 1.56 = \$83.86$$

Hence, we are 98 percent confident that the interval $80.74–$83.86 includes the population mean account balance. In other words, if $82.30 is used as an estimate of μ, the maximum error expected is $1.56 at the .98 confidence level.

Note that it is "safe" to ignore the fpc even if $n/N \geq .05$. In our example, we could compute $s_{\bar{x}}$ ignoring the fpc to obtain $14.10/\sqrt{250} = .89$. Then, we obtain confidence limits as follows:

$$\text{LL} = 82.30 - 2.33(.89) = \$80.23$$
$$\text{UL} = 82.30 + 2.33(.89) = \$84.37$$

Certainly, if we are 98 percent confident of the previous interval ($80.74–$83.86), we are at least 98 percent confident that the wider interval $80.23–$84.37 includes μ. Of course, a narrower interval provides more information, so that it is worthwhile to take the fpc into account in the computations when the sampling ratio is not negligible ($n/N \geq .05$).

Summary: Confidence limits may be computed for a population mean or a population proportion using the methods presented earlier in this chapter, even if the sampling ratio n/N is .05 or greater. Narrower and more useful confidence intervals can be obtained if the standard errors are adjusted by multiplying by the fpc when the sampling ratio is .05 or greater.

9.6 Sample Size

When sampling, a decision must be made on the size of sample to be selected. Sample size determination is often complicated since many factors must be considered, including the costs involved. We have developed sufficient statistical theory to permit a general consideration of how to determine sample size in a given sampling situation.

When a sample is to be selected we must first determine the purpose, since computation of sample size depends on whether it is intended to estimate a population mean, a population proportion, or some other population parameter. It must also be determined how accurate an estimate is required. That is, how much error is acceptable (since perfect estimates are not to be expected) and how much confidence is required that the maximum error associated with a point estimate does not exceed the amount specified.

When $\overline{X}$ is used as an estimate of μ, we know (Equation 9.2.14) that the *maximum error* to be expected (m) at a stated level of confidence is

$$m = z \frac{\sigma}{\sqrt{n}} \tag{9.6.1}$$

where we substituted σ for its estimated value s. Solving for n, we obtain

$$n = \left(\frac{z\sigma}{m}\right)^2 \tag{9.6.2}$$

Equation 9.6.2 provides a basis for determining sample size when it is required to estimate a population mean based on sample data. For example, how large a sample is required to estimate the mean weight of a certain large supply of fittings, if it is specified that the estimate should be off (either way) by no more than 2 oz at the 95 percent level of confidence? We apply Equation 9.6.2, with $m = 2$ oz and $z = 1.96$. What should we use for σ, the unknown population standard deviation? Actually, we do not need to know σ precisely for the purpose at hand. Often, a fair estimate will be available from one source or another (for example, from a previous sample). If no information on the size of σ is available, it will be necessary to collect a small pilot sample and compute s as an estimate of σ.

Suppose σ is estimated as 18 oz. Then, entering the appropriate values into Equation 9.6.2, we obtain

$$n = \left(\frac{(1.96)(18)}{2}\right)^2 = 311.17 \text{ fittings}$$

Therefore, a sample of 311 fittings is required to produce the required estimate. This result is not, of course, a precise measure of the sample size required, since σ is typically an estimated value. The general practice is to use a sensibly rounded figure for sample size. A sample of 300 or 325 fittings may be decided upon.

When X/n is used as a point estimate of a population proportion p, we know (Equation 9.4.5) that the *maximum error* to be expected at a stated level of confidence is

$$m = z \sqrt{\frac{p(1 - p)}{n}} \tag{9.6.3}$$

where p has been substituted for the point estimate X/n. Solving for n, we obtain

$$n = p(1 - p)\left(\frac{z}{m}\right)^2 \tag{9.6.4}$$

Equation 9.6.4 provides a basis for estimating sample size when it is required to estimate a population proportion; however, to use this equation, an estimate of p is required. Such an estimate may be available from various sources, such as previous studies. If no information is available on which to base an estimate of p, we may modify Equation 9.6.4 to provide a measure of the maximum sample size needed. Notice that $p(1 - p)$ appears in Equation 9.6.4 as a multiplier. It can be verified that $p(1 - p)$ is a maximum when $p = 1/2$, so that $(1/2)[1 - (1/2)] = 1/4$ is a maximum value for $p(1 - p)$. Substituting $1/4$ for $p(1 - p)$ in Equation 9.6.4, we obtain an equation for the *maximum sample size* needed for a stated level of confidence and for a specified amount of error:

$$\max n = \frac{1}{4}\left(\frac{z}{m}\right)^2 = \left(\frac{z}{2m}\right)^2 \qquad (9.6.5)$$

Suppose it is desired to estimate the proportion of housewives who prefer detergent A and it is specified that the estimate should not be off by more than .05 at the .99 level of confidence. If no information is available to provide an estimate of p, we may use Equation 9.6.5 to obtain

$$\max n = \left(\frac{2.58}{(2)(.05)}\right)^2 = 665.64$$

Therefore, the largest sample needed is 666 housewives. On the other hand, suppose a fair estimate of p is found to be .30. Then, applying Equation 9.6.4 we obtain

$$n = (.30)(1 - .30)\left(\frac{2.58}{.05}\right)^2 = 559.14$$

Therefore, if p is about .30, a sample of only 559 housewives is needed.

Note: If it is intended to construct a confidence interval for a population proportion using the procedure in Section 9.4, the sample size must meet the requirements specified in that section. This is true even if Equation 9.6.4 or 9.6.5 indicates a smaller sample size.

Study Problems

1.   Define and illustrate: (a) point estimate, (b) confidence interval, (c) confidence coefficient, and (d) confidence limits.

2.   Sampling without replacement from a large finite population (with $n/N < .05$) may be conceptualized as sampling from an infinite population. Explain.

3.   When sampling without replacement from a large finite population (sampling ratio less than .05) and constructing a confidence interval for μ, why does Model B provide an acceptable description of the sampling distribution of the mean for large samples and not for small samples?

4.   Explain the difference between a confidence interval for the mean at the .95 level of confidence and at the .99 level of confidence.

5.   If a confidence interval for the mean is constructed at the .98 confidence level, what is the probability the interval does not include the population mean?

6.   What z value would you use, if you wanted to construct a confidence interval for the mean based on a large sample, if the confidence coefficient to use is (a) .99, (b) .9836, and (c) .9774?

7.   A sample of 225 men selected from a large population has a mean right-arm skinfold of 3.25 cms, with a standard deviation of .75 cms. Construct a .95 confidence interval for the population mean skinfold.

8.   In Problem 7:
        a.   Determine the point estimate for the population mean skinfold.
        b.   Construct a .98 confidence interval for μ.
        c.   What is the maximum error you would expect, if $\overline{X}$ is used as an estimate of μ (at the .98 confidence level)?
        d.   Construct a .98 confidence interval, if $n = 100$ and the same sample results were obtained.

9.   A sample of 144 checks selected from a large population of checks handled by New York banks has a mean amount of $736.25 per check and a standard deviation of $95.50.
        a.   Estimate the mean amount per check for the population.
        b.   Construct a .90 confidence interval for the population mean check amount.
        c.   What is the maximum error you could expect to make at .90 level of confidence, if $\overline{X}$ is used as an estimate of μ?

10.   In Problem 9 the population includes 10,000 checks.
        a.   Estimate the total check amount for the population.
        b.   Construct a .90 confidence interval for the total population check amount.

11.   In Problem 9:
        a.   Construct a confidence interval at the .99 level of confidence.
        b.   How would you interpret this interval estimate?
        c.   What is the probability that this interval does not include the population mean?

12.   A random sample of 2,500 students in a large state university worked on part-time jobs a mean of 170 hours during a certain period, with a standard deviation of 40 hours.
        a.   Estimate the mean number of hours worked per student in the population.
        b.   What is the maximum error you would expect to make in your estimate, at the .80 level of confidence?
        c.   Construct a .99 confidence interval for the population mean number of hours worked per student.

13.   In Problem 12, construct the confidence interval for the population mean number of hours worked per student at the .99 level of confidence assuming the same sample results but with $n = 625$.

14.   In Problem 12, the total number of students in the state university is 60,000.
        a.   Estimate the total amount of time spent on part-time employment during the period by all the students in the university.
        b.   What is the maximum error you would expect to make in your estimate, at the .80 level of confidence?

   c.   Construct a .99 confidence interval for the total amount of time spent on part-time employment for all the students in the university.

15.   A random sample of 16 young women (20 to 24 years of age) selected from a large population obtained a mean score of 75 on a psychological test, with a standard deviation of 5 points. It is known that the score distribution for the population is approximately normal. Construct a .95 confidence interval for the population mean score.

16.   In Problem 15, construct a .98 confidence interval for the population mean score.

17.   A sample of 9 salesmen selected from a large population of salesmen wrote a mean of 30.2 orders during a given week, with a standard deviation of 4.7 orders. The number of orders written per salesmen in the population is approximately normal in distribution. Construct a .99 confidence interval for the population mean number of orders written per salesmen.

18.   In Problem 17, construct the interval estimate at the .95 confidence level.

19.   In Problem 17, construct the interval estimate at the .90 level of confidence.

20.   In Problem 17 the population includes 1,000 salesmen. Construct a confidence interval for the total number of orders written by the population of salesmen at the following confidence level: (a) .99, (b) .95, and (c) .90.

21.   The diameters of round tablets produced by a press is approximately normal in distribution. A random sample of 10 tablets has a mean diameter of .246 cms and a standard deviation of .035 cms. Construct a .98 confidence interval for the mean diameter of tablets produced by the press.

22.   In Problem 21, compute the confidence interval assuming the same sample results but with sample size (a) 20, (b) 40, and (c) 100.

23.   A sample of 25 time-reactions selected from a large normally distributed population of time-reactions (in seconds) has a mean of 35.16 sec and a standard deviation of 4.15 sec.

   a.   If $\overline{X}$ is used as an estimate of μ, what is the maximum error you can expect to make at the .95 level of confidence?

   b.   Construct a .95 confidence interval for μ.

   c.   What is the probability that the confidence interval does not include μ?

24.   A survey of 500 commuters showed that 150 opposed a certain schedule change.

   a.   What would you estimate as the percentage of all commuters who oppose the schedule change?

   b.   Construct a .95 confidence interval for the proportion of all commuters opposing the schedule change.

25.   In Problem 24, compute the confidence interval using .98 as the confidence coefficient.

26.   In Problem 24 the total number of commuters is 20,000.

   a.   Estimate the number of commuters opposing the schedule change.

   b.   Construct a .95 confidence interval for the number of commuters opposing the schedule change.

   c.   Construct the confidence interval at the .98 level of confidence.

27.   A survey of 1,000 residents in a large city showed that 750 favored a civil rights bill under consideration by the municipal legislature. Construct a .99 confidence interval for:

   a.   The proportion of all city residents who favor the bill.

b. The number of city residents who favor the bill. (The total number of residents in the city is 150,000.)

28. In Problem 27, if .75 is used as an estimate of the population proportion favoring the legislation, what is the maximum error you would expect to make at the
    a. .99 level of confidence
    b. .95 level of confidence

29. A sample of 250 married women in a city showed that 50 are not happily married. Construct a .98 confidence interval for:
    a. The proportion of the population of married women who are not happily married.
    b. The number of the 50,000 married women in the city who are not happily married.

30. In Problem 29:
    a. What would you estimate as the population proportion of married women who are not happily married?
    b. How large an error would you expect in this estimate, at most, at the .98 confidence level?

31. A sample of 200 voters indicate that 10 percent prefer Smith as president of the city Board of Higher Education. Construct a .80 confidence interval for the proportion of all voters who prefer Smith.

32. In Problem 31, construct the interval estimate at the .98 level of confidence.

33. When sampling from a finite population, why must we take into consideration the effects of sampling without replacement if the sampling ratio is appreciable (.05 or more)?

34. When sampling without replacement from a finite population, what adjustment should be made to the standard error when constructing a confidence interval?

35. What is the effect on the finite population correction factor if the sample size is increased?

36. Why is it "safe" to ignore the fpc even if $n/N \geq .05$ when constructing confidence intervals?

37. A random sample of 25 out of a population of 250 production workers produced a mean of 136.15 units, with a standard deviation of 26.09 units. Compute:
    a. The sampling ratio
    b. The finite population correction factor
    c. The appropriate standard error of the mean
    d. The .95 confidence interval for the population mean number of units produced. Assume the sampled population can be approximated by a normal curve.

38. A sample of 225 students selected from a population of 1,500 showed that 50 are physics majors. Compute:
    a. The sampling ratio
    b. The fpc
    c. The appropriate standard error of a proportion
    d. A point estimate of the population proportion of physics majors
    e. A .98 confidence interval for the population proportion of physics majors

39.   A survey of legal typists is being planned to estimate average weekly earnings in a large metropolitan area. How large a sample is required, if it is specified that the estimate should not be off by more than $2.50 at the .95 confidence level? Based on earlier surveys, σ is estimated to be $30.

40.   How large a sample of social case workers should be selected in a large city to estimate the mean case load per social worker within three cases, at the .99 level of confidence? It is estimated that $\sigma = 25$ cases.

41.   It is desired to estimate the proportion of housewives who prefer butter to margarine. How large a sample is required if it is specified that the estimate be within .05 of the true population proportion at the .98 confidence level?

42.   In Problem 41, suppose it is estimated that p is about .60. What sample size is required?

43.   How large a sample of voters at most should be selected in a large city to estimate the proportion who favor candidate A? It is desired to produce an estimate which is off by no more than .02 at the .95 confidence level.

44.   How large a sample should be selected to estimate the mean weight of a product for a large supply, at the .80 level of confidence? It is required that the estimate be off by no more than 5 lbs. (σ is believed to be about 36 lbs.)

45.   In Problem 44, how would n be affected if the estimate of σ is (a) too large, or (b) too small?

Chapter 10

Optional:
Confidence Intervals
for Differences

10.1   Estimation

The need to estimate differences between two populations is frequently encountered. For example, what is the difference between the average typing speed for typewriter A and typewriter B? What is the difference between average weekly sales of gasoline when gifts are offered to customers and when no gifts are offered? What is the difference between the proportion of voters favoring candidate Smith in city A and in city B?

Estimates of differences may be point estimates or interval estimates, just as in the case of estimates of parameters discussed in the previous chapter. In this chapter, we will consider the estimation of differences between two population means and between two population proportions. We will speak of population 1 and population 2, with corresponding means μ_1 and μ_2, standard deviations σ_1

175

and σ_2, and proportions p_1 and p_2. The *difference parameters* we will estimate are $\mu_1 - \mu_2$ and $p_1 - p_2$.

Estimates of differences involving two populations are based on two samples, one selected from each population. These may be *independent samples* where the sample selected from one population is in no way influenced by the sample selected from the other population or they may be *dependent samples*. Suppose a sample of males are selected from a population of male cab drivers and a sample of females are selected from a population of female cab drivers in order to estimate the difference in the mean number of accidents per driver for the two populations. If each sample is a random sample from its population, then the two samples are *independent*. On the other hand, if the samples are selected in such a way that the males and females pair off as brother and sister, then they are *dependent samples*. "Before" and "after" samples are a frequently encountered type of dependent samples (e.g., weight before diet and after diet for a group of subjects).

We will denote the samples as n_1 from population 1 and n_2 from population 2, with corresponding sample means $\overline{X}_1$ and $\overline{X}_2$, sample standard deviations s_1 and s_2, and sample proportions X_1/n_1 and X_2/n_2. We will be concerned with the *difference statistics* $\overline{X}_1 - \overline{X}_2$ and $X_1/n_1 - X_2/n_2$.

Point estimates for the difference parameters are the corresponding difference statistics. More specifically, $\overline{X}_1 - \overline{X}_2$ is a point estimate of $\mu_1 - \mu_2$ and $X_1/n_1 - X_2/n_2$ is a point estimate of $p_1 - p_2$. Confidence intervals for difference parameters are discussed in the subsequent sections of this chapter.

10.2   Means

In order to construct confidence intervals for the difference between two population means, we need to describe the *sampling distribution of the statistic* $\overline{X}_1 - \overline{X}_2$. Let us use our imagination and develop this sampling distribution concept. Imagine that a sample of size n_1 is selected from population 1 (infinite) and $\overline{X}_1$ computed and an *independent sample* of size n_2 is selected from population 2 (infinite) and $\overline{X}_2$ computed. Then, compute the difference $\overline{X}_1 - \overline{X}_2$. Now, imagine that such pairs of samples are selected repeatedly and $\overline{X}_1 - \overline{X}_2$ computed each time. The $\overline{X}_1 - \overline{X}_2$ differences are organized into a table similar to Table 7.4.2, presenting the different values obtained for this statistic, the corresponding frequencies, and the corresponding probabilities. We have now constructed, in our imagination, the *sampling distribution of the difference between sample means*. Model F describes this sampling distribution for large independent samples.

Model F

Let population 1 and population 2 represent infinite populations and n_1 and n_2 large independent samples. Then the *sampling distribution of* $\overline{X}_1 - \overline{X}_2$ is approximately normal, with mean $\qquad \mu_{(\overline{X}_1 - \overline{X}_2)} = \mu_1 - \mu_2$

standard error
of the difference
between means $\qquad \sigma_{(\overline{X}_1 - \overline{X}_2)} = \sqrt{\sigma_{\overline{X}_1}^2 + \sigma_{\overline{X}_2}^2} = \sqrt{\dfrac{\sigma_1^2}{n_1} + \dfrac{\sigma_2^2}{n_2}}$

Model F is applicable to problems which satisfy the following conditions for each population and each sample: population large; sample size large; and sampling ratio less than .05. These are the same conditions specified for application of Model B to construct confidence intervals for the mean based on large samples (Section 9.2).

A procedure for construction of confidence limits for $\mu_1 - \mu_2$ is developed in a manner similar to the procedure in Section 9.2 and results in the following limits.

$$(\overline{X}_1 - \overline{X}_2) \pm z s_{(\overline{X}_1 - \overline{X}_2)} \quad \text{or} \quad (\overline{X}_1 - \overline{X}_2) \pm z \sqrt{\dfrac{s_1^2}{n_1} + \dfrac{s_2^2}{n_2}} \quad \textbf{(10.2.1)}$$

where $\qquad s_{(\overline{X}_1 - \overline{X}_2)} = \sqrt{\dfrac{s_1^2}{n_1} + \dfrac{s_2^2}{n_2}} \qquad \textbf{(10.2.2)}$

is the *estimated standard error of the difference between means.*

Suppose the following data is obtained relating to the number of hours of life for light bulbs produced by firm A (population 1) and firm B (population 2): $n_1 = 50$ bulbs, $\overline{X}_1 = 1{,}040$ hrs, $s_1 = 30$ hrs, $n_2 = 40$ bulbs, $\overline{X}_2 = 920$ hrs, $s_2 = 25$ hrs. Construct a .98 confidence interval for $\mu_1 - \mu_2$, the difference in mean number of hours of life for the bulbs produced by the two firms. Noting that $z = 2.33$ for a .98 confidence interval, apply Equation 10.2.1 as follows:

$$(1{,}040 - 920) \pm 2.33 \sqrt{\dfrac{(30)^2}{50} + \dfrac{(25)^2}{40}} = 120 \pm 13.5$$

Then, .98 confidence limits are

$$\text{LL} = 120 - 13.5 = 106.5 \text{ hrs}$$
$$\text{UL} = 120 + 13.5 = 133.5 \text{ hrs}$$

We estimate that the difference in population mean life for these two types of bulbs is 120 hrs more for firm A bulbs, with a maximum expected error in this estimate of 13.5 hrs at the .98 level of confidence. In other words, we are 98 percent confident that the interval 106.5 hrs–133.5 hrs includes the true difference in mean bulb lives for the two firms.

When samples are small, we must take into consideration that the sampling distribution of $(\overline{X}_1 - \overline{X}_2)/s_{(\overline{X}_1 - \overline{X}_2)}$ is approximated by the t distribution. Model G describes this sampling situation.

Model G

Let population 1 and population 2 represent normal populations and n_1 and n_2 independent samples. Then, the sampling distribution of the statistic

$$t = \frac{(\overline{X}_1 - \overline{X}_2) - (\mu_1 - \mu_2)}{s_{(\overline{X}_1 - \overline{X}_2)}} = \frac{(\overline{X}_1 - \overline{X}_2) - (\mu_1 - \mu_2)}{\sqrt{\dfrac{s_1^2}{n_1} + \dfrac{s_2^2}{n_2}}}$$

is like the t distribution, with $n_1 - 1$ df for sample n_1 and $n_2 - 1$ df for sample n_2. Let t_1 for $n_1 - 1$ df and t_2 for $n_2 - 1$ df denote the corresponding t values for a stated level of confidence. Then, the corresponding t value to be used in constructing a confidence interval is approximated as

$$t = \frac{\dfrac{s_1^2}{n_1} t_1 + \dfrac{s_2^2}{n_2} t_2}{\dfrac{s_1^2}{n_1} + \dfrac{s_2^2}{n_2}}$$

Model G is applicable to problems where each population and each sample satisfies the following conditions: population large and approximately normal in distribution, and sampling ratio less than .05. Notice that there is no sample size restriction, so that Model G is applicable for any sample size; however, it is usually applied where small samples are involved, since Model F is generally applied when samples are large.

It is desired to estimate the difference in average driving speed between female and male drivers on a certain road. A sample of $n_1 = 8$ female drivers and $n_2 = 10$ male drivers was selected and the following statistics computed: $\overline{X}_1 = 31.6$ mph (miles per hour), $s_1 = 5.3$ mph, $\overline{X}_2 = 38.1$ mph, $s_2 = 4.9$ mph. Construct a .95 confidence interval for the difference in mean driving speed between male and female drivers. Assume that the distributions of driving speed are approximately normal for male and female drivers. Since n_1 and n_2 are less than 30 and the population distributions are approximately normal, Model G is applicable.

Referring to Table B.5, column headed $t_{.05}$ (for confidence level .95), we see (for df $= 8 - 1 = 7$) 2.365 as the value for t_1 and (for df $= 10 - 1 = 9$) 2.262 as the value for t_2. Compute $t_{.05} = t$ for a .95 confidence interval according to the second equation in Model G as follows:

$$t_{.05} = \frac{\dfrac{(5.3)^2}{8}(2.365) + \dfrac{(4.9)^2}{10}(2.262)}{\dfrac{(5.3)^2}{8} + \dfrac{(4.9)^2}{10}} = 2.33$$

Confidence limits are constructed as follows:

$$(\overline{X}_1 - \overline{X}_2) \pm ts_{(\overline{X}_1 - \overline{X}_2)} \qquad \text{or} \qquad (\overline{X}_1 - \overline{X}_2) \pm t\sqrt{\frac{s_1^2}{n_1} + \frac{s_2^2}{n_2}} \qquad (10.2.3)$$

Applying the foregoing result ($t_{.05} = 2.33$) and Equation 10.2.3 to our problem, we obtain

$$(31.6 - 38.1) \pm 2.33 \sqrt{\frac{(5.3)^2}{8} + \frac{(4.9)^2}{10}} = -6.5 \pm 5.7$$

Then, .95 confidence limits are

$$\text{LL} = -6.5 - 5.7 = -12.2 \text{ mph}$$
$$\text{UL} = -6.5 + 5.7 = -.8 \text{ mph}$$

Based on the sample data we estimate that mean driving speed is 6.5 mph less for female drivers than for male drivers. Furthermore, we are 95 percent confident that this estimate is off by no more than 5.7 mph. We may also state, at the .95 level of confidence, that the interval .8 mph–12.2 mph includes the true difference "male driving speed minus female driving speed."

10.3   Means: Equal Variances

It is often appropriate to assume that two populations have equal variances σ^2, so that $\sigma_1 = \sigma_2 = \sigma$. Where this is appropriate, the procedures for construction of confidence intervals for $\mu_1 - \mu_2$ presented in the previous section are applicable. A more efficient procedure, however, is to pool the data from the two samples to estimate the common population variance σ^2 by computing s^2 as follows:

$$s^2 = \frac{\sum (X_i - \overline{X}_1)^2 + \sum (X_i - \overline{X}_2)^2}{n_1 + n_2 - 2} \qquad (10.3.1)$$

The computation of the *pooled variance estimate* s^2 according to Equation 10.3.1 is conceptually similar to s^2 computed for a single sample (Equation 5.2.2). Recall that in Equation 5.2.2 s^2 is computed by dividing the sum of the squared deviations from the mean, $\sum (X_i - \overline{X})^2$, by the number of degrees of freedom $n - 1$. Similarly, computing s^2 according to Equation 10.3.1, the numerator represents the pooled sum of squared deviations from each sample mean. The first part of the numerator is the sum of squared deviations of the n_1 items from $\overline{X}_1$ and the second part is the sum of the squared deviations of the n_2 items from $\overline{X}_2$. The denominator is obtained by adding $n_1 - 1$ df from sample n_1 and $n_2 - 1$ df from sample n_2 to obtain $(n_1 - 1) + (n_2 - 1) = n_1 + n_2 - 2$.

Consequently, when it is appropriate to assume that two populations have a common variance, σ^2 estimated by s^2 (based upon the pooled sums of squares with $n_1 + n_2 - 2$ df) is a better estimate than the individual variances s_1^2 and s_2^2 (computed from the individual samples with smaller numbers of degrees of freedom, $n_1 - 1$ and $n_2 - 1$). In some problems the individual variances s_1^2 and s_2^2 are available. In such problems, a more convenient form of Equation 10.3.1 is

$$s^2 = \frac{(n_1 - 1)s_1^2 + (n_2 - 1)s_2^2}{n_1 + n_2 - 2} \qquad (10.3.2)$$

Equation 10.2.2 for $s_{(\overline{X}_1 - \overline{X}_2)}$ may be rewritten for populations which are considered to have equal variances as follows:

$$s_{(\overline{X}_1 - \overline{X}_2)} = \sqrt{s^2 \left(\frac{1}{n_1} + \frac{1}{n_2}\right)} \qquad (10.3.3)$$

where s^2 is computed according to Equation 10.3.1 or 10.3.2, whichever is more convenient. Confidence intervals for $\mu_1 - \mu_2$ are computed according to Equation 10.2.1 for large samples, with $s_{(\bar{X}_1 - \bar{X}_2)}$ determined according to Equation 10.3.3. When small samples are used, Equation 10.2.3 should be applied, with $s_{(\bar{X}_1 - \bar{X}_2)}$ determined according to Equation 10.3.3 and the appropriate t value taken from Table B.5 for $n_1 + n_2 - 2$ df for the specified level of confidence. The appropriate t value is not to be computed as shown in the second equation in Model G when equal population variances are assumed and $s_{(\bar{X}_1 - \bar{X}_2)}$ is computed according to Equation 10.3.3.

10.4   Means: Dependent Samples

Various problems arise where *dependent samples* are used to estimate the difference between two population means. For example, a sociologist wishes to estimate the difference in the mean score on a social attitudes test for a population of college students before and after an intensive series of discussions and film-viewing. A sample of students is selected to take the social attitudes test before the discussions and film-viewing, and then the test is administered again to the same sample of students after the discussions and film-viewing. Construction of a confidence interval for the difference in the "before" and "after" mean test scores must take into account that the two samples of test scores are *dependent samples* since there is a score in each sample for each subject.

Matched samples are a frequently encountered type of dependent samples. For instance, suppose a manufacturer is considering which of two types of stamping machines to purchase, and he needs to determine which will result in higher productivity. He may decide to select two samples of workers *matched on productivity* (based on production records for the past year). If there is a worker with a certain level of productivity in one sample, then there is a worker with the same level of productivity in the other sample. Each sample is assigned to use one type of machine and the productivity of the two samples compared. Matched samples are used for the purpose of eliminating the effects of *extraneous factors* when the sample data are compared so that it will be more likely that observed differences between the samples are caused by the *experimental variable* (in our example, the machines used).

Generally speaking, dependent samples are composed of *paired sample items*, one of each pair in each sample. The statistical theory underlying confidence interval construction for the difference parameter $\mu_1 - \mu_2$ when dependent samples are involved is based on the *differences between sample pairs* as illustrated in Table 10.4.1. We will denote the individual sample differences as X_i as shown in the table, the mean difference as $\bar{X}$, and the standard deviation of the differences as s. The population mean difference μ is our difference parameter $\mu_1 - \mu_2$. The sample size is $n = $ number of pairs. In Table 10.4.1, $n = 10$. A sample of 30 or more pairs satisfies our definition of a large sample. When $n \geq 30$, confidence intervals for μ are constructed according to Equation 9.2.14 (Section 9.2). If $n < 30$, Equation 9.3.3 (Section 9.3) is appropriate.

TABLE 10.4.1

Dependent samples: IQ scores for a sample of ten married couples.

IQ score		Difference
Husband	Wife	X_i
109	100	9
110	113	−3
100	96	4
96	103	−7
114	102	12
103	108	−5
112	106	6
118	110	8
102	95	7
98	96	2

10.5 Proportions

Construction of confidence intervals for the difference between two population proportions is based on the *sampling distribution of the statistic* $X_1/n_1 - X_2/n_2$. Model H describes this sampling distribution for independent samples.

Model H

Let population 1 and population 2 represent infinite populations and n_1 and n_2 independent samples, each sufficiently large. Then, the *sampling distribution of* $X_1/n_1 - X_2/n_2$ is approximately normal, with

mean

$$\mu_{(X_1/n_1 - X_2/n_2)} = p_1 - p_2$$

standard error of
the difference between
population proportions

$$\sigma_{(X_1/n_1 - X_2/n_2)} = \sqrt{\frac{p_1(1 - p_1)}{n_1} + \frac{p_2(1 - p_2)}{n_2}}$$

Model H is applicable to problems which satisfy the following conditions for each population and each sample: population large; sampling ratio n/N less than .05; population proportion not too close to 0 or 1; and sample sufficiently large ($n \geq 100$). These are the same conditions specified for application of Model D for construction of confidence intervals for a population proportion (Section 9.4).

Confidence limits for $p_1 - p_2$ are determined in a manner similar to the method used in Section 9.4 for a population proportion:

$$\left(\frac{X_1}{n_1} - \frac{X_2}{n_2}\right) \pm zs_{(X_1/n_1 - X_2/n_2)} \qquad \text{or} \qquad (10.5.1)$$

$$\left(\frac{X_1}{n_1} - \frac{X_2}{n_2}\right) \pm z \sqrt{\frac{\frac{X_1}{n_1}\left(1 - \frac{X_1}{n_1}\right)}{n_1} + \frac{\frac{X_2}{n_2}\left(1 - \frac{X_2}{n_2}\right)}{n_2}} \qquad \begin{array}{c}(10.5.1)\\ (contd.)\end{array}$$

where $\qquad s_{(X_1/n_1 - X_2/n_2)} = \sqrt{\dfrac{\frac{X_1}{n_1}\left(1 - \frac{X_1}{n_1}\right)}{n_1} + \dfrac{\frac{X_2}{n_2}\left(1 - \frac{X_2}{n_2}\right)}{n_2}}$ $\qquad$ (10.5.2)

is the estimated standard error of the difference between proportions.

In a large city, 52 out of a sample of $n_1 = 350$ teen-age boys stated that they prefer soft drinks in cans, whereas 41 out of a sample of $n_2 = 300$ teen-age girls indicated such a preference. Construct a .95 confidence interval for $p_1 - p_2$, the difference in the population proportions of teen-age boys and girls who prefer soft drinks in cans. Noting that $z = 1.96$ for a .95 confidence interval, apply Equation 10.5.1 as follows:

$$\left(\frac{52}{350} - \frac{41}{300}\right) \pm 1.96 \sqrt{\frac{\frac{52}{350}\left(1 - \frac{52}{350}\right)}{350} + \frac{\frac{41}{300}\left(1 - \frac{41}{300}\right)}{300}} = .012 \pm .053$$

Then, .95 confidence limits are

$$LL = .012 - .053 = -.041$$
$$UL = .012 + .053 = .065$$

Hence, we estimate that the proportion of teen-age boys in the city who prefer soft drinks in cans is .012 greater than the proportion of teen-age girls who have such a preference. Also, we are 95 percent confident that the difference in population proportions is included in the interval $(-.041)$–.065. Notice that this interval includes zero as a possible value of the difference between the two proportions. When a confidence interval includes zero, we should conclude that the parameter estimated is equal to zero. Consequently, we conclude that there is no difference in the proportions of teen-age boys and girls who prefer soft drinks in cans.

Study Problems

1.   Which statistic is used as a point estimate of the difference between two population (a) means and (b) proportions?

2.   Distinguish between (and illustrate) independent samples and dependent samples.

3.   A sample of 40 college men (n_1) obtained a mean of 120 on a research aptitude test, with a standard deviation of 10; whereas, a sample of 50 college women (n_2) obtained a mean of 100 on this test, with a standard deviation of 20. Construct a .95 confidence interval for the difference in population mean test scores for college men and women.

4.   The mean height of a sample of 60 plants grown under condition A is 36.2 cms, with a standard deviation of 5.0 cms. A sample of 50 plants (same variety) grown under condition B has a mean height of 40.4 cms, with a standard deviation of 6.0 cms. Construct a .98 confidence interval for the difference in mean height of the plants grown under the two conditions.

5.   A sample of $n_1 = 100$ sheets produced by firm A showed a mean of 11.3 defects per sheet, with a standard deviation of 2.5 defects. A sample of $n_2 = 90$ sheets produced by firm B showed a mean of 8.6 defects per sheet, with a standard deviation of 3.0 defects. Construct a .99 confidence interval for the difference in the mean number of defects per sheet for the two firms.

6.   A sample of 12 corporate preferred stocks showed a mean price rise of $15 per stock and a standard deviation of $2. A sample of 9 corporate common stocks showed a mean price rise of $22 and a standard deviation of $4. Construct a .98 confidence interval for the difference in the population mean price rise for corporate preferred and common stock prices. (Assume that the populations of preferred and common stock price rises are approximately normal in distribution.)

7.   Determine a point estimate of the difference between population means for
   a.   Problem 5
   b.   Problem 6

8.   Determine the maximum error you would expect when using $\overline{X}_1 - \overline{X}_2$ as a point estimate of $\mu_1 - \mu_2$ for
   a.   Problem 5, at the .99 level of confidence
   b.   Problem 6, at the .98 level of confidence

9.   A sample of 10 men in the 55–60 year age group has a mean chest girth of 47.3 inches, with a standard deviation of 4.0 inches; whereas, a sample of 15 men in the 40–45 year age group has a mean chest girth of 38.7 inches, with a standard deviation of 5.0 inches. Chest girths for men in each age group are approximately normal in distribution.
   a.   Determine a point estimate of the difference in mean chest girth for the two age groups.
   b.   What is the maximum error you would expect in the point estimate at the .95 level of confidence?
   c.   Construct a .95 confidence interval for the difference in mean chest girth for the two age groups.

10.   A test of racial tolerance showed a mean score of 70.5 for a sample of 20 boys (standard deviation 5.0) and a mean score of 80.3 for a sample of 18 girls (standard deviation 4.0). Test scores are normally distributed for boys and girls. Construct a .99 confidence interval for the difference in mean test scores for boys and girls.

11.   A sample of 34 construction workers in region A showed a mean of 22.9 days of unemployment last year, with a standard deviation of 2.0 days. A sample of 40 construction workers in region B showed a mean of 28.4 days of unemployment last year, with a standard deviation of 2.5 days.
   a.   Determine a point estimate of the difference in mean number of days of unemployment for regions A and B.
   b.   What is the maximum error in this point estimate you would expect at the .98 level of confidence?
   c.   Construct a .98 confidence interval for the difference in the mean number of days of unemployment for construction workers in the two regions.

12.   A mechanical aptitude test was administered to a sample of 33 college juniors who are chemistry majors and showed a mean of 18.9 and a standard deviation of 2.0. The same test was administered to a sample of 35 college juniors who are physics majors and showed a mean of 25.4 and a standard deviation of 3.0. Determine a .98 confidence

interval for the difference in test mean scores for chemistry and physics college juniors. Assume equal population variances.

13.　A town is located between two shopping centers—A and B. A sample of 45 housewives in the town indicated that it took a mean of 27.5 minutes to travel to shopping center A, with a standard deviation of 4.0 minutes. A second sample of 36 housewives in the town indicated that it took a mean of 22.3 minutes to travel to shopping center B, with a standard deviation of 3.0 minutes. The standard deviations of the populations of travel times to the two shopping centers are considered to be equal. Construct a .95 confidence interval for the difference in the two population mean travel times for the housewives in the town.

14.　A large food supermarket has a number of check-out counters. A sample of 8 customers waited a mean of 15.2 minutes (with a standard deviation of 1.2 minutes) in the line for check-out counter I. A sample of 10 customers in the line for check-out counter II waited a mean of 10.7 minutes (with a standard deviation of 1.5 minutes). The check-out waiting time per customer is approximately normal in distribution for the two counters, with equal variances. Construct a .95 confidence interval for the difference in mean check-out waiting time for the two counters.

15.　A sample of 10 pieces of cotton cloth produced by firm A showed a mean breaking strength of 173 pounds (standard deviation 20 pounds). A sample of 15 pieces o cloth produced by firm B showed a mean breaking strength of 100 pounds (standard deviation 25 pounds). The breaking strength per piece of cloth is approximately normal in distribution for the two firms and variances are considered to be equal. Construct a .90 confidence interval for the difference in mean breaking strength of the cloth from the two firms.

16.　A sample of 9 subjects was selected for a weight reducing plan. The "before" and "after" weights for each subject are as follows:

Subject	Weight (pounds) Before	After
A	160	162
B	175	170
C	170	163
D	168	171
E	184	180
F	195	188
G	200	192
H	185	176
I	198	186

Construct a .95 confidence interval for the mean population difference in "before" and "after" weights for subjects following this weight reducing plan.

17.　A sample of eight pairs of plants was selected, with each pair matched on height and other characteristics. One sample of eight plants was replanted in specially treated soil whereas the other sample of eight plants was replanted in soil of standard formulation. After a two-week period the following plant heights (in cms) were recorded:

Pair	Standard soil	Special soil
A	22	24
B	18	20
C	25	29
D	20	27
E	19	25
F	24	31
G	26	34
H	30	38

Construct a .90 confidence interval for the mean plant height difference due to the two soils.

18. A shock stimulant was administered to a sample of six pairs of brother-sister twins in the 20–21 year age category. The following pairs of reaction-times (seconds) were obtained:

Pair	Brother	Sister
1	10	12
2	9	9
3	12	16
4	8	14
5	13	18
6	9	13

Construct a .95 confidence interval for the mean difference in reaction time between brother and sister (20–21 year age group) to the shock stimulant.

19. A sample of 60 married couples was selected in a city and the husband and wife were each asked to estimate the driving time from home to a new shopping center during the evening hours. The mean of the differences between wife and husband estimates was 10 minutes more for the wives than for the husbands, with a standard deviation of 2.1 minutes. Construct a .98 confidence interval for the mean difference in driving time estimates for a man and his wife.

20. A sample of 500 male college juniors includes 120 who are foreign students compared with 110 out of 650 female college juniors who are foreign students. Construct a .95 confidence interval for the difference in the proportions of male and female college juniors who are foreign students.

21. A sample of 150 brass fittings purchased from company A includes 90 which are perfectly made. A sample of 200 brass fittings purchased from company B includes 150 which are perfectly made. Construct a .95 confidence interval for the difference in the proportions of fittings which are perfectly made for the two companies.

22. A sample of 1,000 voters in city A includes 350 who oppose candidate X. A sample of 1,500 voters in city B includes 450 who oppose this candidate. Construct a .99 confidence interval for the difference in the proportions of voters in the two cities opposing candidate X.

Chapter 11

Tests of Hypotheses

11.1 Statistical Decision Making

In every field of endeavor, we are faced with the need for decision making. Generally, this requires consideration and evaluation of myriad facts and opinions and usually must be performed on the basis of limited information. We will be concerned with decision-making problems where the methods of statistical inference can offer assistance in the decision-making process. This is the area of statistical inference called *tests of hypotheses*. Let us consider two examples of the kinds of decision-making problems we will study.

A producer of livestock feeds claims that a newly developed feed concentrate will result in the same mean weight gain for pigs as a standard feed, but is much lower in price. A farmers' cooperative, interested in switching to the new feed, selects a sample of pigs for feeding with the new concentrate and plans to make a decision after comparing the mean weight gain per pig for this sample with the well-established mean gain for the standard feed. We know that sample means may be expected to vary around the population mean (the sampling distribution of the mean). That is, even if the population mean weight gain for the new feed concentrate is the same as for the standard feed, a particular sample mean could be far different. How could a decision be made based on sample results?

If the sample mean weight gain for the new feed $\overline{X}$ is found to be only slightly different from the population mean weight gain for the standard feed, should the cooperative conclude that the new feed is just as effective as the standard feed? If $\overline{X}$ is moderately different, what should the conclusion be? Suppose $\overline{X}$ is considerably different, should it be concluded that the new feed and the standard feed are not equally effective, even if such an $\overline{X}$ value could be obtained if the new feed actually is just as effective as the standard feed?

A drug manufacturer is advised by his chemist that a new product will be effective in treating a certain ailment 75 percent of the time; however, before using this percentage (cure rate) in advertising he wants to test its validity by trying it out on a sample of subjects who have the ailment. He plans to decide on the validity of .75 (or 75 percent) as a cure rate based on the sample proportion of subjects cured (X/n). Here, as in the previous example, we must take into consideration the sampling distribution of a statistic (in this case, X/n). More specifically, if the sample shows a cure rate different from 75 percent, should the chemist's advice be disregarded, even if such a sample percentage could be obtained if the cure rate is actually 75 percent?

Generally speaking, decision making involves the evaluation of alternative courses of action and the selection of one of these alternatives. In statistical decision making, the decision to be made relates to the value of a *population parameter*. For example, in the first of the two previously presented illustrations the parameter involved was the mean pig weight gain in the population and, in the second, it was the proportion of cures in the population. We call these the *test parameters*. In some problems, other test parameters, such as the population standard deviation, are involved.

Typically, in the statistical decision-making problems we are considering, we evaluate the validity of a claim relating to a population parameter with respect to a *standard* or a *reference value*. For example, in the pig weight gain problem, we evaluate the validity of the claim that the new feed results in the same mean weight gain as the standard feed. In the cure rate problem, we set up the chemist's estimate (75 percent) as a reference value (assume it is true) and evaluate the chemist's claim that the cure rate of the new product equals this reference value.

Then, we express the alternative courses of action as hypotheses relating to the test parameter. First, we formulate the hypothesis that there is no difference between the claim and the standard or reference value. For example, we hypothesize that the mean gain in pig weight is the same for the new feed as for the standard feed. We hypothesize that the chemist's claim is true (no difference between his claim and the true cure rate which is assumed to be 75 percent). We formulate such a hypothesis of no difference, whether or not we actually believe it, since information is not available to formulate any other *specific* hypothesis. Such a hypothesis is called a *null hypothesis* and is denoted as H_0. Statistical decision making involves testing the acceptability of the null hypothesis H_0.

We also formulate an *alternative hypothesis*, denoted as H_1, which is a *logical* course to follow *if the null hypothesis is rejected*. For example, if the null hypothesis is:

H_0: mean weight gain is the same for the new feed as for the standard feed.

Then the logical alternative hypothesis is:

H_1: mean weight gain for the new feed is less than for the standard feed.

This is the logical alternative to H_0 in this problem, since it is reasonable to believe that the producer of the new feed is not likely to understate the effectiveness of his product and if it may be expected to result in a higher mean weight gain he would certainly claim this. On the other hand, if the claim that the new feed is just as effective as the standard feed were considered to be an unprejudiced opinion and if the purpose of the test was to determine whether or not this claim is true, then the logical alternative hypothesis would be H_1: mean weight gain for the new feed is not the same as for the standard feed (it may be more or less). In the problem of the drug manufacturer, assume that the chemist offered an honest and unbiased opinion of the cure rate for the manufacturer's use and the manufacturer wishes to determine whether or not to use this claim in advertising. The logical alternative hypothesis is H_1: the cure rate is not 75 percent (it may be higher or lower, we do not know which). We will be concerned with problems involving only two hypotheses; this is the simplest type of a test of an hypothesis.

Testing an hypothesis H_0 leads to one of the following actions:

(1) If H_0 is rejected, accept H_1.
(2) If H_0 is not rejected, accept it.
(3) If H_0 is not rejected, make no decision.

Notice that in actions (1) and (2) we expose ourselves to the risk of making an incorrect decision. In action (1), we may make the error of rejecting the hypothesis we are testing H_0, when it is actually true. We call this a *Type I error*. In action (2), we may make the error of accepting H_0 when it is actually false. We call this a *Type II error*. In action (3), it is not possible to make an error since no decision is made (the problem may be studied further). Before conducting a test of hypothesis, we must decide whether actions (1) and (2) or (1) and (3) will be incorporated in the test procedure.

An important feature of statistical decision making is the ability it provides for *controlling* the risks of making Type I and Type II errors. Controlling the risk of committing an error means to *predetermine the probability of making the error*. For example, in the drug manufacturer's problem the hypothesis being tested is H_0: the chemist's claim is true or $p = .75$, where p is the cure rate for the population of people who have the ailment. How can he control the risk of committing a Type I error? According to Model D, the sampling distribution of X/n is approximately normal, with mean p and standard error $\sigma_{X/n} = \sqrt{p(1-p)/n}$. Suppose the manufacturer states that he is willing to take a risk of 5 in 100 (or .05) of committing a Type I error (*rejecting H_0 if it is true*). He refers to Table B.4 for normal curve areas to find z values which mark off half of .05 or .025 in each tail of the sampling distribution of X/n. Such z values are ± 1.96. Then he selects a sample of subjects who have the ailment, tries out the new drug on the sample, and computes X/n, the proportion cured. If X/n is *different enough* from $p = .75$ (sufficiently smaller or sufficiently larger) he will reject H_0. What is different enough? This is determined by the risk he is willing to take of making a Type I error, .05, as indicated in the following paragraph.

Transform X/n computed from the sample to z (Equation 8.6.6) and compare this z value with ± 1.96. If this z value is less than -1.96 or more than 1.96, then the corresponding sample proportion X/n is considered to be *different enough* to lead to rejection of H_0: $p = .75$, even though such an X/n may be expected to occur 5 times in 100 if p actually is .75. Clearly, the manufacturer is taking a risk of .05 of committing a Type I error. A similar approach is used to predetermine the probability of committing a Type II error. This will be discussed in Section 11.5. We will be concerned mostly with problems where only the probability of committing a Type I error is predetermined. We will consider the problem of controlling both Type I and Type II errors in Section 11.5. The probability of committing a Type I error is called the *level of significance* of the test and is denoted as α (the Greek letter alpha). The probability of committing a Type II error is denoted as β (the Greek letter beta).

The general approach of tests of hypotheses and the way in which such tests are conducted will become clearer in discussions throughout the remaining sections of this chapter.

Summary: Tests of hypotheses is an application of statistical decision making. In the tests we are considering, the decision to be made is expressed in terms of evaluating the value of a population parameter, called the test parameter. Tests of hypotheses may be looked upon as a three-phase procedure, as follows:

Phase A: Formulate an appropriate hypothesis relating to the test parameter, the null hypothesis H_0. This is the hypothesis of no difference between the claimed value of the test parameter and a standard or reference value. Formulate an alternative hypothesis H_1 which indicates a logical course to follow, if H_0 is rejected. Specify α, the level of significance for the test (the probability of committing a Type I error).

Phase B: Construct a decision model which involves determining the appropriate sampling distribution for the test (based on an appropriate sampling model) and specify when the sample data indicate that there is enough difference to warrant rejection of H_0. In other words, the decision model provides a decision rule which leads to a decision based on the sample data. The meaning and construction of a decision model and a decision rule will be discussed in the balance of the chapter.

Phase C: Evaluate the sample data according to the decision model and make the decision. (We ignore, for the sake of simplicity, the important phase of sample data collection which precedes Phase C. We assume sample data are available when Phase C is to be carried out.) This will be discussed and illustrated in the balance of the chapter.

11.2   Proportions

In this section, we will study tests of hypotheses where the test parameter is a *proportion* and where the following conditions are met (so that Model D is the appropriate sampling model): (1) population large; (2) sampling ratio n/N less than .05; (3) population proportion not too close to 0 or 1; and (4) sample size sufficiently large ($n \geq 100$).

A leading builder in a large city claims that 60 percent of the families whose head of the household is past age 65 prefer to live in an apartment instead of a one-family home. The builders' association in the city wants to test whether or not this claim is acceptable, with a risk of no more than 2 in 100 that this claim is rejected if it is true.

The *test parameter* in this problem is p, which is the proportion of all families in the city where the head of the household is past age 65 who prefer to live in apartments. We formulate the hypotheses:

$H_0: p = .60$

$H_1: p \neq .60$ (p not equal to .60 means that, if $p = .60$ is not acceptable, the association has no basis for concluding that p is less than .60 or greater than .60.)

The level of significance of the test is $\alpha = .02$, as specified by the builders' association. We construct a *decision model* based on the sampling distribution of the proportion X/n. According to Model D, the sampling distribution is approximately normal with mean p and standard error $\sigma_{X/n} = \sqrt{p(1 - p)/n}$. This sampling distribution is presented in Figure 11.2.1 in terms of the standard normal curve, where the variable is z. Notice that the sampling distribution is divided into regions: a *region of rejection*, also called a *critical region*, in each tail (reject H_0) and a *region of acceptance* (accept H_0) in the center of the distribution. These regions are determined by the level of significance of the test and the way in which the alternative hypothesis is formulated, as explained in the following paragraph.

Notice that H_1 states that $p \neq .60$, so that p could be either below or above .60, and $\alpha = .02$ indicates that the risk of a Type I error is predetermined as 2 chances in 100. Accordingly, we determine (from Table B.4) that $z = \pm 2.33$ cuts off .01 in each tail or .02 ($= \alpha$) in both tails combined. Then, if sample data results in an X/n which, after transformation to standard z units, falls in *either* region of rejection, the difference between .60 and the obtained X/n is considered to be different enough to require rejection of H_0 and acceptance of H_1.

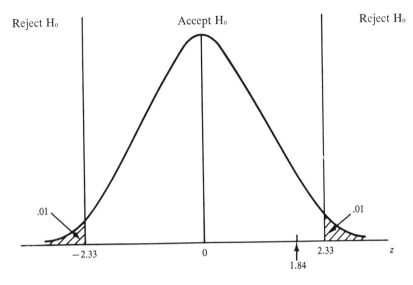

FIGURE 11.2.1

In other words, an X/n different enough from .60 to fall in the lower-tail region of rejection (too small) or in the upper-tail region of rejection (too large) is possible if $p = .60$; however, it is so different from .60 that it is *improbable* (only 2 chances in 100 or α). Hence, H_0 is to be rejected and the logical alternative H_1 accepted.

Figure 11.2.1 represents a decision model and it includes a *decision rule* as follows. *Transform the sample proportion X/n to z (Equation 8.6.6). If $z < -2.33$ or > 2.33, reject H_0 and accept H_1, otherwise, accept H_0.*

Suppose that a sample of $n = 100$ families includes $X = 69$ who prefer to live in an apartment. We compute $X/n = 69/100 = .69$. Transforming to standard units (Equation 8.6.6), we obtain

$$z = \frac{.69 - .60}{\sqrt{\dfrac{.60(1 - .60)}{100}}} = 1.84$$

Referring to our decision rule (Figure 11.2.1), we find that $z = 1.84$ falls in the region of acceptance. Hence, H_0 is accepted and the association concludes that the builders' claim of .60 is acceptable. (Of course, this does not prove that .60 is the correct proportion. The sample result .69 is close enough to .60 to make it highly probable that .60 is correct.)

Notice that the alternative hypothesis $p \neq .60$ requires that we provide for the possibilities that the sample proportion X/n may be extreme (different enough) in either direction (too small or too large). Such an alternative hypothesis, called a *two-sided alternative*, always results in a *two-tail test*, where there is a critical region in each tail of the sampling distribution. Some tests have a *one-sided alternative*, resulting in a *one-tail test* (critical region in one tail of the sampling distribution, either tail). Let us consider such an example.

A typist in a large insurance company complained to her supervisor that she typed as many letters but made fewer errors than the other typists, and therefore, should be paid a higher salary. The supervisor promised to consider this matter. The supervisor is concerned that if she decided to raise this typist's salary, the others would complain unless there is clear evidence that this typist is better (makes fewer errors). So the supervisor selects a random sample of 600 letters from those typed by this typist during the past year (carbon copies are retained in the files). If a letter was retyped because of typing errors, this is indicated by a code on the carbon copy. The supervisor knows that on the average about 5 percent of the letters are retyped because of typing errors. She decides to take only a small risk (.01) of concluding erroneously that this typist has a smaller proportion of letters retyped. She does not plan to use this sample of letters to decide that this typist is not better then average in typing accuracy. If this test does not indicate that this typist is more accurate, she will review her work again later on.

The test parameter is p, the proportion of the typist's letters which must be retyped. We formulate the hypotheses:

H_0: $p = .05$ (This is the null hypothesis of no difference between the typist's retyping rate and the average rate. The formulation of this hypothesis does not mean, necessarily, that the supervisor believes it is true.)

H_1: $p < .05$ (This is the appropriate alternative, since the supervisor wishes to decide whether the typist is average in her retyping rate or lower than average. Whether the typist is more than average in this rate is not pertinent to this test.)

The specified level of significance for this test is .01. The appropriate sampling distribution is the sampling distribution of X/n, as before, which is presented in Figure 11.2.2. Notice that the alternative hypothesis in this problem, $p < .05$, takes into consideration only the possibility of an extreme X/n below $p = .05$. This illustrates a one-sided alternative. As a result, there is the need for only one region of rejection (in the lower tail), as indicated in Figure 11.2.2. Thus, this is a one-tail test.

Since the supervisor plans not to make a decision based on this test, if the test does not lead to rejection of H_0, there is no region of acceptance shown in Figure 11.2.2. Instead, there is a *region of no decision*. The lower-tail region of rejection was determined by finding a z which cuts off $\alpha = .01$ in the lower tail. Such a z (called a *critical value*) is -2.33, as shown in the figure.

Suppose the sample of 600 letters showed that 16 were retyped because of typing errors. We compute $X/n = 16/600 = .027$. Transforming to z, we obtain

$$z = \frac{.027 - .05}{\sqrt{\dfrac{.05(1 - .05)}{600}}} = -2.56$$

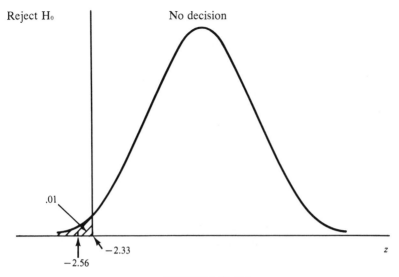

FIGURE 11.2.2

Referring to the decision rule (Figure 11.2.2), we find that $z = -2.56$ falls in the region of rejection. Therefore, reject H_0 and accept H_1. The supervisor decides to raise the typist's salary.

11.3 Means

In this section we are concerned with tests of hypotheses where the test parameter is the mean and where the following conditions hold (so that Model B is the appropriate sampling model for large samples and Model E for small samples): (1) population large, and (2) sampling ratio n/N less than .05. Let us first consider problems involving large samples.

A filling mechanism is set to fill cleaning powder boxes with a mean fill of 35 ozs. A sample of 50 filled boxes is selected to test whether or not the filling mechanism is operating as required. The mean quantity of cleaning powder per box for the sample was found to be 36.42 ozs and the standard deviation was computed as $s = 6.30$ ozs. Use the .05 level of significance for the test.

The *test parameter* in this example is μ, the mean quantity of cleaning powder per box. Formulate the hypotheses:

> H_0: $\mu = 35$ ozs
> H_1: $\mu \neq 35$ ozs (Why is a two-sided alternative appropriate in this
> problem?)

The decision model is constructed on the basis of the sampling distribution of the mean $\overline{X}$. According to Model B, the sampling distribution is approximately normal with mean μ and estimated standard error $s_{\bar{x}} = s/\sqrt{n}$. Since the alternative hypothesis is two-sided, we mark off a critical region (a region of rejection) in each tail of the sampling distribution (a two-tail test) by finding the critical z values which cut off $(1/2)\alpha = (1/2)(.05)$ or $.025$ in each tail. Such z values are ± 1.96. We write the *decision rule* as follows. *Transform the sample mean $\bar{X}$ to z (Equation 8.4.2). If $z < -1.96$ or > 1.96, reject H_0 and accept H_1, otherwise, accept H_0.* (It is a good exercise for the reader to express the decision rule in chart form, similar to Figure 11.2.1, as well.)

Referring to the sample data, transform $\bar{X} = 36.42$ to z to obtain

$$z = \frac{36.42 - 35}{\dfrac{6.30}{\sqrt{50}}} = 1.60$$

Then, according to the decision rule, $z = 1.60$ falls in the region of acceptance. Therefore, conclude that the mean fill is 35 ozs as required. (Of course, this does not prove that the mean fill is 35 ozs. How do you interpret the finding that $z = 1.60$ falls in the region of acceptance?)

If it were suspected that the mean fill was too high, the appropriate alternative hypothesis would have been $\mu > 35$ ozs, a one-sided alternative. Then, a critical region would appear only in the upper tail of the sampling distribution, determined by the critical value $z = 1.64$ which cuts off $\alpha = .05$ in the upper tail.

In tests of hypotheses where the test parameter is μ and the sample size is small, Model E is the appropriate sampling model; however, this model requires that the population have a normal or approximately normal distribution.

The owner of a large gambling casino in Las Vegas claims that the average loss of players who lose money at the gambling tables is $250 or less. An investigating committee doubts this claim and selects a sample of 25 losers to test the claim at the .05 level of significance. The sample shows a mean loss of $263.10 and a standard deviation of $35.25. (Assume that gambling losses are approximately normal in distribution.)

Formulate the hypotheses:

H_0: $\mu = \$250$ (Even though the claim of the casino owner is $250 or less, we must formulate a specific hypothesis to be tested.)

H_1: $\mu > \$250$

The decision model for small samples is constructed on the basis of the t distribution, with $n - 1$ df (Model E). Since the alternative hypothesis is one-sided and hypothesizes that μ is greater than specified in H_0, mark off a critical region in the upper tail of the t distribution by finding the critical $t_{.10}$ value for $25 - 1 = 24$ df. This value $t_{.10} = 1.711$ cuts off $\alpha = .05$ in the upper tail. This t value is found in Table B.5, column headed .10 (and represents a value which cuts off .10 in both tails or .05 in each tail). The decision rule is as follows. *Compute the t statistic based on the sample data (according to Model E). If $t > 1.711$, reject H_0 and accept H_1, otherwise, accept H_0.* This decision rule is also presented in Figure 11.3.1.

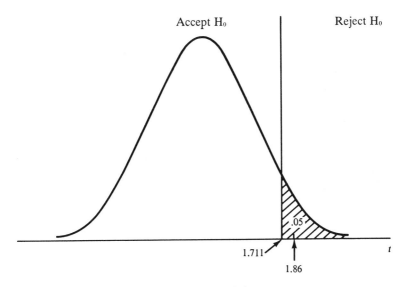

FIGURE 11.3.1

Referring to the sample data, compute

$$t = \frac{263.10 - 250}{\dfrac{35.25}{\sqrt{25}}} = 1.86$$

According to the decision rule (and as shown in Figure 11.3.1) $t = 1.86$ falls in the region of rejection so that H_0 is rejected and H_1 accepted. Hence, the claim of the casino owner is rejected and it is concluded that the average loss of gambling losses is more than \$250. (Of course, this does not prove that the claim is false. Since $t = 1.86$ falls in the critical region, it indicates that it is improbable that the claim is true. However, is it possible that the claim is true? Why?)

11.4   Optional: Variances

In some problems, the *variance* σ^2 (or the *standard deviation* σ) is the test parameter. The sampling situation in such problems may be represented by Model I.

Model I

If samples of size n are selected from a normal population with variance σ^2, estimated by s^2 computed from sample data, the sampling distribution of the statistic

$$\chi^2 = \frac{(n - 1)s^2}{\sigma^2}$$

is like the x^2 distribution, with $n - 1$ df. (x^2 is the square of the Greek letter chi, often written as *chi-square*.)

Notice that Model I, like Models A and E, requires that the population sampled be normal (or approximately normal) in distribution and that the sample size is not restricted. Furthermore, the x^2 distribution, like the t distribution, depends on the number of degrees of freedom involved with df $= n - 1$. A x^2 distribution is presented in Figure 11.4.1 and indicates that the distribution is skewed to the right, continuous, and can take on only positive values (0 to infinity). Notice that Figure 11.4.1 also presents the regions of acceptance and rejection for a one-tail test, with the critical region in the upper tail.

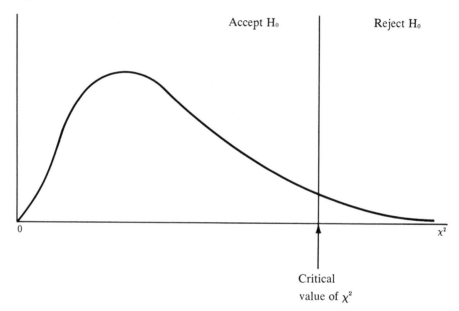

FIGURE 11.4.1

A chi-square distribution.

Table B.6 presents critical x^2 values for various levels of significance and for 1 to 30 df. The critical values in this table cut off upper-tail critical regions only. For example, the column headed ".05" presents x^2 values for various df which cut off .05 of the area under the x^2 curve in the upper tail. What do we do about lower-tail tests? For example, suppose $\alpha = .05$ and we need to determine a x^2 value which marks off .05 in the lower tail. This means that to the right of the required x^2 value lies $1 - .05$ or .95 of the area under the x^2 curve. Referring to Table B.6, column headed .95, we see the x^2 value corresponding to the appropriate df. This x^2 ($x_{.95}^2$) is the critical value for a one-tail test with $\alpha = .05$

and the critical region in the lower tail. Similarly, the critical χ^2 values may be determined for each tail in a two-tail test.

The manager of a large jewelry store claims that the number of watches sold on a Saturday during the fall and winter months has a standard deviation of $\sigma = 10$ watches. Test this claim by using the .05 level of significance and assume that the manager is offering his best and most honest opinion. (Assume that the distribution of the number of watches sold on a Saturday during the specified months is approximately normal.) A sample of 18 Saturdays showed $s = 15$ watches.

The test parameter is σ^2, the variance for the distribution of the number of watches sold on a Saturday during the specified months. Formulate the hypotheses:

H_0: $\sigma^2 = (10)^2 = 100$
H_1: $\sigma^2 \neq 100$

The decision model is constructed on the basis of the χ^2 distribution (Model I). Since H_1 is two-sided, we mark off a region of rejection in each tail of the chi-square distribution by finding the χ^2 values which cut off $(1/2)\alpha = (1/2)(.05)$ or .025 in each tail. Notice that df $= 18 - 1 = 17$. Hence, we see from Table B.6 that $\chi_{.975}^2 = 7.564$ and $\chi_{.025}^2 = 30.191$. The decision rule is as follows. *Compute the χ^2 statistic based on the sample data (Model I). If $\chi^2 < 7.564$ or > 30.191, reject H_0 and accept H_1, otherwise, accept H_0.*

Referring to the sample data compute:

$$\chi^2 = \frac{(18 - 1)(15)^2}{(10)^2} = 38.25$$

According to the decision rule, $\chi^2 = 38.25$ falls in the region of rejection. Therefore, reject H_0 and accept H_1. Conclude that σ is not equal to 10 watches (or that $\sigma^2 \neq 100$).

11.5 Optional: Controlling Decision Risks

So far, we have shown how an acceptable risk of rejecting H_0 if it is true (Type I error) can be incorporated into a test procedure by specifying α, the level of significance. What about Type II errors (accepting H_0 if it is false)? Let us consider this.

The president of a large university claims that no more than 15 percent of the applicants accepted for admission do not enroll in the university. The trustees of the university doubt this claim and believe the percentage not enrolling is much larger. It is decided to conduct a test of the president's claim based on a sample of $n = 500$ accepted applicants, with H_0: $p = .15$, H_1: $p > .15$ and $\alpha = .05$. The sampling distribution of X/n is shown in Figure 11.5.1, with the regions of acceptance and rejection marked off. This sampling distribution assumes H_0 is true, so that the mean of the sampling distribution is $z = 0$ or $X/n = p = .15$. The alternative hypothesis $p > .15$ is not specific. Let us formulate the specific alternative $p = .21$. *Only when a specific alternative hypothesis is formulated is it possible to compute* β, *the probability of committing a Type II error.* Figure 11.5.2 presents the sampling distribution of X/n assuming that $p = .21$ is true, so that the mean of this distribution is $z = 0$ or $X/n = p = .21$.

Notice that in Figure 11.5.1 the critical value $z = 1.64$ marks off the region of rejection in the upper tail. We can determine the X/n value corresponding

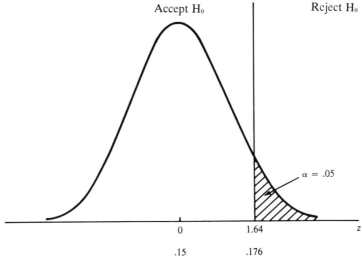

FIGURE 11.5.1

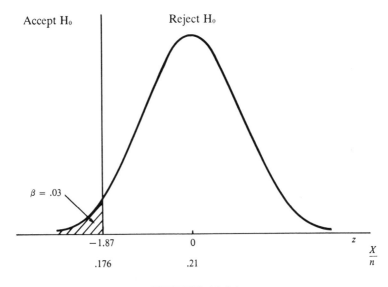

FIGURE 11.5.2

to $z = 1.64$ by first substituting in Equation 8.6.6 for z all the known values ($z = 1.64$, $p = .15$, $n = 500$) as follows:

$$1.64 = \frac{\frac{X}{n} - .15}{\sqrt{\frac{.15(1 - .15)}{500}}}$$

Then, solving for X/n, we obtain $X/n = .176$. Therefore, in Figure 11.5.1, $z = 1.64$ and $X/n = .176$ represent two ways to identify the critical value which marks off the region of rejection (as shown in the figure).

The critical value $X/n = .176$ is also marked off in Figure 11.5.2 and the regions of acceptance and rejection identified to correspond with these regions in Figure 11.5.1. Of course, the corresponding z value is not 1.64 for this distribution. The corresponding z value for the distribution in Figure 11.5.2 is computed by applying Equation 8.6.6 and substituting $n = 500$, $X/n = .176$, and $p = .21$ to obtain

$$z = \frac{.176 - .21}{\sqrt{\frac{.21(1 - .21)}{500}}} = -1.87$$

This z value is shown in Figure 11.5.2. We may determine by the use of the normal curve table (Table B.4) that .03 of the total area under the sampling distribution curve in Figure 11.5.2 is in the lower tail (to the left of $X/n = .176$ or $z = -1.87$), as shown in the figure. This area lies in the region of acceptance, so that the probability is .03 that if $p = .21$ is true, H_0 (which states $p = .15$) will be accepted and a Type II error committed. The probability that a Type II error will occur ($\beta = .03$) is shown in Figure 11.5.2.

The probability of committing a Type II error (β) can be computed in the same way for any specific alternative hypothesis. For example, for the alternative $p = .20$, $\beta = .09$; for $p = .19$, $\beta = .21$; etc. If β values are computed for possible alternative values of p and plotted on the vertical axis, with p on the horizontal axis, we have an *operating characteristic (OC) curve. This curve shows the probability of accepting H_0 for possible alternative values of the test parameter.* Often, $1 - \beta$ is computed, which is called the power of the test and represents the probability of rejecting H_0 if it is false. If $1 - \beta$ is plotted (vertical axis) with alternative possible test parameter values on the horizontal axis, we have a *power curve.* This curve shows the probability of rejecting H_0 if an alternative value of the test parameter is true.

It is not always possible to formulate a *specific* alternative hypothesis (which is required if β is to be determined). Then α is predetermined and not β. In some problems there is a basis for formulating a specific alternative. For example, in our illustration, suppose the trustees of the university feel that changes in university policies will have to be considered if the proportion of accepted applicants who do not enroll in the university is .21 or more. Then, $p = .21$ is an important alternative to be considered. It would be necessary to conduct the test so that the risk of accepting H_0: $p = .15$ is small if p is actually .21 or more. Assume that the trustees do not wish to take a risk of more than $\beta = .02$ that a Type II error will be committed if $p = .21$ (or more). How can a test of hypothesis be conducted with *both* α and β predetermined?

First, let us see how α and β are related. In Figure 11.5.3, we placed the sampling distribution of Figure 11.5.1 on the X/n scale of values with the mean at $p = .15$ and the sampling distribution of Figure 11.5.2 with the mean at $p = .21$. Then, an ordinate (vertical line) was constructed at the point $X/n = .176$ and the regions of acceptance and rejection identified. The probabilities of committing a Type I error ($\alpha = .05$) and a Type II error ($\beta = .03$) are shown in the figure. Notice that we may move the vertical line up and down the X/n scale of values, thus, shifting the critical value of X/n and changing the regions of acceptance and rejection and the size of α and β. Notice that α is increased and β is decreased if the critical value of X/n is reduced. If the critical value of X/n is increased, α is reduced and β is increased. In other words, if the test is conducted with a reduced risk (probability) of committing a Type I error (α), the risk of a Type II error (β) is increased. If the risk of committing a Type II error (β) is reduced, then the risk of a Type I error (α) is increased. How can both types of error risks be reduced? *The risks of committing both types of errors may be predetermined and incorporated into a test of hypothesis if the proper sample size is used.*

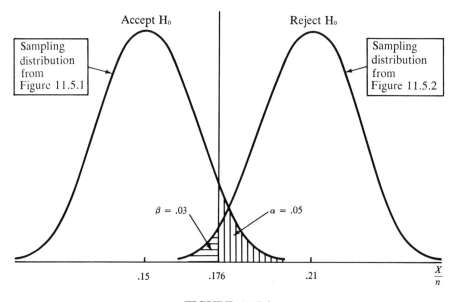

FIGURE 11.5.3

Suppose we design a one-tail test (critical region in the upper tail) with $H_0: p = p_0$ and the specific alternative $p = p_1$. Then, the critical z value corresponding to a specified α may be denoted as z_0 and the critical z value corresponding to a specified β may be denoted as z_1. As indicated in Figure 11.5.3, these critical values are identical when expressed in terms of X/n. According to Equation 8.6.6, we have

$$z = \frac{\frac{X}{n} - p}{\sqrt{\frac{p(1 - p)}{n}}} \qquad (11.5.1)$$

Solving for X/n, we obtain

$$\frac{X}{n} = p + z\sqrt{\frac{p(1 - p)}{n}} \qquad (11.5.2)$$

Applying the foregoing equation, we may express the critical value X/n in terms of z_0 and p_0 (based on H_0) as well as in terms of z_1 and p_1 (based on the alternative hypothesis), as follows:

$$\frac{X}{n} = p_0 + z_0\sqrt{\frac{p_0(1 - p_0)}{n}} \qquad (11.5.3)$$

and

$$\frac{X}{n} = p_1 + z_1\sqrt{\frac{p_1(1 - p_1)}{n}} \qquad (11.5.4)$$

Equating the right-hand sides of Equations 11.5.3 and 11.5.4, we obtain

$$p_0 + z_0\sqrt{\frac{p_0(1 - p_0)}{n}} = p_1 + z_1\sqrt{\frac{p_1(1 - p_1)}{n}} \qquad (11.5.5)$$

Solving for n, we obtain

$$\sqrt{n} = \frac{(z_1\sqrt{p_1(1 - p_1)} - z_0\sqrt{p_0(1 - p_0)})}{(p_0 - p_1)} \qquad (11.5.6)$$

and

$$n = \frac{(z_1\sqrt{p_1(1 - p_1)} - z_0\sqrt{p_0(1 - p_0)})^2}{(p_0 - p_1)^2} \qquad (11.5.7)$$

Equation 11.5.7 provides the basis for determining sample size for a test when p is the test parameter. Negative or positive values are to be substituted for z_0 and z_1 as appropriate. In a one-tail test, z_0 is positive if the critical region is in the upper tail and negative if the critical region is in the lower tail. In a one-tail test with the critical region in the upper tail, z_1 is negative if $\beta < .50$ and positive if $\beta > .50$. In a one-tail test with the critical region in the lower tail, z_1 is positive if $\beta < .50$ and negative if $\beta > .50$. If $\beta = .50$, then $z_1 = 0$.

Some interesting relationships may be noted from an examination of Equation 11.5.7. The closer together p_0 and p_1, the smaller is the difference and the larger is the required sample size. Notice that we may predetermine any two of the three α, β, n, but not all three. In the previously presented testing procedure, n was given and α predetermined. Consequently, it was not possible to control β. If both α and β are predetermined, then Equation 11.5.7 will indicate the sample size needed for a given specific alternative hypothesis.

The manager of a department store claims that the exchange and refund policy of the store is adequate since at most 18 percent of the purchases made are exchanged or returned for a refund. This is well below the national average of 25 percent determined by a trade association. The president of the firm doubts the manager's claim and requests his assistant to have the claim evaluated. The president states that he wishes to take a risk of no more than .05 that the claim will be rejected if it is true. On the other hand, he is willing to take only a small risk (1 chance in 100) that the claim will be accepted if 25 percent or more of purchases made are exchanged or returned for refund. How large a sample of purchases is required if it is decided to conduct a statistical test of the manager's claim?

Noting that a one-tail test (upper tail critical region) is to be conducted, apply Equation 11.5.7 using $p_0 = .18$, $z_0 = 1.64$ (since $\alpha = .05$), $p_1 = .25$, and $z_1 = -2.33$ (since $\beta = .01$), to obtain

$$n = \frac{(-2.33\sqrt{(.25)(.75)} - 1.64\sqrt{(.18)(.82)})^2}{(.18 - .25)^2} = 538.96$$

Therefore, a sample of 539 or 540 customers is required. The test is conducted in the same manner as previously discussed.

Equation 11.5.7 may be used to determine sample size for a two-tail test if $p_0 = .50$ (or very close to .50). Suppose we have $H_0: p = .50$, $H_1: p \neq .50$, $\alpha = .05$, and it is specified that if the true value of p differs from .50 by as much as .05 either way then $\beta = .01$ is the degree of protection required against a Type II error. Then, we are concerned with the specific alternative hypotheses $p = .50 - .05 = .45$ and $p = .50 + .05 = .55$. We need to consider only one of these alternatives (either one) to apply Equation 11.5.7 to determine sample size. Assume we use $H_1: p = 55$. This equation is applied with $p_0 = .50$, $p_1 = .55$, $z_0 = 1.96$ (since $\alpha = .05$ and H_1 is two-sided), and $z_1 = -2.33$ (since $\beta = .01$ and the full probability of .01 must be represented by the area in the tail).

If the test parameter is the *mean*, $H_0: \mu = \mu_0$, z_0 corresponds to a specified α, μ_1 is a specific alternative value hypothesized for μ, and z_1 corresponds to a specified β, then sample size may be determined as follows:

$$n = \frac{(z_1 - z_0)^2\sigma^2}{(\mu_0 - \mu_1)^2} \tag{11.5.8}$$

where σ denotes the population standard deviation. This equation was developed following the same approach as was used for Equation 11.5.7. If σ is not known (which is usually the case), an estimate must be used (based upon previous studies or a pilot sample). Equation 11.5.8 may be used for a two-tail test or a one-tail test. In the case of a two-tail test, z_0 corresponds to $(1/2)\alpha$ and z_1 corresponds to β.

Study Problems

1. A grooming and beauty aids firm claims that 20 percent of men aged 18–24 prefer a new after-shave lotion. A random sample of men in the specified age group showed that 16 percent prefer the new product.

        a. Does the sample information prove that the claim is false? Why?

        b. What is the probability that a sample proportion would equal .16 or less, if the true population proportion is actually $p = .20$? Assume $n = 500$.

2. What is a test parameter? Name three.

3. What is a null hypothesis (H_0)? Give an illustration.

4.   How is the null hypothesis used in a test of an hypothesis?

5.   What is the alternative hypothesis and the basis for its formulation?

6.   The superintendent of a large school system claims that the mean reading level of sixth grade pupils in his schools is 100 (on a specified reading test). A research organization wants to test this claim and administers the specified reading test to a sample of the sixth grade pupils.

      a.   What is the appropriate test parameter?

      b.   Formulate the appropriate hypothesis to be tested.

      c.   Formulate the appropriate alternative hypothesis if the research organization has no preconceived idea that the superintendent's claim is not true.

      d.   Formulate H_1, assuming that it is believed that the claim is an overstatement of the true population mean test score.

7.   Explain the meaning of Type I error and Type II error.

8.   What do we mean by controlling the risk of making an error?

9.   In Problem 1, assume that the claim made by the firm is completely objective and unprejudiced and that it is desired to test the validity of this claim.

      a.   What is the appropriate test parameter?

      b.   Formulate the appropriate hypothesis to be tested.

      c.   Formulate the logical alternative hypothesis.

      d.   Assume $\alpha = .02$. In terms of z, how much different from .20 must the sample proportion be (in either direction) to warrant rejection of H_0?

      e.   In terms of X/n (the sample proportion), how much different from .20 must the sample proportion be (in either direction) to warrant rejection of H_0? (Assume $\alpha = .02$, $n = 500$.)

10.   What is meant by the level of significance of a test?

11.   Examine the statements below and indicate whether a Type I or a Type II error is possible (or both or neither):

      a.   In Problem 1, it is decided to accept the claim.

      b.   In Problem 1, no decision on the validity of the claim is made.

      c.   In Problem 6, the claim is rejected and H_1 is accepted.

12.   What is meant by (a) region of acceptance and (b) critical region (or region of rejection)?

13.   What is meant by (a) one-sided alternative and (b) two-sided alternative?

14.   What is a (a) one-tail test and (b) two-tail test?

15.   Explain the meaning of a decision rule.

16.   Suppose a test of hypothesis is conducted with $\alpha = .01$.

      a.   What is the predetermined risk of committing a Type I error?

      b.   If H_1 is one-sided, what is the proportion of the area under the appropriate sampling distribution which lies in the critical region?

      c.   If H_1 is two-sided, what is the proportion of the area (see part (b)) in each region of rejection?

      d.   What is the proportion of the area (see part (b)) in the region of acceptance?

17.   What is a critical value?

18.   Suppose a test of hypothesis is conducted with $n = 1,000$. What is the critical

z value (or values) if (a) $\alpha = .01$ and H_1 is one-sided (upper tail) and (b) $\alpha = .01$ and H_1 is two-sided?

19.   A public figure states that he will allow himself to be drafted for the governorship if no more than 12 percent of the voters oppose him. A random sample of 1,000 voters includes 144 who oppose him. Should he decide to accept a draft for the governorship? Conduct the test at the .01 level of significance.

20.   In Problem 19, conduct the test on the assumption that the public figure wishes to determine whether or not 12 percent is an acceptable estimate of the proportion of the voters who oppose him.

21.   A sample of 500 teen-agers indicated that 350 prefer soft drinks in cans. Would you accept the claim of a can manufacturer that at least three-fourths of the teen-agers prefer soft drinks in cans, at the .05 level of significance?

22.   In Problem 21, conduct the test on the assumption that you want to test the impartial claim that 75 percent of the teen-agers prefer soft drinks in cans and you have no preconceived basis to support or reject the claim.

23.   A philanthropic foundation wishes to endow a new social studies department in a large university if there is sufficient assurance that at least 30 percent of the student body will take one or more courses in this department. If this does not seem likely at this time, the matter of the endowment will be further considered by the foundation. A survey of 800 students shows that 230 would take at least one course in the new department. What decision should the foundation make? Use the .02 level of significance.

24.   A machine is set to produce rods which have a mean length of 12 inches. A sample of 36 rods was found to have a mean length of 12.35 inches, with a standard deviation of one inch. Using the .01 level of significance, would you conclude that the machine needs readjustment?

25.   In Problem 24, conduct the test at the .05 level of significance.

26.   In Problem 24, if it is very important that the rods produced have a mean length of 12 inches and if checking the adjustment of the machine and readjusting is simple and inexpensive, which level of significance .01 or .05 is preferable? Why? (Assume that the machine will be tested at regular intervals.)

27.   In Problem 24, suppose checking the adjustment of the machine is time-consuming and expensive. Is .01 or .05 a more desirable level of significance to use? Why? (Assume that the machine will be tested at regular intervals.)

28.   Third-grade pupils in a large school system obtained a mean score of 30 on a reading test. Has reading ability improved for these pupils after a month of reading practice using a new system? Base your conclusion on the information that a sample of 33 pupils obtained a mean of 34 on a different form of the same reading test (standard deviation 8 points) after the month of additional reading instruction. Conduct the test at the .05 level of significance.

29.   In Problem 28, assume that test scores are approximately normal in distribution and that only 9 pupils were in the sample.

30.   It was determined that the mean time required by the patrons of a shopping center to travel from home to the center is 22 minutes. After the designation of certain streets as one-way thoroughfares, it was claimed that the mean travel time was reduced by five minutes. Would you accept this claim, at the .01 level of significance, if a sample of 16 patrons required a mean travel time of 21 minutes with a standard deviation of

five minutes? Assume that travel time is approximately normal in distribution. (Conduct a two-tail test.)

31. In Problem 30, assume $n = 9$.

32. In Problem 30, assume $n = 49$.

33. The head of a suburban hospital claims that hospital admissions average 18 patients per day and requests an increased budget. The hospital trustees doubt that average daily admissions is that high. A sample of 40 days showed a mean of 17.2 patients admitted per day, with a standard deviation of 3.4 patients. Would you accept the claim of the hospital head, if the trustees state that they will take a risk of only 2 in 100 that the claim will be falsely rejected?

34. In Problem 33, suppose $n = 16$, $\alpha = .01$, and assume that hospital admissions per day are approximately normal in distribution.

35. A recreation center in a small city claims that attendance averages 13 youngsters per day and so the center should not be closed. A sample of 22 days showed a mean daily attendance of 12.50 youngsters, with a standard deviation of .75 youngsters. Would you conclude, at the .05 level of significance, that the mean daily attendance claimed by the recreation center is acceptable? (Assume that daily attendance is approximately normal in distribution.)

36. A large university claims that entering freshmen score a mean of 35 on a racial tolerance test. It is desired to test this claim, even though it is generally agreed that it may be true. A sample of 60 entering freshmen obtained a mean score of 37, with a standard deviation of 3. Would you accept the claim based on the sample data? Use the .05 significance level.

37. Given $\sigma = 3$, $s = 4$, $n = 12$. Compute the chi-square statistic according to Model I.

38. Determine the critical χ^2 value (or values) for a test of hypothesis with
    a. $\alpha = .01$, $n = 12$, H_1 one-sided (upper-tail test)
    b. $\alpha = .01$, $n = 10$, H_1 one-sided (lower-tail test)
    c. $\alpha = .05$, $n = 14$, H_1 two-sided

39. Scores on a learning ability test are claimed to have a standard deviation of .50 by an impartial research worker. Would you accept this claim if a sample of 22 scores has a standard deviation of .55? Use the .05 significance level. Assume the sampled population is approximately normal in distribution.

40. In Problem 39, assume that the test is conducted because it is felt that the claim is an understatement.

41. An office manager states that the time required to handle customer complaints has a standard deviation of 15 minutes. A sample of 27 complaints shows a standard deviation of 13.8 minutes. Suppose you feel the office manager is overstating the standard deviation in order to have an excuse to reassign office personnel. If you are willing to take a risk of 5 in 100 of making a Type I error, would you accept the office manager's claim? Assume the sampled population is approximately normal in distribution.

42. In Problem 41, assume you considered the office manager's claim unbiased.

43. A national fraternity claims that the ages of its members have a standard deviation of 3.5 years. A sample of 25 members has a standard deviation of 4.3 years. If you suspect the claim was too low and wish to take a risk of .01 of committing a Type I error, would you accept the claim? Assume the sampled population is approximately normal in distribution.

44. Is it possible to compute β and why in a test of an hypothesis where
    a. $H_0: p = .30, H_1: p > .30$
    b. $H_0: p = .09, H_1: p = .04$

45. How is an operating characteristic (OC) curve constructed and what does it show?

46. In a test of an hypothesis with $H_0: p = .25; H_1: p = .30;$ and $\beta = .06;$ compute the power of the test and explain what it means.

47. How is a power curve constructed and what does it show?

48. The seller of a very large inventory of steel rings is willing to sell the entire inventory at a sharp discount, if no less than 30 percent of the rings are defective (out of shape). He wants to take a risk of only 2 in 100 of making a Type I error. What is the probability of committing a Type II error, given $H_1: p = .25$? Assume $n = 1,000$.

49. In Problem 48, determine the probability of a Type II error given $H_1: p = .29$.

50. It is claimed that the average number of traffic tickets written per day in a city is 100, with a standard deviation of 3, and it is desired to select a sample of days to test this claim. There is a feeling that the average is somewhat higher. How large a sample of days should be selected if the significance level is set at .02 and it is specified that the probability of concluding that the mean number of tickets written per day is 100, if it is actually 102, should be .01?

51. Computing clerks in a research laboratory are required to complete 150 computations per hour to meet work standards. The supervisor suspects that one of the clerks is producing below standard. How large a sample of hourly computation counts for this clerk should be selected, if he wishes to test this clerk's productivity using the .05 level of significance? It is indicated that a risk of only 2 in 100 is acceptable in concluding that this clerk is meeting the standard, if his mean computation rate is as low as 145 per hour. The standard deviation of computations is estimated as 5 per hour.

52. In Problem 50, suppose it is desired to test whether or not the mean number of tickets written daily is 100. How large a sample of days should be selected if the level of significance is set at .01 and the probability of a Type II error is set at .01 if the mean is off by 2 tickets either way?

53. The majority leader in a state legislature is willing to support a pending bill if it can be established that no more than 30 percent of the voters are opposed. How large a sample of voters is required to conduct a test of the percent opposed, if the level of significance is set at .05? It is specified that the risk of deciding to support the bill should be only .01, if the percentage opposed is as much as 35 percent.

54. A drug firm is warned by its chemist that 50 percent of those who take a new medication will suffer minor side effects. The firm wishes to determine whether or not this percentage claim is acceptable based on sample results. How large a sample of subjects should be selected for administration of the new product, if the probability of a Type I error is to be .02? It is specified that the probability of a Type II error should be .05, if the percentage suffering side effects is off by ten percentage points (either way) from the chemist's claim.

Chapter 12

Optional:
Tests of Hypotheses
for Comparisons

12.1   Proportions

In the previous chapter we studied tests of hypotheses involving a standard or a reference value for a parameter. In some problems, the hypothesis to be tested involves a comparison of the values of a parameter in two populations. For example, how do boys and girls compare in mean learning speed in mathematics? How do two machines compare in the proportion of defective units produced? How do two stores compare in the variance of daily sales? In tests of hypotheses relating to comparisons we will be concerned with two populations and a sample selected from each population. We will use the same notation and, generally, the same sampling models as in Chapter 10.

In this section, we will study tests of hypotheses comparing two population proportions when the following conditions are met by each population and each sample: population large; sampling ratio n/N less than .05; population proportion not too close to 0 or 1; and sample sufficiently large ($n \geq 100$). Model H which relates to independent samples is the appropriate sampling model. The test parameter is $p_1 - p_2$, the difference between two population proportions. The decision model is constructed on the basis of the sampling distribution of $X_1/n_1 - X_2/n_2$, the difference between sample proportions. According to Model H, this sampling distribution is approximately normal and has the mean $p_1 - p_2$ and standard error

$$\sigma_{(X_1/n_1 - X_2/n_2)} = \sqrt{[p_1(1 - p_1)/n_1] + [p_2(1 - p_2)/n_2]}$$

A sample of 350 wire rings (n_1) was taken from the production stream of machine 1 and 300 (n_2) from the production stream of machine 2. The first sample showed that 52 rings (X_1) were out-of-round (not a perfect circle) and the second sample showed that 41 (X_2) were similarly defective. Would you conclude, at the .05 level of significance, that the proportion of out-of-round rings is the same for the two machines?

Formulate the hypotheses:

H_0: $p_1 - p_2 = 0$ (This is the null hypothesis of no difference in the proportions for the two machines.)

H_1: $p_1 - p_2 \neq 0$ (A two-sided alternative is appropriate in this problem, since there is no basis for expecting either p_1 or p_2 to be larger. In some problems, the appropriate alternative hypothesis is one-sided.)

As we have learned in the previous chapter, the decision model is constructed on the assumption that H_0 is true. Consequently, if $p_1 - p_2 = 0$, so that $p_1 = p_2$, we may denote the common population proportion as p and we may rewrite the standard error equation in Model H, substituting p for p_1 and p_2, as follows:

$$\sigma_{(X_1/n_1 - X_2/n_2)} = \sqrt{\frac{p(1 - p)}{n_1} + \frac{p(1 - p)}{n_2}}$$

$$= \sqrt{p(1 - p)\left(\frac{1}{n_1} + \frac{1}{n_2}\right)} \tag{12.1.1}$$

We estimate the common proportion p by *pooling* the sample information to obtain

$$\frac{X}{n} = \frac{X_1 + X_2}{n_1 + n_2} \tag{12.1.2}$$

Then, we compute the *estimated* standard error of the difference between proportions by substituting X/n for p in Equation 12.1.2 to obtain

$$s_{(X_1/n_1 - X_2/n_2)} = \sqrt{\frac{X}{n}\left(1 - \frac{X}{n}\right)\left(\frac{1}{n_1} + \frac{1}{n_2}\right)} \tag{12.1.3}$$

The difference statistic $X_1/n_1 - X_2/n_2$ is transformed to standard units using the following equation.

$$z = \frac{\left(\dfrac{X_1}{n_1} - \dfrac{X_2}{n_2}\right) - (p_1 - p_2)}{s_{(X_1/n_1 - X_2/n_2)}} = \frac{\dfrac{X_1}{n_1} - \dfrac{X_2}{n_2}}{s_{(X_1/n_1 - X_2/n_2)}} \tag{12.1.4}$$

The second form for computing z is obtained by assuming that $p_1 - p_2 = 0$, according to the null hypothesis.

The decision model, based on the sampling distribution of $X_1/n_1 - X_2/n_2$ and the assumption that H_0 is true, leads to the following decision rule. *Transform the sample statistic $X_1/n_1 - X_2/n_2$ to z (Equation 12.1.4). If $z < -1.96$ or > 1.96, reject H_0 and accept H_1, otherwise, accept H_0.*

Applying the sample data, we compute

$$\frac{X_1}{n_1} = \frac{52}{350} = .149$$

$$\frac{X_2}{n_2} = \frac{41}{300} = .137$$

$$\frac{X}{n} = \frac{52 + 41}{350 + 300} = .143$$

The difference statistic $\dfrac{X_1}{n_1} - \dfrac{X_2}{n_2} = .149 - .137 = .012$ is transformed to z (Equation 12.1.4) to obtain

$$z = \frac{.012}{\sqrt{.143(1 - .143)\left(\frac{1}{350} + \frac{1}{300}\right)}} = .44$$

Applying the decision rule, $z = .44$ falls in the region of acceptance. Therefore, accept H_0 and conclude that the proportion of out-of-round wire rings produced by the two machines is the same.

12.2 Means

Tests of hypotheses comparing two population means, using data from independent samples, may be based on Model F for large samples and Model G for small samples. The following conditions must be met by each population and each sample: population large and sampling ratio n/N less than .05. The decision model for large samples is constructed on the basis of the sampling distribution of $\overline{X}_1 - \overline{X}_2$, which is approximately normal (Model F). The mean of the sampling distribution is $\mu_1 - \mu_2$ and the standard error is $\sigma_{(\overline{X}_1 - \overline{X}_2)} = \sqrt{\sigma_1^2/n_1 + \sigma_2^2/n_2}$.

An agricultural testing station wants to compare the effect on plant growth of two fertilizers. A sample of $n_1 = 75$ plants using an experimental fertilizer showed a mean height of 21.6 cms after a certain period of time, with a standard deviation of 8.9 cms. A control sample of $n_2 = 80$ plants using a well-established fertilizer showed a mean height of 18.2 cms after an equal period of time, with a standard deviation of 10.1 cms. Would you conclude at the .01 level of significance that the mean height for plants using the two fertilizers is the same? It is expected that the experimental fertilizer will lead to a higher mean height than the well-established fertilizer. If the test does not lead to this conclusion, further testing will take place.

Formulate the hypotheses:

H_0: $\mu_1 - \mu_2 = 0$ (We formulate the null hypothesis even though we believe $\mu_1 > \mu_2$.)

H_1: $\mu_1 - \mu_2 > 0$

The decision rule may be stated as follows. *Transform the difference statistic* $\overline{X}_1 - \overline{X}_2$ *to z. If z > 2.33, reject H_0 and accept H_1, otherwise, make no decision.* Computation of z is as follows.

$$z = \frac{\overline{X}_1 - \overline{X}_2}{\sqrt{\dfrac{s_1^2}{n_1} + \dfrac{s_2^2}{n_2}}} \tag{12.2.1}$$

Applying the sample data, compute the difference statistic $\overline{X}_1 - \overline{X}_2 = 21.6 - 18.2 = 3.4$ cms. Transforming to z, we obtain

$$z = \frac{3.4}{\sqrt{\dfrac{(8.9)^2}{75} + \dfrac{(10.1)^2}{80}}} = 2.22$$

According to the decision rule, $z = 2.22$ falls in the region of no decision. It is decided to conduct further tests comparing the two fertilizers.

When small samples are the basis for a test, the applicable sampling model is Model G which requires that each of the populations sampled be normal (or approximately normal) in distribution. The appropriate sampling distribution is the t distribution, with $n_1 - 1$ df for sample n_1 and $n_2 - 1$ df for sample n_2.

A sociologist selected a sample of $n_1 = 8$ male adults in a neighborhood where the residents are mostly foreign-born and a sample of $n_2 = 10$ adults in a neighborhood where the residents were mostly born in this country. The same social adjustment test was administered to each sample. The following results were computed based on the test scores: $\overline{X}_1 = 31.6$, $s_1 = 5.3$, $\overline{X}_2 = 38.1$, and $s_2 = 4.9$. Use the .05 level of significance and test whether or not the mean test score is the same for the two neighborhoods.

Formulate the hypotheses:

H_0: $\mu_1 - \mu_2 = 0$

H_1: $\mu_1 - \mu_2 \neq 0$

The decision model is constructed on the basis of the t distribution. The t statistic is computed according to Model G, with $\mu_1 - \mu_2 = 0$ (since H_0 is assumed to be true). The decision rule is as follows. *Compute the t statistic based on the sample data. If t < $-t_{.05}$ or > $t_{.05}$, reject H_0 and accept H_1, otherwise, accept H_0.* $t_{.05}$ is the critical value of t for a two-tail test with $\alpha = .05$ and is computed according to the second equation in Model G. Applying this equation, with $t_1 = 2.365$ for $8 - 1 = 7$ df and $t_2 = 2.262$ for $10 - 1 = 9$ df compute

$$t_{.05} = \frac{\frac{(5.3)^2}{8}(2.365) + \frac{(4.9)^2}{10}(2.262)}{\frac{(5.3)^2}{8} + \frac{(4.9)^2}{10}} = 2.33$$

Based on the sample data, compute $\bar{X}_1 - \bar{X}_2 = 31.6 - 38.1 = -6.5$. Then,

$$t = \frac{-6.5}{\sqrt{\frac{(5.3)^2}{8} + \frac{(4.9)^2}{10}}} = -2.77$$

Referring to the decision rule, since $t = -2.77$ is less than $-t_{.05} = -2.33$, reject H_0 and accept H_1. Conclude that the mean test score is not the same for the two neighborhoods.

12.3 Means: Equal Variances

In tests which compare two population means, it is often appropriate to assume that $\sigma_1^2 = \sigma_2^2$. When the two populations may be considered to be approximately normal in distribution and $\sigma_1^2 = \sigma_2^2$, the null hypothesis $\mu_1 - \mu_2 = 0$ amounts to the hypothesis that the two populations are one and the same population. For example, suppose it is desired to test whether a specified treatment for cotton cloth improves the breaking strength of the cloth. Suppose strength tests are conducted on a sample of untreated cloth and also on a sample of treated cloth and a number of breaking strength measurements (in pounds per square inch) are obtained for each sample. If breaking strength measurements are considered to be approximately normal in distribution for each type of cloth and the variances of these distributions are assumed to be equal, then the hypothesis that the mean breaking strength is the same for the two types of cloth amounts to the hypothesis that the population distributions of breaking strength measurements are identical (or that the treatment has no effect on the breaking strength of the cloth).

When it is appropriate to assume that the population variances (or standard deviations) are equal, the procedures for conducting a test of hypothesis in the previous section are applicable. A more efficient testing procedure, however, is based upon a pooling of the information from the two samples to estimate the common population variance, as discussed in Section 10.3. This *pooled variance estimate* s^2 is computed according to Equation 10.3.1 or Equation 10.3.2, whichever is more convenient, as discussed in Section 10.3. For a *large sample* test, compute z as follows.

$$\frac{\bar{X}_1 - \bar{X}_2}{\sqrt{s^2\left(\frac{1}{n_1} + \frac{1}{n_2}\right)}} \tag{12.3.1}$$

And conduct the test as described in Section 12.2. If the test is based on *small samples*, the t statistic is also computed as in the foregoing equation and the critical value of t is found in Table B.5 for $n_1 + n_2 - 2$ df. The critical t value is not to be computed according to the second equation in Model G when equal population variances are assumed and the t statistic for the samples is computed according to Equation 12.3.1.

12.4   Means: Dependent Samples

Tests of hypotheses which compare population means may be based on dependent samples. Dependent samples were discussed in Section 10.4. We will use the same notation as presented in that section. When dependent samples are involved, the test is conducted on the basis of a *single population of differences*, X_i = the difference for a pair. In other words, the test comparing population means μ_1 and μ_2 based on dependent samples becomes a test of the *population mean difference μ.*

Suppose it is desired to investigate the productivity of two presses. A sample of 14 workers are selected and each is assigned to work for one hour on each press. Then, there are two samples of production amounts, $n = 14$ for each press, and these are dependent samples. The difference in amount produced per worker for the two presses is denoted as X_i; the mean difference is $\overline{X}$; and the standard deviation of the differences is s. The mean population difference is μ.

If n (the number of sample pairs) is 30 or more, the test procedure to be used is the same as discussed in Section 11.3 for large samples. If n is less than 30, follow the test procedure for small samples in Section 11.3.

12.5   Variances

Tests of the equality of variances (or standard deviations) of two populations usually involve testing $H_0: \sigma_1^2 = \sigma_2^2$ against the alternative $H_1: \sigma_1^2 \neq \sigma_2^2$. Model J is the appropriate sampling model.

Model J

Given two normal populations, with standard deviations σ_1 and σ_2. Let n_1 and n_2 represent independent random samples from these populations, with standard deviations s_1 and s_2. If $\sigma_1 = \sigma_2$, the sampling distribution of the statistic

$$F = \frac{s_1^2}{s_2^2}$$

is like the F distribution, with $n_1 - 1$ df and $n_2 - 1$ df.

Notice that Model J requires that the sampled populations be normal (or approximately normal) in distribution, there is no restriction on sample size, and independent samples are specified. Notice that the F distribution depends on two numbers of degrees of freedom, the df associated with the variance in the numerator of the F statistic and the df associated with the variance in the denominator. There is a different F distribution for each pair of df.

An F distribution is presented in Figure 12.5.1 and indicates that this distribution is skewed to the right, continuous, and that F can take on only positive values (0 to infinity). Figure 12.5.1 also presents the regions of acceptance and rejection for a one-tail test, with the critical region in the upper tail. Clearly, critical values of the F statistic are always positive.

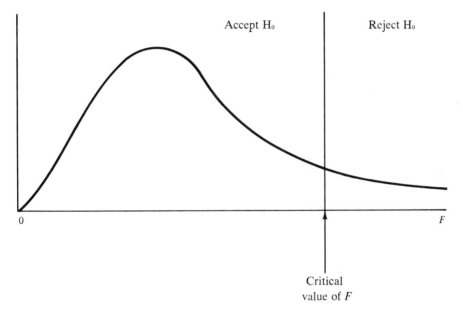

FIGURE 12.5.1

An F distribution.

Tables B.7 and B.8 present critical F values for various pairs of degrees of freedom. The critical values in these tables cut off upper-tail critical regions only. In Table B.7 the proportion of the total area marked off in the upper tail by the F values is .05, so that this table presents $F_{.05}$ critical values for indicated pairs of df. Table B.8 presents $F_{.01}$ critical values, since they mark off .01 of the total area in the upper tail. In each table, the column headings show the df for the *numerator variance* of the F statistic and the stub (the first column) shows the df for the *denominator variance*. Suppose $n_1 = 8$ and df $= n_1 - 1 = 7$, $n_2 = 15$, and df $= n_2 - 1 = 14$. Then, for $\alpha = .05$ and an upper-tail critical region test, we see $F_{.05} = 2.76$ from Table B.7 as the appropriate critical value. If $\alpha = .01$ for this test, we would read $F_{.01} = 4.28$ from Table B.8 as the appropriate critical value.

Tests to evaluate whether an observed difference between two sample variances (s_1^2 and s_2^2) is consistent with the hypothesis $H_0: \sigma_1^2 = \sigma_2^2$ are usually referred to as *tests of the homogeneity of variances*. The alternative hypothesis in such tests is usually $\sigma_1^2 \neq \sigma_2^2$, but it may also be $\sigma_1^2 > \sigma_2^2$ or $\sigma_1^2 < \sigma_2^2$. If a two-sided alternative is used there is a critical region in each tail of the F distribution, with the critical F values determined so that the proportion of the total area in each tail is equal to $\alpha/2$. We will, however, follow the procedure of always computing the F statistic by placing the larger of the two sample variances in the numerator. Consequently, we will need to consider specifically only the critical region in the upper tail of the distribution. Hence, we will need to determine only the upper-tail critical F value. This is particularly convenient, since F tables (B.7 and B.8) show upper-tail critical F values only. Then, the F statistic will be computed in one of the following two ways, as appropriate:

1. If $s_1{}^2 \geq s_2{}^2$, then $F = s_1{}^2/s_2{}^2$ with $n_1 - 1$ df (numerator) and $n_2 - 1$ df (denominator).
2. If $s_1{}^2 < s_2{}^2$, then $F = s_2{}^2/s_1{}^2$ with $n_2 - 1$ df (numerator) and $n_1 - 1$ df (denominator).

A sample of $n_1 = 22$ cans of corn filled by an old machine showed a variation of $s_1 = .03$ ozs in drained weight. A sample of $n_2 = 25$ cans filled by a new machine showed a variation of $s_2 = .07$ ozs in drained weight. Would you conclude that the variation in drained weight (as measured by the standard deviation) is the same for cans filled by the old and new machines? Use the .02 level of significance. Assume that the weight distributions per can are approximately normal.

Formulate the hypotheses:

$H_0: \sigma_1 = \sigma_2$
$H_1: \sigma_1 \neq \sigma_2$

The decision model is constructed based on the F distribution, with the F statistic computed according to Model J. The decision rule is as follows. *Compute the F statistic (with the larger sample variance in the numerator). If F > $F_{.01}$ (= 2.80), reject H_0 and accept H_1, otherwise, accept H_0.* $F_{.01} = 2.80$ was found in Table B.8 for $n_2 - 1 = 24$ df (numerator) and $n_1 - 1 = 21$ df (denominator), since $s_2 > s_1$.

Applying the sample data, compute

$$F = \frac{(.07)^2}{(.03)^2} = 5.44$$

Referring to the decision rule, $F = 5.44$ falls in the region of rejection. Therefore, reject H_0 and accept H_1. Conclude that the variation in drained weight of fill per can is not the same for the two machines.

Study Problems

1.   A sample of 300 young married women of ethnic group A in a large region included 15 who had babies under one year of age; whereas a sample of 200 young married women of ethnic group B in the same region included 12 who had babies under one year of age. Would you conclude, at the .02 level of significance, that the proportion of young married women with babies under one year of age is the same for the two ethnic groups?

2.   In Problem 1, conduct the test assuming it is desired to evaluate the belief that ethnic group B has a higher proportion of young married women with babies under one year of age.

3.   A sample of 500 voters from urban areas indicated that 350 favored a certain candidate for state senator. A sample of 300 voters from rural areas indicated that 250 favored this candidate. Would you conclude that the proportion of voters favoring the candidate is the same in urban and rural areas? Use the .05 level of significance.

4.   In Problem 3, conduct the test assuming it is desired to evaluate the claim that the proportion favoring the candidate is smaller in urban areas.

5.   Fifty out of a sample of 200 college women stated that they traveled abroad at least once. A sample of 300 college men included 50 who traveled abroad at least once. Would you conclude that the proportion of college women who traveled abroad at least once is the same as for college men? Use the .01 significance level.

6.   In Problem 5, conduct the test assuming it is desired to evaluate the claim that a larger proportion of college women traveled abroad at least once.

7.   A sample of 40 small-sized northern cities reported a mean of 4.50 crimes per day, with a standard deviation of .50 crimes per day. A sample of 50 small-sized western cities reported a mean of 4.90 crimes per day, with a standard deviation of .40 crimes per day. Would you conclude that the mean crime rate per day is the same for small-sized northern and western cities, at the .05 level of significance?

8.   In Problem 7, assume it is desired to test whether western small-sized cities have a higher mean number of crimes per day.

9.   In Problem 7, assume that the sample size is 15 for the northern cities and 10 for the western cities and that the sampled populations are approximately normal in distribution.

10.   A sample of 30 workers selected in a large factory required a mean of 40 minutes per worker with a standard deviation of 8 minutes to construct a certain product using method A. A second independent sample of 40 workers from the same factory required a mean of 35 minutes per worker with a standard deviation of 9 minutes to construct the same product using method B. Using the .02 level of significance, would you conclude that the same mean time per worker is required for methods A and B?

11.   In Problem 10, suppose sample size is 20 for method A and 25 for method B and the sampled populations are approximately normal in distribution.

12.   A sample of 35 pieces of wrapping paper produced by firm A had a mean strength of 2.4 pounds per square inch, with a standard deviation of .3 pound per square inch. A similar type of paper produced by firm B had a mean strength of 2.8 pounds per square inch, with a standard deviation of .5 pound per square inch, for a sample of 45 pieces. Would you conclude, at the .01 level of significance, that mean strength is the same for the wrapping paper produced by the two firms?

13.   In Problem 12, suppose the sample size was 12 pieces for firm A and 15 pieces for firm B and the sampled populations are approximately normal in distribution.

14.   Two diets, A and B, were developed for pig feeding. A sample of 40 pigs were fed diet A for a certain period of time resulting in a mean weight gain of 10.5 pounds, with a standard deviation of 2.2 pounds. A sample of 50 pigs were fed diet B for the same time period resulting in a mean weight gain of 14.1 pounds, with a standard deviation of 2.5 pounds. Test, at the .01 level of significance, whether diets A and B

result in equal weight gains for pigs, on the average. Assume that the population standard deviation of weight gains per pig is the same for both diets.

15. In Problem 14, assume it is desired to test whether diet A results in less weight gain per pig, on the average.

16. In Problem 14, assume the sample size was 10 pigs for diet A and 15 pigs for diet B and the sampled populations are approximately normal in distribution.

17. The number of nonfatal accidents occurring during the winter months reported in a large city for a sample of eight weeks showed a mean of 1,030 accidents per week, with a standard deviation of 50 accidents. The number of accidents occurring during the summer months in this city for a sample of ten weeks showed a mean of 990 accidents per week, with a standard deviation of 45 accidents. Assume that the standard deviation of the number of weekly accidents is the same for the winter and summer months. Would you conclude, at the .05 level of significance, that the mean weekly number of accidents is the same for winter and summer months? Assume that the sampled populations are approximately normal in distribution.

18. A sample of 60 plants from type A seed showed a mean height of 10.3 cms after an eight-week period of growth, with a standard deviation of .9 cm. A sample of 50 plants from type D seed for the same plant variety showed a mean height of 11.8 cms during the same time, with a standard deviation of 1.3 cms. Assume that the standard deviation of plant height is the same for both seed types. Would you conclude, at the .02 level of significance, that the mean height per plant from type D seed is greater than for type A seed, as claimed by its developer?

19. In Problem 18, assume that the sample size was 12 for type A seed and 10 for type D seed. Conduct the test using a two-sided alternative. Assume that the sampled populations are approximately normal in distribution.

20. Two forms of a leadership test were constructed for young executives. Form I was administered to a sample of nine young executives. About three weeks later form II was administered to the same sample of executives. The scores obtained were:

Executive	Form I	Form II
1	93	96
2	99	103
3	105	101
4	103	94
5	96	83
6	85	90
7	92	97
8	108	106
9	91	100

Would you conclude that the two forms have the same population mean score, at the .05 significance level? Assume that the population of pair differences is approximately normal in distribution.

21. A sample of 12 typists were asked to type a specified report using typewriter A. Two weeks later these same typists were asked to type the same report using typewriter B. The time required to type the report by each typist on each machine was recorded as follows (in minutes):

Typist	Typewriter A	Typewriter B
1	9	12
2	20	17
3	15	10
4	18	16
5	13	18
6	22	18
7	10	14
8	9	10
9	19	13
10	16	17
11	18	15
12	16	18

Would you conclude, at the .05 significance level, that the time required to type the report is the same, on the average, for typewriters A and B? Assume that the population of pair differences is approximately normal in distribution.

22.   A sample of ten college students took a social hostility test before viewing a film (test form A) and again after viewing the film (test form B). The following scores were obtained.

Student	Before film	After film
1	30	11
2	17	7
3	23	7
4	20	25
5	18	20
6	20	5
7	14	5
8	32	20
9	19	23
10	14	17

Would you conclude, at the .05 level of significance, that the mean score is the same before and after the film-viewing? It was expected that the effect of the film would be to reduce the mean test score. Assume that the population of pair differences is approximately normal in distribution.

23.   A sample of eight pieces of cloth produced by firm A showed a standard deviation of 25 lbs per square inch in strength measurement. A sample of ten pieces of similar cloth produced by firm B showed a standard deviation of 20 lbs per square inch in strength measurement. Would you conclude, at the .02 level of significance, that the cloth produced by the two firms have equal strength variation? Assume that the sampled populations are approximately normal in distribution.

24.   A sample of 25 men and a sample of 16 women were asked to perform a specified task. The male sample showed a standard deviation of 15 minutes in time required per man to perform the task; whereas, the female sample showed a standard deviation of 27 minutes. Would you conclude that the variance in time required per person to perform the task is the same for men and women? Use the .10 level of significance. Assume that the sampled populations are approximately normal in distribution.

Part V

Statistical Decision Making— Further Topics

Chapter 13

Tests Relating to Frequencies

13.1   Chi-Square Tests

We encountered the chi-square (χ^2) statistic in Section 11.4 in our discussion of tests relating to variances. In this chapter we will apply the χ^2 statistic to tests relating to *frequencies*. Generally speaking, the χ^2 tests presented in this chapter compare *observed (sample) frequencies* for a set of categories with corresponding *expected frequencies* determined on the basis of an hypothesis or on a theoretical basis.

Several types of chi-square tests will be considered. The sample data used in these tests are usually presented in a "row by column" arrangement, as illustrated in Table 13.1.1. This table contains three rows and three columns and includes $3 \cdot 3 = 9$ *categories* or *cells*. Each cell contains a frequency or *count*. Such a table is often referred to as a 3 x 3 (3 by 3) table.

TABLE 13.1.1

Number of voters in samples from three cities who favor
candidate A, oppose him, or have no opinion.

	A	City B	C
Favor	60	115	70
Oppose	20	80	60
No opinion	15	10	20

We will denote the *sample (observed) frequency* in a category as X_i, $i = 1, 2,$ $\ldots, m$ (for m categories). We will let E_i denote the *expected frequency* corresponding to the observed frequency X_i. The sampling distribution applicable to the tests presented in this chapter is the distribution of the χ^2 statistic, however, for these tests, this statistic is computed differently than in Section 11.4. The determination of the expected frequencies E_i and the computation of χ^2 will be discussed and illustrated in the subsequent sections of this chapter.

The χ^2 statistic is a discrete variable. Its distribution is approximated by the χ^2 distribution, which is a continuous distribution, as discussed in Section 11.4 and illustrated in Figure 11.4.1.

13.2 Comparing Several Proportions

This section presents tests of hypotheses relating to the comparison of several population proportions based on independent samples. Generally, we will speak of c samples $(n_1, n_2, \ldots, n_c)$ selected from c populations. We will denote the population proportions as p_i $(p_1, p_2, \ldots, p_c)$.

A sample of girls in the 13–15 year age bracket was selected in each of three large cities (A, B, and C). The girls in each sample were asked if their parents permitted them to date. The frequency in each response category (yes or no) is shown in Table 13.2.1. Would you conclude, at the .05 level of significance, that the proportion of girls permitted to date is the same in the three cities?

TABLE 13.2.1

Samples of girls aged 13–15 in cities A, B, and C: the
number permitted to date by their parents and the number
not permitted to date.

Permitted to date	A	City B	C	Total
Yes	8	20	14	42
No	202	300	256	758
Total	210	320	270	800

In this problem, we have $c = 3$ populations (girls 13–15 years of age in cities A, B, and C). There are two categories for each sample: the "yes" category, which is the category of interest, and the "no" (other) category. There are in all $m = 2 \cdot c = 2 \cdot 3$ or 6 categories or cells, as shown in Table 13.2.1. We use X_i to denote the sample frequency in a cell and r_i to denote a row total (r_1 represents the total $\sum X_i$ for the first or "yes" row and r_2 represents the total $\sum X_i$ for the second or "no" row). Of course, the total for a column is the sample size n_i. The sum of the sample sizes $\sum n_i$ is equal to n, the combined size of the c samples. Finally, p_i ($i = 1, 2, 3$) denotes the proportion of girls in the category of interest (the "yes" category) in a population.

The following hypotheses are formulated:

$H_0: p_i = p$ (The population proportions are equal; they all equal a common value p.)

$H_1: p_i \neq p$ (The population proportions are not all equal.)

Assuming that H_0 is true, we estimate the common population proportion p by pooling the data from the $c = 3$ samples and computing r_1/n (following the procedure of Equation 12.1.2) as follows:

$$\frac{r_1}{n} = \frac{\sum X_i \text{ (for row 1)}}{\sum n_i} \qquad (13.2.1)$$

Applying this equation to our problem, we obtain (using the data in Table 13.2.1)

$$\frac{r_1}{n} = \frac{8 + 20 + 14}{210 + 320 + 270} = \frac{42}{800} = .0525$$

Hence, we estimate that, if H_0 is true, $r_1/n = .0525$ is the proportion of girls in each population who are permitted to date. Consequently, we would expect that the number of girls in the "yes" category for a sample, for instance for the sample $n_1 = 210$, would equal $(r_1/n)n_1 = (.0525)(210)$ or 11.0. This result (11.0) represents the *expected frequency* E_1 corresponding to the *observed frequency* $X_1 = 8$ for sample $n_1 = 210$. We may write this as

$$E_1 = \left(\frac{r_1}{n}\right) n_1 = \frac{r_1 n_1}{n} \qquad (13.2.2)$$

According to Equation 13.2.2, the expected frequency E_1 corresponding to the sample frequency X_1 is computed by multiplying the total of the row, which contains X_1, r_1, by the total of the column which contains X_1, n_1, and dividing the product by the combined sample size n. Then, in general, the expected frequency E_i corresponding to a sample frequency X_i is computed as

$$E_i = \frac{\text{corresponding row total} \cdot \text{corresponding column total}}{n} \qquad (13.2.3)$$

Applying Equation 13.2.3, the expected frequency corresponding to the sample frequency 202 for city A in Table 13.2.1 is computed as $(758)(210)/800 = 199.0$; the expected frequency corresponding to the sample frequency 20 for city B is computed as $(42)(320)/800 = 16.8$; etc.

Model K is the appropriate sampling model for tests relating to the comparison of population proportions.

Model K

Given independent samples selected from c infinite populations, with p_i representing the proportion in a category of interest for a population. If the p_i are all equal, then the statistic

$$\chi^2 = \sum \frac{(X_i - E_i)^2}{E_i}$$

is distributed approximately like the chi-square distribution, with $c - 1$ degrees of freedom.

Model K is applicable to problems which satisfy the following conditions for each population and each sample: population large, sampling ratio n/N less than .05, population proportion not too close to 0 or 1, and sample size sufficiently large ($n \geq 100$).

Then, according to Model K, the decision model for testing the equality of population proportions is constructed on the basis of the chi-square distribution, with $c - 1$ df. Notice that the χ^2 statistic (Model K) is smaller, the closer X_i and E_i are to each other, because the numerator will be small. Hence, the smaller the χ^2 statistic computed for a problem, the closer the agreement between the sample frequencies (X_i) and the expected frequencies (E_i) and the more likely that H_0 is true. On the other hand, the greater the differences between X_i and E_i, the larger is χ^2 and the less likely that H_0 is true. Consequently, *in testing the equality of proportions, the critical region is always in the upper tail.*

Since the level of significance was specified as .05 and df $= c - 1 = 3 - 1$ or 2, the critical value for our problem is $\chi^2 = 5.991$ (Table B.6). We state the decision rule as follows: *Compute the χ^2 statistic according to Model K. If $\chi^2 >$ 5.991, reject H_0 and accept H_1, otherwise, accept H_0.*

Table 13.2.2 presents a convenient procedure for the computation of the chi-square statistic, using the data in Table 13.2.1 (X_i) and the corresponding

TABLE 13.2.2

Computation of the χ^2 statistic: sample data from Table 13.2.1.

X_i	E_i	$X_i - E_i$	$(X_i - E_i)^2$	$\dfrac{(X_i - E_i)^2}{E_i}$
8	11.0	−3.0	9.00	.82
202	199.0	3.0	9.00	.05
20	16.8	3.2	10.24	.61
300	303.2	−3.2	10.24	.03
14	14.2	−.2	.04	.003
256	255.8	.2	.04	.0001
				$\chi^2 = \overline{1.51}$

expected frequencies (E_i) computed according to Equation 13.2.3. Then, $\chi^2 = 1.51$ falls in the region of acceptance, according to the decision rule. Therefore, accept H_0 and conclude that the proportion of girls aged 13–15 permitted to date by their parents is the same in cities A, B, and C.

13.3 Tests of Independence

Chi-square *tests of independence* are used to evaluate whether or not two characteristics (variables) are independent. Suppose a sample of 300 adults are asked their opinion of a certain issue, and the responses are tabulated in a *two-way frequency table* as in Table 13.3.1. The two characteristics or variables of classification are "attitude toward the issue" and "educational attainment." A two-way table such as Table 13.3.1 is usually called a *contingency table*. Table 13.3.1 has 3 rows and 3 columns so that it is a 3 x 3 contingency table, with $m = 3 \cdot 3$ or 9 cells or joint-classification categories.

TABLE 13.3.1

Cross-tabulation of a sample of 300 adults on two charac-teristics "attitude toward an issue" and "educational attainment."

| Attitude toward issue | Educational attainment | | | |
	No high school A	Some high school B	Some college C	Total
For	20	40	10	70
Against	40	55	50	145
Indifferent	30	30	25	85
Total	90	125	85	300

We will use X_i to denote the observed (sample) frequency in a cell, E_i to denote the expected frequency, and r_i to denote a row total, as in the previous section. Then, c_i will denote a column total, r (without a subscript) will denote the number of rows and c (without a subscript) will denote the number of columns. The sample size will be denoted by n.

In contingency tables, we test the hypothesis that the two variables of classification (the row variable and the column variable) are independent against the alternative hypothesis that they are not independent. In our illustrative problem, we wish to test the hypothesis that attitude toward the issue is independent of the educational attainment of the respondent. Let us consider the joint-classification category "for the issue and educational level A" which shows an observed frequency (an X_i count) of 20 in Table 13.3.1. Assuming that the hypothesis of independence is true, we may use probability Rule E for independent events and write

$$P(\text{For issue and A}) = P(\text{For issue}) \cdot P(\text{A}) \tag{13.3.1}$$

The terms of this probability equation may be interpreted as follows.

P(For issue) denotes the probability that, if an adult is selected at random from the population, he will be for the issue. Using the sample data in the total column of Table 13.3.1, we estimate this probability as (probability Definition 1)

$$P(\text{For issue}) = \frac{r_1}{n} = \frac{70}{300} \tag{13.3.2}$$

$P(A)$ denotes the probability that, if an adult is selected at random from the population, he will have attained educational level A. Using the sample data in the total row of Table 13.3.1, we estimate this probability as (probability Definition 1)

$$P(A) = \frac{c_1}{n} = \frac{90}{300} \tag{13.3.3}$$

P(For issue and A) denotes the probability that, if an adult is selected at random from the population, he will be for the issue and have attained educational level A. Applying Equation 13.3.1 (probability Rule E), we estimate this probability as

$$P(\text{For issue and } A) = \frac{r_1}{n} \cdot \frac{c_1}{n} = \frac{r_1 c_1}{n^2}$$

$$= \frac{70}{300} \cdot \frac{90}{300} = .07 \tag{13.3.4}$$

Recall that a probability has a relative frequency interpretation, so that P(For issue and A) $= .07$ indicates that $.07$ of the adults in the population fall in the joint-classification category "for the issue and educational level A." Hence, the expected frequency for the "For issue and A" cell in Table 13.3.1 is the product of this proportion $r_1 c_1/n^2$ (Equation 13.3.4) and the sample size n. We may write this expected frequency as

$$E_1 = \frac{r_1 c_1}{n^2} \cdot n = \frac{r_1 c_1}{n} \tag{13.3.5}$$

Generalizing, based on the foregoing equation, we may compute the expected frequency E_i for any cell in a contingency table as

$$E_i = \frac{\text{corresponding row total} \cdot \text{corresponding column total}}{n} \tag{13.3.6}$$

According to this equation, the expected frequency E_i corresponding to the observed (sample) frequency X_i in a cell is computed by multiplying the row and column totals corresponding to X_i and dividing the product by the sample size n. For example, the first cell frequency in the "For" row in Table 13.3.1 is 20 and the corresponding expected frequency is $(70)(90)/300 = 21.0$; the second cell frequency in this row is 40 and the corresponding expected frequency is $(70)(125)/300 = 43.5$; the first cell frequency in the "Against" row is 40 and the corresponding expected frequency is $(145)(90)/300 = 29.2$; etc.

Model L is the appropriate sampling model for tests of independence for two variables of classification in a contingency table.

Model L

Given samples of size n selected from an infinite population and two independent variables of classification, one with r categories and one with c categories, so that there are $m = r \cdot c$ joint-classification categories. Then, the statistic

$$\chi^2 = \sum \frac{(X_i - E_i)^2}{E_i}$$

is distributed approximately like the chi-square distribution, with $(r - 1)(c - 1)$ degrees of freedom.

Model L is applicable to problems involving samples selected from large populations and where the sample size is not too small (for instance, $n \geq 50$).
Returning to our problem (Table 13.3.1), we formulate the hypotheses:

H_0: Attitude toward the issue and educational attainment are independent.
H_1: These characteristics of the adults in the population are not independent (they are related).

According to Model L, the decision model for testing for the independence of two variables of classification is constructed on the basis of the chi-square distribution, with $(r - 1)(c - 1)$ df. In such tests the critical region is always in the upper tail (as we found in the previous section). In our problem, df $= (3 - 1)$ $(3 - 1) = 4$. Let us conduct the test at the .01 level of significance. Referring to Table B.6, we see the critical value for our test is $\chi^2 = 13.277$. We state the decision rule as follows. *Compute the χ^2 statistic according to Model L. If $\chi^2 > 13.277$, reject H_0 and accept H_1, otherwise, accept H_0.*

Table 13.3.2 presents the computation of χ^2 using the data from Table 13.3.1 (X_i) and the corresponding expected frequencies (E_i) computed according to Equation 13.3.6. $\chi^2 = 13.22$ falls in the region of acceptance, according to the decision rule. Therefore, accept H_0 and conclude that attitude toward the issue and level of education are independent for the population of adults.

TABLE 13.3.2

Computation of chi-square: sample data from Table 13.3.1.

X_i	E_i	$\dfrac{(X_i - E_i)^2}{E_i}$
20	21.0	.05
40	43.5	.28
30	25.5	.79
40	29.2	3.99
55	60.4	.48
30	35.4	.82
10	19.8	4.85
50	41.1	1.93
25	24.1	.03
	$\chi^2 =$	13.22

If a chi-square test leads to the conclusion that two variables are related (not independent), how should this be interpreted? This test tells us nothing at all about how strong the relationship is or about the direction of the relationship. In our illustrative problem (Table 13.3.1), suppose the test led to the conclusion that attitude toward the issue and educational level are related. Is there a close relationship between these characteristics or is there a weak relationship? Does a higher educational level tend to be associated with the attitude "for the issue" or "against the issue"? The methods of *correlation analysis* (some of which are discussed in Chapters 17–19) are required to measure the strength and direction of relationship between two variables.

The χ^2 test should not be applied if one or more of the expected frequencies E_i are less than 5. In such cases, columns or rows should be combined so that each E_i is 5 or more. For example, if a 3 x 3 table has one or more E_i less than 5, combining two rows will result in a 2 x 3 table and combining two columns will result in a 3 x 2 table. Notice, however, that there are $(3 - 1)(3 - 1) = 4$ df for a 3 x 3 table but only 2 df after two rows or two columns are combined.

If one or more E_i are less than 5 in a 2 x 2 table, it is not possible to combine rows or columns and still conduct a test of independence. In such situations, since the χ^2 statistic is discrete whereas the χ^2 distribution is continuous, apply *Yates' correction for continuity* and compute the χ^2 statistic as follows:

$$\chi^2 = \sum \frac{(|X_i - E_i| - .5)^2}{E_i} \tag{13.3.7}$$

Notice that, in this equation, the absolute deviations $|X_i - E_i|$ are reduced by .5 before squaring, resulting in a smaller value. A desirable alternative is to use a sample large enough so that none of the expected frequencies are less than 5.

13.4   Goodness of Fit

Tests relating to "goodness of fit" involve comparison of a distribution of sample frequencies with a distribution standard determined on a theoretical basis or hypothesized. Often a theoretical distribution, such as the normal or binomial probability distributions, provide the theoretical standard for comparison. Model M is the appropriate sampling model.

Model M

Given an infinite population classified into m categories. Then the χ^2 statistic

$$\chi^2 = \sum \frac{(X_i - E_i)^2}{E_i}$$

is distributed approximately like the chi-square distribution, with $m - k$ degrees of freedom, where k denotes the number of constraints imposed on the computation of the expected frequencies E_i.

As noted in the previous section, adjacent categories should be combined, where an E_i is less than 5. The meaning of "number of constraints" will become clear in the illustrations which follow. The computation of E_i and the determination of k are different in different situations, as will become evident in the illustrations. In "goodness of fit" tests, we deal with categories in a single column or a single row.

An automobile manufacturer asked a sample of 350 university men which of four styles of sports car they prefer. The number who prefer each style (X_i) is shown in Table 13.4.1. Would you conclude, at the .01 level of significance, that the number preferring each style is the same?

Formulate the following hypotheses:

H_0: The number of university men who prefer each style of sports car is the same.

H_1: The number who prefer each style is not the same.

Assuming that the hypothesis to be tested H_0 is true, the expected frequencies E_i are computed by dividing the total sample size equally among the four styles of sports car, as shown in Table 13.4.1. Notice that the expected frequencies were computed subject to the constraint that $\sum E_i = \sum X_i = 350$ and this was the only restriction placed on the E_i. Hence, for this problem, the number of constraints $= k = 1$.

TABLE 13.4.1

Computation of chi-square: distribution of university men with respect to their preference for a style of sports car.

Style	Number in sample who prefer a style of sports car X_i	Expected number E_i	$\dfrac{(X_i - E_i)^2}{E_i}$
A	80	87.5	.64
B	120	87.5	12.07
C	60	87.5	8.64
D	90	87.5	.07
Total	350	350	$\chi^2 = 21.42$

Then, according to Model M, the decision model for testing the "goodness of fit" of the hypothesized distribution of preferences (equal number for each style) is constructed on the basis of the chi-square distribution, with $m - k$ df. In such tests the critical region is in the upper tail (as for the other χ^2 tests presented in this chapter). In our illustrative problem, $m = 4$ categories and $k = 1$, so that df $= 4 - 1 = 3$. Then, since the specified level of significance is .01, $\chi^2 = 11.345$ is the critical value for the test. Accordingly, we may express the decision rule as follows. *Compute the χ^2 statistic according to Model M. If $\chi^2 > 11.345$ reject H_0 and accept H_1, otherwise, accept H_0.*

Table 13.4.1 also shows the computation of chi-square. Then, $\chi^2 = 21.42$ falls in the region of rejection. Therefore, reject H_0 and accept H_1. Conclude that the number of university men who prefer each style of sports car is not the same.

A random sample of 1,500 families in a large city were asked how many times during the past year they went to a restaurant for dinner. The replies are summarized in the frequency distribution in Table 13.4.2. Would you conclude, at the .005 level of significance, that the number of times a family goes to a restaurant for dinner during a year is distributed like the normal curve?

TABLE 13.4.2

Frequency distribution of the number of times during the past year that a sample of 1,500 families went to a restaurant for dinner.

Number of times dinner was had in a restaurant	Number of families	Continuous class limits
		39.5
40–44	70	
		44.5
45–49	160	
		49.5
50–54	350	
		54.5
55–59	550	
		59.5
60–64	300	
		64.5
65–69	50	
		69.5
70–74	20	
		74.5

In this problem, the "distribution standard" is the *normal distribution* with the same mean and standard deviation as computed from the sample data ($\overline{X} = 55.6$ and $s = 6.0$). The 7 classes in Table 13.4.2 represent $m = 7$ categories and the class frequencies represent the observed frequencies X_i. The expected frequencies are computed by fitting a normal curve to the sample data. Table 13.4.3 presents the computation procedure for fitting a normal curve.

First, the continuous class limits are computed (Table 13.4.2) and posted in the first column of Table 13.4.3. These limits are transformed to z_i (Equation 5.4.4) as shown in the second column. Referring to Table B.4, the area under the standard normal curve between each z_i and the mean is determined. Then the area below each z_i is determined using the methods of Section 8.2. The "expected (theoretical) less than cum f_i" is obtained by multiplying the area below each z_i by the sample size ($n = 1,500$). Finally, the expected (theoretical) frequencies

E_i are obtained by subtracting successive pairs of the "less than cum f_i." Notice in Table 13.4.3 that the first and last z_i values are not computed and the corresponding areas between the mean and z_i are shown as .5000. This is done to insure that the E_i total $n = 1,500$.

TABLE 13.4.3

Fitting a normal curve to the data in Table 13.4.2 and computing the expected frequencies.

Continuous class limits	z_i	Normal curve area Between mean and z_i	Below z_i	Expected less than cum f_i	E_i
39.5	—	.5000	0	0	
					48.3
44.5	−1.85	.4678	.0322	48.3	
					182.6
49.5	−1.02	.3461	.1539	230.9	
					412.0
54.5	−.18	.0714	.4286	642.9	
					470.4
59.5	.65	.2422	.7422	1113.3	
					282.6
64.5	1.48	.4306	.9306	1395.9	
					88.8
69.5	2.32	.4898	.9898	1484.7	
					15.3
74.5	—	.5000	1.0000	1500.0	

The E_i were computed with the *three constraints:* the mean and the standard deviation of the fitted normal curve were made to equal $\overline{X} = 55.6$ and $s = 6.0$, respectively, and the E_i were made to total $n = 1,500$. Therefore, in Model M, $k = 3$ and df $= m - k = 7 - 3 = 4$. We formulate the following hypotheses:

H_0: The distribution of the number of times during a year a family goes to a restaurant for dinner is like the normal distribution.
H_1: The distribution is not normal.

We may write the decision rule as follows. *Compute the χ^2 statistic according to Model M. If $\chi^2 > 14.860$ reject H_0 and accept H_1, otherwise, accept H_0.* The critical χ^2 value was determined for 4 df and $\alpha = .005$.

Table 13.4.4 presents the computation of χ^2 based on the data in Tables 13.4.2 and 13.4.3. $\chi^2 = 54.81$ falls in the region of rejection, according to the decision rule. Therefore, accept H_1 and conclude that the number of times during a year a family goes to a restaurant for dinner is not normal in distribution.

TABLE 13.4.4

Computation of χ^2 based on the data in Tables 13.4.2 and 13.4.3.

X_i	E_i	$\dfrac{(X_i - E_i)^2}{E_i}$
70	48.3	9.75
160	182.6	2.80
350	412.0	9.33
550	470.4	13.47
300	282.6	1.07
50	88.8	16.95
20	15.3	1.44
		$\chi^2 = \overline{54.81}$

Study Problems

1.   A sample of 150 customer accounts in branch A of a large department store showed that 25 were behind in payments; in branch B, a sample of 200 customer accounts showed that 35 were behind in payments; and in branch C, a sample of 250 customer accounts showed that 40 were behind in payments. Would you conclude that the proportion of customer accounts behind in payments is the same for the three branches? Use the .05 significance level.

2.   A sample of 500 men in ethnic group I included 50 who were living on welfare payments. A sample of 1,000 men from ethnic group II included 90 in this category. Similarly, 110 men out of a sample of 800 from ethnic group III and 100 men out of a sample of 700 from ethnic group IV were living on welfare payments. Using the .01 level of significance, would you conclude that the proportion of each ethnic group living on welfare payments is the same?

3.   An agricultural research institute experimented with soil types A, B, and C for a certain plant variety. A sample of 200 two-month-old plants were planted in soil type A, 100 in soil type B, 150 in soil type C, and 200 in standard type soil. After a three-month period, 25 percent of the plants in the standard type soil grew to greater than normal height for plants of the given variety; whereas, 28 percent of the plants in soil type A, 32 percent of the plants in soil type B, and 26 percent of the plants in soil type C grew to greater than normal height. Would you conclude at the .005 significance level that the proportion of plants of the given variety which grow to greater than normal height for the plant is the same for all four soil types?

4.   A sample of 1,000 college seniors were asked to identify a fellow student of the same sex who is their best friend. Then, the academic major of each student and each best friend was determined. The number of seniors in the sample fitting into each cross-classification category is as follows:

Major category of senior	Major category of best friend		
	A	B	C
A	170	90	100
B	200	100	130
C	80	60	70

Would you conclude that there is a relationship between a senior's major category and the major category of his best friend? Use the .05 level of significance.

5.   A sample of 500 voters in a city were asked which of four candidates they favor for city manager. The following information was obtained as to the number of voters in the sample falling into each of 12 cross-classification categories as follows:

Candidate preferred	Low	Income category Medium	High
A	20	35	60
B	40	25	70
C	60	50	20
D	30	60	30

Would you accept the claim of a city politician that there is a relationship between candidate preferred and income class? Conduct the test at the .005 level of significance.

6.   A sample of 230 drivers were asked if they would be willing to pay the increased price for automobiles if rear-window defrosters were installed as standard equipment. The number in each response category, by sex, is shown below.

	Response Willing	Not willing
Women	150	10
Men	120	20

Use the .01 level of significance and test whether there is a relationship between response and sex.

7.   A sample of 110 boys between the ages of six and ten were asked if they prefer shooting-type toys to other toys. The number responding by type of toy and age class is as follows:

Age class	Shooting-type	Other toys
6–8 yrs	47	3
9–10 yrs	53	7

Would you conclude, at the .05 level of significance, that there is no relationship between type of toy preferred and age class for boys 6–10 years of age?

8.   The following data appeared in a research journal relating to the labor force.

Occupational areas	Sample from city A (number)	Distribution for the region (proportion)
Professional	200	.30
Managerial	60	.15
Sales	70	.10
Clerical	40	.10
Manual	120	.30
Other	10	.05
Total	500	1.00

Using the .01 significance level, would you conclude that the occupational area distribution of the labor force for city A is the same as for the region?

9.   A, B, and C are the leading detergents on the market. A sample of 800 housewives included 220 who prefer A, 185 who prefer B, 210 who prefer C, and the rest prefer other detergents. Based on these sample results, would you accept the claim that A, B, C, and the other detergents combined share the market equally (one-fourth of the market each)? Use the .05 level of significance.

10.   According to the Mendelian law, a cross between two specified plants should produce four seed types, A, B, C, and D in the proportion $9:3:3:1$. A researcher found that such a cross of the two specified plants produced 50 type A seeds, 20 type B seeds, 25 type C seeds, and 10 type D seeds. Would you conclude, at the .005 significance level, that these results are consistent with expectations according to the Mendelian law?

11.   You are told that two coins were tossed 400 times and resulted in two heads 95 times, only one head 195 times, and two tails 110 times. Are these results consistent with the claim that the outcomes of the tosses represent randomly occurring events for two well-balanced coins? Use the .05 level of significance.

12.   A sample of 100 test scores was selected from a large population of scores and the following frequency distribution constructed:

Scores	f_i
10–14	1
15–19	8
20–24	21
25–29	42
30–34	18
35–39	9
40–44	1

a.   Compute the expected class frequencies by fitting a normal curve to the sample data.

b.   Using the .05 level of significance, would you conclude that the population score distribution is approximately normal?

Chapter 14

Analysis of Variance

14.1 Comparison of Means

Analysis of variance is a powerful statistical procedure. It is used for the evaluation of experimental results which involve outcomes under a variety of *experimental conditions* called *treatments*. For example, is the efficiency of a production process the same if performed under a temperature of 65° F, 75° F, 85° F, or 95° F? The experimental conditions or the treatments are the different temperatures. Are miles per gallon different for gasolines A, B, and C? The three makes of gasoline are the treatments. The experimental condition which varies, the treatment, is called the *criterion of classification*.

The general approach in an analysis of variance is to specify the conditions or treatments to be studied and to try out each treatment on an independently selected random sample selected from the population of interest. The mean outcome is computed for each sample and the sample means compared to determine whether the variation in treatments resulted in differences among the sample outcomes. Clearly, then, analysis of variance tests involve the *comparison of means*.

Analysis of variance tests range from the simpler tests to tests based on highly complex experimental designs. We will study the simpler tests. In Sections 14.2 and 14.3 we will consider tests involving only one criterion of classification (one set of treatments), called *one-way analysis of variance*. In Section 14.4 we will study tests involving two criteria of classification (two sets of treatments), called *two-way analysis of variance*.

14.2 One-Way Analysis of Variance

A large firm must decide which of three methods (A, B, or C) to adopt for the production of precision-cut steel clips. Three independent samples of four workers each are selected from the large work force, and the workers in each sample are asked to produce steel clips by one of the methods. After each worker has become thoroughly familiar with the method he is to use, the methods are put into effect and the number of clips produced during one working day by each worker is observed and recorded as shown in Table 14.2.1. Our interest is in evaluating whether the different production methods (treatments) have differing effects on the number of steel clips produced.

TABLE 14.2.1

Number of precision-cut steel clips produced by three samples of four workers each using methods A, B, and C.

| | Method | |
A	B	C
30	20	30
25	15	20
20	15	10
15	10	10

The three samples of observations (production amounts) in Table 14.2.1 represent three theoretical populations. The four production amounts for method A are a sample of the population of production amounts which would be obtained if the firm used only method A. Similarly for the samples of production amounts for methods B and C. *In conducting an analysis of variance test, it is assumed that the samples were independently selected from normally distributed populations with equal variance σ^2.* Hypotheses are formulated relating to the population means, as follows:

H_0: $\mu_1 = \mu_2 = \mu_3$ (for the three populations relating to methods A, B, and C)
H_1: The μ_i are not all equal.

The hypothesis H_0 implies that the different treatments (production methods) all have the same effect on the number of steel clips produced, since all methods result in the same average amount of production per worker. Alternatively, if the several production methods have different effects on production, then it is to be expected that the population means would not all be equal (H_1).

If H_0 is true, normal populations with equal treatment means μ_i and a common variance σ^2 may be considered to make up a single normal population with mean μ and variance σ^2. We estimate σ^2 by pooling the data for the three samples in Table 14.2.1, using the approach for two samples in Equation 10.3.1, as follows:

$$\frac{\sum_i \sum_j (X_{ij} - \overline{X}_i)^2}{n - c}$$

$$= \frac{\sum_j (X_{1j} - \overline{X}_1)^2 + \sum_j (X_{2j} - \overline{X}_2)^2 + \sum_j (X_{3j} - \overline{X}_3)^2}{(n_1 - 1) + (n_2 - 1) + (n_3 - 1)} \qquad \textbf{(14.2.1)}$$

Where: X_{ij} = jth observation in the ith treatment sample. (In Table 14.2.1: $X_{11} = 30$ is the first observation in the first sample; $X_{21} = 20$ is the first observation in the second sample; $X_{32} = 20$ is the second observation in the third sample; etc.)

$\overline{X}_i$ = mean of the ith treatment sample. ($\overline{X}_1$ = mean of the first sample, etc.)

$\left.\begin{matrix} n_1 \\ n_2 \\ n_3 \end{matrix}\right\}$ = treatment sample size

n = size of the three samples combined ($= 4 + 4 + 4 = 12$ in our illustration)

c = number of treatment samples ($= 3$ in our illustration)

The *double summation* $\sum_i \sum_j (X_{ij} - \overline{X}_i)^2$ denotes the sum of the squared deviations of the production amounts (X_{ij}), each from its sample mean ($\overline{X}_i$), and is obtained by first adding the squared deviations for each sample ($\sum_j$) and then adding these sums across all samples ($\sum_i$). This is indicated more specifically in the expression on the right-hand side of the equality sign in Equation 14.2.1. Let us apply this to the data in Table 14.2.1. These data are also shown in Table 14.2.2 with certain computational results as follows:

$\sum_j X_{ij}$ = sum of the production amounts (observations) for a sample

(For the first sample, this sum is $\sum_j X_{1j} = 90$, etc.)

$\sum_i \sum_j X_{ij}$ = sum of the production amounts across all samples

n_i = sample size

$\overline{X}$ = the *grand mean* (general mean) for the n production amounts

The symbols n and $\overline{X}_i$ have been explained previously.

TABLE 14.2.2

Data for three samples from Table 14.2.1, with certain computational results.

	Method				
	A	B	C		
	30	20	30		
	25	15	20		
	20	15	10		
	15	10	10		
$\sum\limits_{j} X_{ij}$	$\overline{90}$	$\overline{60}$	$\overline{70}$	$\sum\limits_{i}\sum\limits_{j} X_{ij} = 220$	
n_i	4	4	4	$n = 12$	
$\overline{X}_i$	22.50	15.00	17.50	$\overline{X} = 18.33$	

We are now ready to compute the terms in the numerator of Equation 14.2.1 as follows.

$\sum\limits_{j} (X_{1j} - \overline{X}_1)^2$

$= (30 - 22.50)^2 + (25 - 22.50)^2 + (20 - 22.50)^2 + (15 - 22.50)^2 = 125$

$\sum\limits_{j} (X_{2j} - \overline{X}_2)^2$

$= (20 - 15.00)^2 + (15 - 15.00)^2 + (15 - 15.00)^2 + (10 - 15.00)^2 = 50$

$\sum\limits_{j} (X_{3j} - \overline{X}_3)^2$

$= (30 - 17.50)^2 + (20 - 17.50)^2 + (10 - 17.50)^2 + (10 - 17.50)^2 = 275$

Then, the estimate of the population variance according to Equation 14.2.1 is

$$\frac{\sum\limits_{i}\sum\limits_{j} (X_{ij} - \overline{X}_i)^2}{n - c} = \frac{125 + 50 + 275}{3 + 3 + 3} = \frac{450}{9} = 50$$

The numerator of Equation 14.2.1 $\sum\limits_{i}\sum\limits_{j} (X_{ij} - \overline{X}_i)^2$ is called the "sum of squares within treatment samples" and reflects the variation of the X_{ij} (production amounts) in each sample from the sample mean. We would expect that worker production X_{ij} in a particular sample would vary around the sample mean $\overline{X}_i$ only because of chance (reflecting, for example, random selection of workers in a sample). Clearly, since the same production method was used by all the workers in a sample, no part of this variation reflects differing effects among the production methods. Consequently, the variation within treatment samples represents experimental error (chance) variation and is denoted as SSE (sums of squares for error). The denominator of Equation 14.2.1, $n - c = 12 - 3 = 9$ (or $(n_1 - 1) + (n_2 - 1) + (n_3 - 1) = 9$) represents the number of degrees of freedom involved in the variance estimate. In analysis of variance, a variance is called a *mean square*. More specifically, the mean square computed with SSE in the numerator, as in Equation 14.2.1, is called the *mean square for error* and denoted as MSE. Then,

$$\text{MSE} = \frac{\text{SSE}}{n - c} = \frac{\sum_i \sum_j (X_{ij} - \overline{X}_i)^2}{n - c} \qquad (14.2.2)$$

Recall that the sampling distribution of the mean is a distribution of all possible sample means for samples of a given size selected from a population (Section 7.4). If H_0 (previously formulated) is true, then the treatment sample means $\overline{X}_i$ may be considered to be a sample of possible sample means. Referring to the illustrative problem in Table 14.2.2, the three sample means make up a sample of size $c = 3$. Then, following the approach used for computing the variance s^2 (Equation 5.2.2), we estimate the sampling variance of the mean $s_{\overline{x}}^2$ as

$$s_{\overline{x}}^2 = \frac{\sum_i (\overline{X}_i - \overline{X})^2}{c - 1} \qquad (14.2.3)$$

We know from Section 8.4 that $s_{\overline{x}}^2 = s^2/n$. In other words, the estimated sampling variance $s_{\overline{x}}^2$ may be computed from a sample by dividing the estimated population variance by the sample size. Then, since we are denoting sample size as n_i, we may write

$$s_{\overline{x}}^2 = \frac{s^2}{n_i} \qquad (14.2.4)$$

Solving for s^2, we obtain

$$s^2 = n_i s_{\overline{x}}^2 \qquad (14.2.5)$$

Finally, substituting for $s_{\overline{x}}^2$ according to Equation 14.2.3, we have

$$s^2 = \frac{n_i \sum_i (\overline{X}_i - \overline{X})^2}{c - 1} = \frac{\sum_i n_i (\overline{X}_i - \overline{X})^2}{c - 1} \qquad (14.2.6)$$

The numerator of the foregoing equation $\sum_i n_i (\overline{X}_i - \overline{X})^2$ represents the "sum of squares among treatment sample means" or "the sum of squares for treatments" and is denoted as SST. This sum of squares measures the variation of the treatment sample means $\overline{X}_i$ around the grand mean $\overline{X}$. We would expect the treatment sample means to vary around the grand mean partly (perhaps mainly) because of the differing effects of the several production methods and partly because of chance effects (such as random selection of workers for each sample). If the production methods are equally effective, then these methods make no contribution to the variation among sample means. Then SST would reflect solely chance variation (as we found for SSE). The denominator of Equation 14.2.6, $c - 1$, represents the number of degrees of freedom involved in the variance estimate. The variance estimate in this equation (where the numerator is SST) is called the *mean square for treatments* and denoted as MST. We may write

$$\text{MST} = \frac{\text{SST}}{c - 1} = \frac{\sum_i n_i (\overline{X}_i - \overline{X})^2}{c - 1} \qquad (14.2.7)$$

Applying Equation 14.2.7 to the data in Table 14.2.2, compute

$$\text{MST} = \frac{(4)(22.50 - 18.33)^2 + (4)(15.00 - 18.33)^2 + (4)(17.50 - 18.33)^2}{3 - 1}$$

$$= 58.33$$

Then, if H_0 is true so that the treatment means μ_i are all equal and the treatments have the same effect on production, MSE (Equation 14.2.2) and MST (Equation 14.2.7) both represent estimates of the common population variance σ^2 and we would expect them to be about equal. On the other hand, if the three methods have different effects on production (H_1 is true), we would expect MST to be larger than MSE. In our illustration we find that MST ($= 58.33$) is larger than MSE ($= 50$); however, is MST sufficiently larger to warrant rejection of H_0 and acceptance of H_1? Model N is the appropriate sampling model.

Model N

Given c independent samples selected from c normal populations, with equal means μ_i and a common variance σ^2. The sampling distribution of the statistic

$$F = \frac{\text{MST}}{\text{MSE}}$$

is like the F distribution, with $c - 1$ (numerator) and $n - c$ (denominator) degrees of freedom.

It should be clear that the larger MST compared to MSE, or the larger the value of the F statistic, the less likely is it that the population treatment means μ_i are equal and that H_0 is true. Consequently, when the F statistic exceeds a specified critical value, we reject H_0 and accept H_1 (conclude that the treatment means are not all equal). Consequently, *in analysis of variance tests, the critical region is always in the upper tail.* (The F statistic and the F distribution were introduced and discussed in Section 12.5 where tests comparing variances were presented.) Referring to our illustrative problem, we may write the decision rule as follows. *Compute the F statistic. If $F > 4.26$, reject H_0 and accept H_1, otherwise, accept H_0.* The critical value $F_{.05} = 4.26$ was obtained from Table B.7, assuming that the level of significance was specified as $\alpha = .05$, for df $= c - 1 = 3 - 1$ or 2 (numerator) and df $= n - c = 12 - 3$ or 9 (denominator).

We compute F as $58.33/50 = 1.17$ for our problem. Then, according to the decision rule, accept H_0 and conclude that the population treatment means are equal and one production method is just as good as another.

Sometimes, MST is smaller than MSE. In such instances, we do not compute the F statistic. We conclude immediately that there are no differences among the population means μ_i and that any observed differences are the result of chance variation.

14.3   Test Procedure

Referring to our discussion in the previous section, if H_0 is true, we may consider the 12 observations (production amounts) in Table 14.2.2 as making up a single sample of $n = 12$ from a single population of production amounts. We may rewrite Equation 5.2.2 for s^2 in terms of our new symbols as

$$s^2 = \frac{\sum_i \sum_j (X_{ij} - \overline{X})^2}{n - 1} \qquad (14.3.1)$$

The numerator of this equation states that the squared deviations of the 12 production amounts from the grand mean $(X_{ij} - \overline{X})^2$ are added. This sum represents the total variation around the grand mean $\overline{X}$ in the set of n X_{ij} observations and is called the "total sum of squares." It is denoted as SS Total. The denominator of this equation, $n - 1$, represents the number of degrees of freedom involved in the computation of s^2. We would expect the $n = 12$ production amounts to vary around the grand mean because of the different effects on production of the several methods (A, B, and C) as well as because of the effect of chance (experimental error). Of course, if the different production methods all have the same effect on production, then production methods make no contribution to the total variation. In such event, total variation in the X_{ij} observations would reflect solely chance variation.

The fundamental relationship of a one-way analysis of variance partitions the total variation of the n X_{ij} observations into treatment variation and chance variation, as follows:

$$\sum_i \sum_j (X_{ij} - \overline{X})^2 = \sum_i n_i(\overline{X}_i - \overline{X})^2 + \sum_i \sum_j (X_{ij} - \overline{X}_i)^2 \qquad (14.3.2)$$

Or, more briefly as

$$\text{SS Total} = \text{SST} + \text{SSE} \qquad (14.3.3)$$

The total number of degrees of freedom in the denominator of Equation 14.3.1 $(n - 1)$ may be partitioned similarly into df associated with SST $(c - 1)$ and df associated with SSE $(n - c)$. We may write

$$n - 1 = (c - 1) + (n - c) \qquad (14.3.4)$$

Conducting a one-way analysis of variance requires the computation of the F statistic, which is a ratio of the variances MST and MSE; however, computation of these mean squares is laborious if Equations 14.2.7 and 14.2.2 are used. We will present a more convenient procedure for conducting an analysis of variance test. This procedure will be discussed in terms of an illustrative problem.

Table 14.3.1 presents the number of overseas telephone calls handled by each operator in the morning, afternoon, and night shifts of a large telephone answering service during a 24-hour period. There are five operators on the morning shift, four on the afternoon shift, and six on the night shift. Would you conclude, at the .05 level of significance, that the mean number of overseas calls handled per operator is the same for the three shifts?

Table 14.3.1 also presents some of the computations required for a one-way analysis of variance. T_i denotes the sum of the observations in a sample and T represents the sum of the T_i. As before, n_i denotes the sample size, $n = \sum n_i$; $\overline{X}_i$ denotes a sample mean; and $\overline{X}$ is the grand mean. We compute T_i^2/n_i for the first sample (morning shift) as $T_1^2/n_1 = (32)^2/5 = 204.80$. Then, $\sum T_i^2/n_i$ is the sum of these quantities for all the samples. $\sum_j X_{ij}^2$ denotes the sum of the squares of the observations in a sample. For the first sample

$$\sum_j X_{1j}^2 = (8)^2 + (7)^2 + (6)^2 + (4)^2 + (7)^2 = 214$$

Finally, $\sum_i \sum_j X_{ij}^2$ denotes the total of these sums for all the samples.

TABLE 14.3.1

Computations for a one-way analysis of variance test:
the number of overseas calls handled by operators on three
shifts of a telephone answering service.

	Shift			
	Morning	Afternoon	Night	
	8	5	3	
	7	4	2	
	6	6	2	
	4	2	1	
	7		3	
			4	
$T_i = \sum_j X_{ij}$	32	17	15	$T = \sum_i \sum_j X_{ij} = 64$
n_i	5	4	6	$n = 15$
$\overline{X}_i$	6.40	4.25	2.50	$\overline{X} = 4.27$
$\dfrac{T_i^2}{n_i}$	204.80	72.25	37.50	$\sum \dfrac{T_i^2}{n_i} = 314.55$
$\sum_j X_{ij}^2$	214	81	43	$\sum_i \sum_j X_{ij}^2 = 338$

The following *computational equations* are convenient to use.

$$\text{SS Total} = \sum_i \sum_j X_{ij}^2 - \frac{T^2}{n} \tag{14.3.5}$$

$$\text{SST} = \sum \frac{T_i^2}{n_i} - \frac{T^2}{n} \tag{14.3.6}$$

$$\text{SSE} = \text{SS Total} - \text{SST} \tag{14.3.7}$$

Equation 14.3.7 is based on Equation 14.3.3.
Applying these equations to the data in Table 14.3.1, we obtain

$$\text{SS Total} = 338 - \frac{(64)^2}{15} = 64.93$$

$$SST = 314.55 - \frac{(64)^2}{15} = 41.48$$

$$SSE = 64.93 - 41.48 = 23.45$$

The computations are summarized and completed in an analysis of variance (ANOVA) table as presented in Table 14.3.2. The df were determined based on Equation 14.3.4 and the mean squares (MS) based on Equations 14.2.7 and 14.2.2.

TABLE 14.3.2

ANOVA table for the data in Table 14.3.1.

Source of variation	SS	df	MS	F
Treatments	41.48	2	20.74	10.64
Error	23.45	12	1.95	
Total	64.93	14		

The following hypotheses are formulated:

$H_0: \mu_1 = \mu_2 = \mu_3$
$H_1:$ The μ_i are not all equal.

The decision model is based on Model N. The decision rule is as follows. *Compute the F statistic. If $F > 3.89$, reject H_0 and accept H_1, otherwise, accept H_0.* The critical value $F_{.05} = 3.89$ was found in Table B.7 for df = 2 (numerator) and 12 (denominator). The F statistic 10.64 (Table 14.3.2) falls in the region of rejection. Therefore, reject H_0 and accept H_1. Conclude that the mean number of overseas telephone calls is not the same for all three shifts.

14.4 Two-Way Analysis of Variance

In one-way analysis of variance tests, we deal with one variable of classification and partition the total sum of squares into two sources of variation—treatments and error. More advanced tests involve two or more variables of classification and partition the total sum of squares into three or more sources of variation. The experimental design and the statistical analysis become considerably more complex for these tests.

We will consider briefly a simple type of two-way analysis of variance, where two variables of classification will be involved and we will have only one observation per cell. For example, Table 14.4.1 presents the time required (in minutes) to dry a standard load of clothes for three different dryers and four levels of temperature. The two variables of classification are the column treat-

TABLE 14.4.1

Time required (minutes) to dry a standard load of clothes for three different dryer machines and four levels of temperature.

Temperature level	Dryer A	B	C	T_i	$\overline{X}_i$	T_i^2
I (low)	15	10	13	38	12.67	1444
II (low medium)	12	7	11	30	10.00	900
III (medium)	9	8	6	23	7.67	529
IV (high)	8	5	7	20	6.67	400
T_i	44	30	37	$111(=T)$		$3273(=\sum T_i^2)$
$\overline{X}_i$	11.00	7.50	9.25			
T_i^2	1936	900	1369	$4205(=\sum T_i^2)$		
$\sum_j X_{ij}^2$	514	238	375	$1127(=\sum_i \sum_j X_{ij}^2)$		

ments (dryer machines) and the row treatments (temperature levels). We will use slightly different notation than in the previous section:

c = number of columns (= 3, in our illustration)
r = number of rows (= 4)
rc = total number of observations or total number of cells (= $4 \cdot 3 = 12$)
T_i = a column total
$\overline{X}_i$ = a column mean
$\sum_j X_{ij}^2$ = sum of the squared observations in a column (For the first column, $\sum_j X_{1j}^2 = (15)^2 + (12)^2 + (9)^2 + (8)^2 = 514$.)
T_j = a row total
$\overline{X}_j$ = a row mean

In a two-way analysis of variance, with one observation per cell, we partition the total sum of squares into three sources of variation: column treatments, row treatments, and error. Two tests of hypotheses are conducted. One is a test of the effects of the column treatments. The following hypotheses are formulated:

H_0: The effects of the column treatments (dryers) are equal.
H_1: These effects are not all equal.

The other is a test of the effects of the row treatments. The following hypotheses are formulated:

H_0: The effects of the row treatments (temperature levels) are equal.
H_1: These effects are not all equal.

The partitioning of the total sum of squares may be written as

$$\text{SS Total} = \text{SST}_{\text{columns}} + \text{SST}_{\text{rows}} + \text{SSE} \qquad (14.4.1)$$

where:

$$\text{SS Total} = \sum_i \sum_j X_{ij}^2 - \frac{T^2}{rc} \qquad (14.4.2)$$

$\text{SST}_{\text{columns}} = $ sum of squares for columns

$$= \frac{\sum T_i^2}{r} - \frac{T^2}{rc} \qquad (14.4.3)$$

$\text{SST}_{\text{rows}} = $ sum of squares for rows

$$= \frac{\sum T_j^2}{c} - \frac{T^2}{rc} \qquad (14.4.4)$$

$$\text{SSE} = \text{SS Total} - \text{SST}_{\text{columns}} - \text{SST}_{\text{rows}} \qquad (14.4.5)$$

The available degrees of freedom are $rc - 1$ for SS Total, $c - 1$ for $\text{SST}_{\text{columns}}$, $r - 1$ for SST_{rows}, and $(r - 1)(c - 1)$ for SSE. The df are related as follows:

$$rc - 1 = (c - 1) + (r - 1) + (r - 1)(c - 1) \qquad (14.4.6)$$

Returning to the data in Table 14.4.1, let us specify the level of significance as .05. Table 14.4.1 presents certain computational results to be used in determining the various sums of squares and mean squares. Applying the foregoing equations, we obtain

$$\text{SS Total} = 1{,}127 - \frac{(111)^2}{(4)(3)} = 100.25$$

$$\text{SST}_{\text{columns}} = \frac{4{,}205}{4} - \frac{(111)^2}{(4)(3)} = 24.50$$

$$\text{SST}_{\text{rows}} = \frac{3{,}273}{3} - \frac{(111)^2}{(4)(3)} = 64.25$$

$$\text{SSE} = 100.25 - 24.50 - 64.25 = 11.50$$

$\text{SST}_{\text{columns}}$ is the sum of squares for dryers and SST_{rows} is the sum of squares for temperature levels. Table 14.4.2 presents the ANOVA table.

TABLE 14.4.2

ANOVA table for the data in Table 14.4.1.

Source of variation	SS	df	MS	F
Dryers	24.50	2	12.25	$\frac{12.25}{1.92} = 6.38$
Temperature levels	64.25	3	21.42	$\frac{21.42}{1.92} = 11.16$
Error	11.50	6	1.92	
Total	100.25	11		

The critical value for the .05 level of significance is $F_{.05} = 5.14$ for 2 and 6 df for dryers and $F_{.05} = 4.76$ for 3 and 6 df for temperature levels. We may write the decision rules as follows:

For dryers: *If F statistic > 5.14, reject H_0 and accept H_1, otherwise, accept H_0.*

For temperature levels: *If F statistic > 4.76, reject H_0 and accept H_1, otherwise, accept H_0.*

Referring to the ANOVA table, $F = 6.38$ for dryers falls in the region of rejection. Therefore, reject H_0 and conclude that drying time is not the same for the three dryers. $F = 11.16$ for temperature levels also falls in the region of rejection. Therefore, conclude that drying time is not the same for the different temperature levels.

It is interesting to note the difference in the analysis of the data in Table 14.4.1, if we conducted a one-way analysis of variance instead of a two-way analysis of variance. Suppose we were interested in testing whether the dryers have different effects on drying time, without removing the effects of the different temperature levels. In such an analysis, the total sum of squares is partitioned into a sum of squares for dryers and a sum of squares for error. SS Total and SST$_{columns}$ (or SS dryers) would be the same as in Table 14.4.2 and with the same df.

SSE, however (computed according to Equation 14.3.7 in the previous section), equals the total of SSE and SS temperature levels in Table 14.4.2 or $64.25 + 11.50 = 75.75$ and the df is the sum of the corresponding df or $3 + 6 = 9$. Then, MSE becomes $75.75/9$ or 8.41. Consequently, the F statistic to test the hypothesis of equal mean drying times for the three machines is MS dryers/MSE $= 12.25/8.41 = 1.46$. Comparing this with the critical value for 2 and 9 df, $F_{.05} = 4.26$, leads to acceptance of H_0. Hence, it would be concluded for a one-way analysis of variance that there is no difference in mean drying time among the three dryers. This shows that it is important to conduct the analysis of variance test in a manner consistent with the experimental design. The data in Table 14.4.1 resulted from a two-criteria-of-classification experimental design. Hence, a one-way analysis of variance is not the appropriate test procedure for these data.

Study Problems

1.   Three different texts are being considered for a course in statistics. A sample of five students were assigned textbook A; a second sample of five students were assigned textbook B; and a third sample of five students were assigned textbook C. After the same topics in statistical inference were assigned for study to the three samples of students, a quiz was given. The following scores were obtained:

	Sample	
A	B	C
10	6	10
8	8	12
12	10	10
9	5	14
10	4	13

Would you conclude, at the .01 level of significance, that the three textbooks are equally effective for the study of statistics? (Construct a one-way analysis of variance table.)

2. A land developer experimented with four different sales approaches in selling small homesites in a large, newly developed beach area. He assigned five salesmen to use sales approach A, five to use sales approach B, six to use sales approach C, and four to use sales approach D. The number of homesites sold by each salesman during a specified period of time is as follows:

	Sales approach used		
A	B	C	D
50	60	100	100
60	80	90	110
40	70	90	120
50	70	100	110
50	80	90	
		80	

Would you conclude, at the .05 significance level, that the different sales approaches are equally effective? (Construct a one-way analysis of variance table.)

3. Skinfold measurements were made for a sample of seven men 18–24 years of age, a sample of five men 25–34 years of age, and a sample of six men 35–44 years of age. These measurements (in cms) are as follows:

Skinfold measurements by age group		
18–24	25–34	35–44
3	7	2
4	2	8
2	5	6
5	4	2
6	3	1
4		4
5		

Would you conclude, at the .05 level of significance, that mean skinfold measurements are the same for males in the three age groups? (Construct an ANOVA table.)

4. A chemical process was performed a number of times under five different temperatures. The time for a complete reaction (in seconds) was recorded as follows:

		Degrees (Fahrenheit)		
110°	100°	90°	80°	70°
16	20	50	70	90
18	32	60	80	110
10	35	55	75	120
20	25	65	90	100
12	48			95
	45			105

Would you conclude that time for a complete reaction is affected by temperature? Use the .05 level of significance. (Construct an ANOVA table.)

5. Quiz grades earned by three students in four subjects were recorded as follows:

		Subject		
Student	A	B	C	D
1	8	7	9	6
2	6	5	7	6
3	5	3	4	3

Conduct a two-way analysis of variance to test whether quiz grades are about the same, on the average, for the four subjects; also, whether it may be concluded that the three students have about the same ability (as measured by their quiz grades). Use the .05 level of significance. (Construct the appropriate ANOVA table.)

6. The number of units produced by five workers with different amounts of experience and each worker using three different machines are as follows:

Worker	Months of experience	Machine I	II	III
A	Under 1	2	3	5
B	1–1.9	4	6	5
C	2–3.9	6	5	10
D	4–4.9	6	9	12
E	5 or more	9	10	15

Would you conclude that the number of units produced is not affected by length of experience nor by the machine used? Use the .05 significance level. (Construct the appropriate ANOVA table.)

Chapter 15

Nonparametric Tests

15.1 Parametric vs. Nonparametric

For the most part, statistical test procedures presented so far are applicable only if certain conditions pertaining to the population or populations involved are satisfied. For example, a condition may be that the population is at least approximately normal or that a parameter (such as the mean or standard deviation) is equal for two populations. Tests of this type are called *parametric tests*. Another type of statistical test, useful in a variety of situations, is the *nonparametric test*. Such tests, generally, are applicable to a wider group of test situations since they impose fewer and looser restrictions.

Nonparametric tests involve simpler computations than parametric tests and are easier to apply and understand. Nonparametric tests are also called *distribution-free tests* since they are applicable to situations where the population or populations involved have unknown distributions. An important characteristic of nonparametric tests is their applicability to ranking data, ordinal-scale data generally, and nominal-scale or classificatory data. (This type of data was discussed in Section 3.3.) Such data cannot be tested by parametric test procedures. For example, nonparametric tests may be used to compare the ranking of ten applicants on suitability for a certain job by two interviewers, the preferences of teen-age girls relating to certain cosmetics in two high schools, and the distributions by type of job held by employed males aged 30 to 35 in two communities.

We have already encountered an important group of nonparametric tests, the chi-square tests presented in Chapter 13. We will consider certain additional nonparametric tests in this chapter. It should be noted, however, that there are a large number of such tests in the literature.

A serious weakness of nonparametric tests is that they are generally wasteful of information. That is, they do not always use all the information in the sample data. For example, if a sample of weights is used in a nonparametric test, only the weight rankings, not the weights themselves, may be used in the test. It is usually true that a nonparametric test has less power ($1 - \beta$, the probability of rejecting H_0 if it is false, is smaller) than the comparable parametric test. This may lead to the use of a larger sample when a nonparametric test is applied than would be used if a comparable parametric test were applied instead.

In many situations a nonparametric test is used even if a suitable parametric test is available because the nonparametric test is quicker and easier to apply or because there is not enough assurance that the conditions relating to the parametric test are adequately met. Nevertheless, a parametric test is preferable to a nonparametric test, if it is applicable to a test situation.

15.2 Sign Test for Paired Data

One of the simplest nonparametric tests to apply is the *sign test*, so-called because it is based on the direction (sign) of a difference for a pair of observations. Suppose a sample of 24 male drivers recorded the number of miles per gallon obtained when using a certain gasoline, and then were asked to drive the same cars and determine the number of miles per gallon obtained when using the same gasoline with an additive introduced. Suppose 16 drivers reported increased mileage (positive (+) differences), 4 drivers reported reduced mileage (negative (−) differences), and 4 drivers reported no difference in mileage. Would you conclude, at the .05 level of significance, that the additive increased gasoline mileage?

In conducting the sign test, ignore the cases with no change (zero difference). Therefore, we are concerned with 16 "+'s" and 4 "−'s" and a sample size of $n = 20$. If the additive had no effect on mileage, we would expect that the mileage would be about the same with or without the additive. In such case, mileages for the gasoline with and without the additive may differ for each driver solely because of chance. Then, we would expect that the number of

"+'s" would be about the same as the number of "−'s." Hence, formulate the hypotheses:

H_0: $p = .5$ (The proportion of "+'s" is .5 or the additive has no effect on mileage.)

H_1: $p > .5$ (The proportion of "+'s" is greater than .5 or the additive increases mileage.)

The decision model is constructed on the basis of Model D (Section 8.6) which describes the sampling distribution of a proportion and also the sampling distribution of the frequency of successes. It is more convenient to state the decision rule in terms of the frequency of successes. According to Model D, the sampling distribution of a frequency X is approximately normal with mean np and standard error $\sqrt{np(1 - p)}$. We may state the decision rule as follows. *Compute z. If z > 1.64, reject H_0 and accept H_1, otherwise, accept H_0.*

Returning to our sample data, we have $X = 16$ (the number of + differences), $n = 20$, and $np = 10$. Since X is a discrete variable, transform 15.5 (not 16) to z (Equation 8.6.3) to obtain

$$z = \frac{15.5 - 10}{\sqrt{20(.5)(.5)}} = 2.46$$

Then, $z = 2.46$ falls in the region of rejection. Accordingly, reject H_0 and accept H_1. Conclude that the gasoline with the additive results in higher mileage.

The effect of the additive on gasoline mileage could also be tested statistically by the parametric test procedure for small samples discussed in Section 12.4. This test is based on the amount of difference in mileage for the gasoline with and without the additive for the $n = 24$ drivers. The appropriate sampling model for this test is Model E which is applicable only if the mileage differences are considered to be approximately normal in distribution.

15.3 Mann-Whitney Test for Independent Samples

Suppose it is claimed that entrance salaries paid to male programmers are the same as for female programmers, on the average. Table 15.3.1 presents the entrance salaries paid to a sample of $n_1 = 10$ male programmers and $n_2 = 12$ female programmers. Would you accept the claim based on these data? Use the .05 level of significance.

The *Mann-Whitney U test* provides a test of the difference between means based on *rank sums*. Applying the sample data in Table 15.3.1, the mean for the n_1 sample is $\overline{X}_1 = \$5,100$ and for the n_2 sample, it is $\overline{X}_2 = \$4,466.67$. Do these means result from a difference in population means, or do they differ merely due to chance? Combining the two samples into one of size $n = n_1 + n_2$ or $10 + 12 = 22$, rank the salaries from lowest (rank 1) to highest (rank 22). These ranks are shown in Table 15.3.1. The n_1 sample ranks total $R_1 = 136$ and the n_2 sample ranks total $R_2 = 117$.

TABLE 15.3.1

Entrance salaries paid to a sample of $n_1 = 10$ male programmers and a sample of $n_2 = 12$ female programmers: ranks and rank sums for the Mann-Whitney U test.

Sample		Ranks based on combined sample $n = n_1 + n_2$	
n_1	n_2	n_1	n_2
5,000	6,300	13	22
3,500	3,200	4	2
4,800	4,300	11	8
6,000	5,600	20	17
4,500	4,200	9	7
5,500	5,900	16	19
5,100	3,000	14	1
4,600	3,600	10	5
6,200	3,800	21	6
5,800	3,400	18	3
	4,900		12
	5,400		15
Total 51,000	53,600	136	117

If the population distributions of entrance salaries for males and females are the same, we would expect that for samples of equal size the rank sums would be about equal. If the mean salary is lower for one of the populations, we would expect that the sample ranks and the rank sum would tend to be smaller for that population. In our illustration, the rank sum R_2 (for female programmers) is smaller than R_1 even though it is based on a larger sample. This does not appear to support the claim that, on the average, male salaries are the same as female salaries. Let us conduct the test to determine whether the sample differences reflect only chance effects.

The Mann-Whitney test is based on the U statistic which is computed as

$$U = n_1 n_2 + \frac{n_1(n_1 + 1)}{2} - R_1 \qquad (15.3.1)$$

or

$$U = n_1 n_2 + \frac{n_2(n_2 + 1)}{2} - R_2 \qquad (15.3.2)$$

Even though U has a different value when computed by these two equations, the outcome of the test will be the same no matter which one is used in conducting the test, as will be noted later in the discussion.

Formulate the hypotheses:

H_0: $\mu_1 = \mu_2$ (The populations of male and female entrance salaries are identical in distribution, so that the means of the male salaries μ_1 and female salaries μ_2 are equal.)

H_1: $\mu_1 \neq \mu_2$

If H_0 is true, then the combined sample of n observations may be considered as one sample selected from one population and the sampling distribution of the U statistic has

$$\text{mean} \qquad \mu_U = \frac{n_1 n_2}{2} \qquad \text{(15.3.3)}$$

$$\text{standard error} \qquad \sigma_U = \sqrt{\frac{n_1 n_2 (n_1 + n_2 + 1)}{12}} \qquad \text{(15.3.4)}$$

If n_1 and n_2 are each at least of size 10 then the sampling distribution of U is approximately normal. Hence, we may state the decision rule in terms of the z (normal) distribution, as follows. *Compute z. If z < −1.96 or > 1.96, reject H_0 and accept H_1, otherwise, accept H_0.* The equation for z is

$$z = \frac{U - \mu_U}{\sigma_U} \qquad \text{(15.3.5)}$$

Returning to the data in Table 15.3.1, compute

$$\mu_U = \frac{(10)(12)}{2} = 60$$

$$\sigma_U = \sqrt{\frac{(10)(12)(10 + 12 + 1)}{12}} = 15.17$$

Using Equation 15.3.1,

$$U = (10)(12) + \frac{(10)(10 + 1)}{2} - 136 = 39$$

and $\qquad z = \dfrac{39 - 60}{15.17} = -1.38$

Finally, $z = -1.38$ falls in the region of acceptance. Hence, accept H_0 and conclude that male and female programmers obtain the same entrance salary, on the average. (The reader should compute U using Equation 15.3.2 and conduct the test. It will be seen that the resulting z value will only differ in sign, 1.38 instead of -1.38. Hence, the conclusion reached will be the same.)

The Mann-Whitney test is not as wasteful of information as the sign test, since it uses the ranking information contained in the data. If the combined sample of n observations has ties, ranks should be assigned to the tied observations as if they can be ranked and then assign the mean of the ranks so obtained to each of the tied observations. For example, suppose the n observations are 20, 22, 22, 30, 34, 34, 34, 40. Assigning ranks, we have 1, 2, 3, 4, 5, 6, 7, 8. The two observations of 22 have ranks 2 and 3 which have the mean $(2 + 3)/2 = 2.5$, and the three observations of 34 have ranks 5, 6, 7 which have the mean $(5 + 6 + 7)/3 = 6$. Then, the ranking becomes 1, 2.5, 2.5, 4, 6, 6, 6, 8.

If samples n_1 and n_2 are not large enough to permit the use of the normal distribution in conducting the test, special tables will have to be used, such as those in *Nonparametric Statistics for the Behavioral Sciences*, Sidney Siegel, 1956, McGraw-Hill Book Company, Inc.

15.4   Wilcoxon Test for Paired Data

The *Wilcoxon rank-sum test for paired data* is more powerful than the sign test (Section 15.2) because it takes into account the *magnitudes* of pair differences as well as the signs. (The Mann-Whitney U test is an equivalent test and may also be applied to such data.)

Table 15.4.1 presents the heights of pairs of plants for 15 varieties of the plant where one of each pair was treated according to method A and the other according to method B. Would you conclude, at the .01 level of significance, that the distribution of plant heights is the same for both methods?

TABLE 15.4.1

Heights of 15 varieties of a plant for method A *and method* B: *pair differences, ranks, and rank sums for the Wilcoxon test.*

Height (cms)		Difference	Rank	
A	B	$d = A - B$	$+d$	$-d$
40	35	5	3	
15	17	−2		1
30	22	8	6.5	
20	14	6	4	
45	35	10	9	
50	30	20	15	
42	31	11	10	
36	43	−7		5
25	29	−4		2
48	36	12	11	
20	12	8	6.5	
18	9	9	8	
37	24	13	12	
48	32	16	13	
50	32	18	14	
Total 524	401		112	8

The Wilcoxon test begins by computing the pair-differences in height and ranking these differences according to amount of difference, without regard to sign. The smallest difference magnitude is assigned rank 1. In Table 15.4.1, the smallest difference magnitude 2 is assigned rank 1, the next higher difference magnitude 4 is assigned rank 2, the third smallest difference magnitude 5 is assigned rank 3, etc. (Treat ties as explained in Section 15.3.) These ranks are shown in the table separately for the positive differences $(+d)$ and the negative differences $(-d)$. Zero differences are ignored, so that the sample size involved in a test is $n =$ the number of pairs with non-zero differences. The rank sums of the $(+d)$ and $(-d)$ ranks are also presented in Table 15.4.1. Clearly, if the two methods A and B have the same effect on plant height, we would expect the number of ranks as well as the rank sums for $(+d)$ and $(-d)$ to be about equal. On the other hand, if one of the methods produced taller plants, for instance method A, we would expect that more than half the A $-$ B differences would be $(+d)$'s and the rank sum for $(+d)$ would exceed the rank sum for $(-d)$.

The Wilcoxon test is based on the T *statistic* which is the smaller of the two rank sums $(+d$ sum or $-d$ sum). In our problem $T = 8$, the $(-d)$ sum as shown in Table 15.4.1. When the number of pairs n is for example 10 or more (25 or more is better), the T statistic is approximately normal in distribution with

$$\text{mean} \qquad \mu_T = \frac{n(n + 1)}{4} \qquad (15.4.1)$$

$$\text{standard error} \qquad \sigma_T = \sqrt{\frac{n(n + 1)(2n + 1)}{24}} \qquad (15.4.2)$$

Returning to the data in Table 15.4.1, where $n = 15$, compute

$$\mu_T = \frac{(15)(16)}{4} = 60$$

$$\sigma_T = \sqrt{\frac{(15)(16)(31)}{24}} = 17.6$$

Formulate the hypotheses:

H_0: Methods A and B have the same effect on plant height.
H_1: Methods A and B have different effects on plant height.

The decision rule is as follows. *Compute z. If z < -2.58 or > 2.58, reject H_0 and accept H_1, otherwise, accept H_0.* The equation for z is

$$z = \frac{T - \mu_T}{\sigma_T} \qquad (15.4.3)$$

Applying this equation to our illustration, we obtain

$$z = \frac{8 - 60}{17.6} = -2.95$$

Since $z = -2.95$ falls in the region of rejection, reject H_0 and accept H_1. Conclude that methods A and B have different effects on plant height.

Special tables are required if n is small, such as the tables in Siegel (noted in Section 15.3).

15.5 Kruskal-Wallis One-Way Analysis of Variance

Table 15.5.1 presents the amounts deposited during a year by three samples of depositors for savings plans A, B, and C. Would you conclude that the mean amount deposited during the year is the same for the three plans? Use the .05 level of significance.

TABLE 15.5.1

Deposits during a year by three samples of depositors for savings plans A, B, and C: ranks and rank sums for the Kruskal-Wallis one-way analysis of variance.

	Savings plans				Ranks	
A	B	C	A	B	C
$ 370	$ 170	$ 195	21	2	7
206	150	188	8	1	6
175	184	182	3	5	4
215	246	308	9	13	19
248	235	258	14	11	16
226	236	265	10	12	17
250		282	15		18
520			20		
Total $2,010	$1,221	$1,678	100	44	87

The appropriate parametric test for this problem is a one-way analysis of variance as presented in Section 14.2. A comparable nonparametric test is the *Kruskal-Wallis test* which is based on the *H statistic* as follows:

$$H = \frac{12}{n(n+1)} \sum \frac{R_i^2}{n_i} - 3(n+1) \qquad (15.5.1)$$

where: n_i = size of each sample (in our illustration $n_1 = 8$, $n_2 = 6$, $n_3 = 7$).
n = $\sum n_i$ = combined sample size.
R_i = sum of the ranks assigned to the observations in a sample. Ranks are determined by ordering the combined sample of n observations from low (rank 1) to high and handling ties in the same way as for the Mann-Whitney test in Section 15.3.

Formulate the hypotheses:

H_0: $\mu_1 = \mu_2 = \mu_3$ (The populations of deposits for savings plans A, B, and C are identical in distribution, so that the means for plan A μ_1, plan B μ_2, and plan C μ_3 are equal.)
H_1: The means are not all equal.

If H_0 is true, then the H statistic has approximately the chi-square distribution with df = $c - 1$ (c = the number of samples or treatments); however, each n_i must be of size 5 or more. We may determine from Table 15.5.1 that the mean amount deposited for plan A is $\overline{X}_1 = \$251.25$; for plan B, $\overline{X}_2 = \$203.50$; and for plan C, $\overline{X}_3 = \$239.71$. Do these means result from savings plans with equal mean deposit amounts? The decision rule may be stated as follows. *Compute the H statistic. If $H > \chi_{.05}^2 = 5.991$, reject H_0 and accept H_1, otherwise, accept H_0.*
Applying the data in Table 15.5.1, compute

$$H = \frac{12}{(21)(22)} \left[\frac{(100)^2}{8} + \frac{(44)^2}{6} + \frac{(87)^2}{7} \right] - (3)(22) = 3.00$$

Therefore, since $H = 3.00$ falls in the region of acceptance, accept H_0 and conclude that the mean amount deposited is the same for the three savings plans.

Special tables are required if each sample size is not at least 5, such as those in Siegel (noted in Section 15.3).

15.6 A Test for Randomness

Sometimes it is a question whether a sample selection procedure resulted in a random sample. There is a simple *runs test* which may be applied.

Table 15.6.1 presents a sample of 32 weights. The weight observations are presented in the order selected, a prerequisite for conducting the runs test. The sample median was determined to be $M = 58$ lbs. Each weight in the sample is identified as B (below the median) or A (above the median), as shown in the table. Each sequence of A's and B's is bracketed and each sequence is called a *run*.

TABLE 15.6.1

A sample of 32 weights X_i (in pounds) listed in the order selected: identification of each observation as A (above the median) or B (below the median) and runs of A's and B's for use in a runs test.

X_i		X_i	
19	B	98	A
6	B	15	B
64	A	43	B
27	B	62	A
13	B	8	B
23	B	74	A
70	A	82	A
99	A	83	A
94	A	41	B
69	A	88	A
2	B	66	A
23	B	34	B
60	A	74	A
39	B	37	B
71	A	74	A
13	B	54	B

If there was a tendency to select heavier weights first and then to compensate by selecting some lighter weights later, or if there were some other departure from a random selection procedure, the number of runs would be too small or, under some conditions, too high. Therefore, the test of randomness is based on the *statistic r, the number of runs.* Let n_1 denote the number of A's and n_2, the number of B's. Then, if n_1 and n_2 are each at least 10, the sampling distribution of r is approximately normal with

$$\text{mean} \qquad \mu_r = \frac{2n_1n_2}{n_1 + n_2} + 1 \qquad (15.6.1)$$

$$\text{standard error} \qquad \sigma_r = \sqrt{\frac{2n_1n_2(2n_1n_2 - n_1 - n_2)}{(n_1 + n_2)^2(n_1 + n_2 - 1)}} \qquad (15.6.2)$$

We formulate the hypotheses:

H_0: The sequence of weight selection indicates that the sample is random.
H_1: The sample cannot be considered to be a random sample.

If $\alpha = .05$, we may state the decision rule as follows. *Compute z. If z < −1.96 or > 1.96, reject H_0 and accept H_1, otherwise, accept H_0.* The equation for z is

$$z = \frac{r - \mu_r}{\sigma_r} \qquad (15.6.3)$$

Table 15.6.1 indicates 21 runs of A's or B's, so that $r = 21$. Then, $n_1 = 16$ (the number of A's) and $n_2 = 16$ (the number of B's). If a sample observation is exactly equal to the median, it is disregarded for this test. Compute

$$\mu_r = \frac{(2)(16)(16)}{16 + 16} + 1 = 17$$

$$\sigma_r = \sqrt{\frac{(2)(16)(16)[(2)(16)(16) - 16 - 16]}{(16 + 16)^2(16 + 16 - 1)}} = 1.90$$

$$z = \frac{21 - 17}{1.90} = 2.11$$

Hence, $z = 2.11$ falls in the region of rejection. Conclude that the sample is not a random one.

The runs test is a versatile test and may be used in various types of problems. Sometimes the appropriate alternative hypothesis is one-sided. This is a useful test procedure to determine whether a time series contains a trend (number of runs too small) or a series of short cycles (number of runs too high). The test procedure just presented is similar to the *Wald-Wolfowitz runs test*, a nonparametric test to compare two independent samples. This test as well as special tables to conduct the runs test if n_1 or n_2 is less than 10 may be found in Siegel (noted in Section 15.3).

Study Problems

1.   The number of units produced per hour was recorded for a sample of 25 workers. Then, hourly productivity was recorded again for this sample, after an improved machine was used by the workers. Hourly productivity increased for 14 workers, decreased

for 8 workers, and remained unchanged for 3 workers. Would you conclude that the new machine increased hourly productivity? Use the .05 level of significance.

2.   The following pairs of scores were obtained by a sample of 30 college seniors on a social acceptance test before and after film-viewing.

Before	After	Before	After
20	18	18	16
25	20	20	21
19	18	15	15
20	20	17	15
23	25	20	22
18	20	19	17
24	24	23	20
22	25	17	13
18	16	25	20
15	12	19	18
17	19	17	15
16	16	13	13
13	10	18	16
24	20	20	16
21	19	23	20

Apply the sign test, using the .01 level of significance, to test whether it may be concluded that film-viewing reduces test scores.

3.   Average daily rainfall was recorded for the two months June and July for a sample of 35 cities in a certain region. The average was higher in July than in June for 16 cities, lower for 10 cities, and the same in the two months for 9 cities. Would you accept the claim that average daily rainfall in the region is about the same in June and July? Use the .05 significance level.

4.   A sample of 13 brass fittings (n_1) was selected from bin A and a sample of 12 brass fittings (n_2) from bin B. The weights of the selected fittings (in ounces) are as follows:

Sample weights

n_1	n_2
20	61
22	30
28	26
29	52
43	57
63	42
36	47
65	35
53	58
49	32
38	59
55	69
33	

    a.   Conduct a Mann-Whitney U test to test the claim that the mean weight per fitting is the same in bins A and B. Use the .05 level of significance. (Compute U according to Equation 15.3.1.)

    b.   Conduct the test again with U computed according to Equation 15.3.2.

5.   The number of motor vehicle accidents was recorded for a sample of 12 winter days (n_1) in city K and for a sample of 14 winter days (n_2) in city L as follows:

Number of accidents

n_1	n_2
6	13
23	38
11	43
29	8
31	39
10	4
27	46
32	33
15	44
24	7
30	49
20	53
	41
	35

It is claimed the number of motor vehicle accidents per winter day is the same, on the average, in city K and city L. Would you accept this claim, at the .05 level of significance? Apply the Mann-Whitney U test.

6.   A psychological test was administered to samples of 11 men (n_1) and 13 women (n_2) selected at a large state university. The following scores were obtained:

Test scores

n_1	n_2
115	151
165	111
131	126
163	138
107	151
111	172
142	117
132	122
157	137
145	177
138	138
	174
	166

Would you conclude that men and women attending the state university would obtain equal test scores, on the average, if this test were administered to all men and women at the university? Use the .05 significance level and apply the Mann-Whitney U test.

7. A sample of 13 typists were asked to type a set of letters before the morning coffee break. They were asked to type a similar set of letters in the afternoon when they were more tired, but after the coffee break. The average number of errors per letter was recorded for each typist as follows:

Morning	Afternoon
2.1	3.0
3.0	3.3
4.3	2.1
1.9	2.0
4.3	2.9
4.7	3.4
5.0	3.1
4.9	2.9
2.0	3.0
2.8	1.2
3.3	1.6
5.0	3.9
2.8	4.6

Would you conclude, at the .05 level of significance, that the distribution of average number of errors per typist is the same for both typing situations? Apply the Wilcoxon rank-sum test.

8. A sample of 16 common stocks was selected for study from the New York stock exchange list. Average monthly prices (rounded to whole dollars) were determined for May and October of last year, as follows:

May	October
$31	$13
22	18
32	32
31	20
36	26
24	25
20	25
38	18
27	27
37	31
14	16
19	22
20	27
34	21
19	10
33	18

Would you accept the claim that the distribution of common stock prices on the New York stock exchange was the same for the two months? Apply the Wilcoxon rank-sum test and use the .05 significance level.

9.   A sample of 17 brother-sister junior high school student pairs was selected and an arithmetic skill test was administered to the students in the sample. The following test scores were obtained:

Brother	Sister
145	210
120	120
95	90
70	100
130	80
115	170
90	90
110	105
120	130
115	110
140	160
155	180
135	150
87	145
70	130
65	80
100	95

Would you conclude, at the .05 significance level, that the distribution of arithmetic skill test scores is the same for junior high school boys and girls? Conduct a Wilcoxon rank-sum test.

10.   The number of weekly sales made by each of three salesmen (L, M, N) were recorded for a sample of nine weeks for L, a sample of nine weeks for M, and a sample of eight weeks for N, as follows:

L	M	N
49	31	19
45	67	27
57	38	22
48	13	17
16	37	24
39	54	12
46	63	10
15	42	33
34	51	

Would you conclude, at the .05 level of significance, that the mean weekly number of sales is the same for the three salesmen? Conduct a Kruskal-Wallis one-way analysis of variance test.

11.   Three firms' cloth sheets of a given type and grade were tested for blemishes. A sample of ten sheets was tested for firm I, a sample of six sheets for firm II, and a sample of seven sheets for firm III. The following numbers of blemishes per sheet tested were recorded:

Firm I	Firm II	Firm III
42	37	20
19	36	18
41	22	16
40	20	20
25	23	17
33	36	15
39		28
44		
47		
38		

It is claimed that the mean number of blemishes per sheet is the same for the three firms. Would you accept this claim at the .01 significance level? Conduct a Kruskal-Wallis test.

12. A sample of IQ scores was obtained for each of four ethnic groups in a large city, as follows:

Ethnic group			
A	B	C	D
79	144	110	137
117	93	87	134
93	121	74	90
114	145	129	129
125	142	106	84
124	83	131	122
		93	

Would you conclude that the mean IQ score is the same for the four ethnic groups? Conduct a one-way analysis of variance test (Kruskal-Wallis) and use the .05 level of significance.

13. A special evaluation examination was administered to entering freshmen in a large university. The scores obtained by a sample of 24 entering freshmen, in the order of selection, are: 11, 70, 67, 20, 45, 12, 13, 19, 85, 39, 14, 10, 62, 78, 60, 15, 51, 58, 32, 16, 82, 87, 27, and 37. Would you accept the claim that this sample was selected by a random selection procedure? Use the .05 significance level.

14. The volume of weekly imports of a commodity (in tons) for a period of 38 weeks were as follows:

Week	Tons	Week	Tons
1	60	20	118
2	63	21	95
3	65	22	73
4	131	23	133
5	92	24	139
6	81	25	108
7	93	26	144

Week	Tons	Week	Tons
8	110	27	163
9	129	28	58
10	100	29	54
11	135	30	165
12	96	31	143
13	70	32	169
14	93	33	56
15	111	34	51
16	125	35	142
17	75	36	147
18	94	37	152
19	120	38	158

Would you conclude, at the .05 level of significance, that weekly volume of imports indicates a growth trend? Apply the runs test (to evaluate the number of runs above and below the median weekly volume of imports).

15.   A sample of 32 heads of households were selected from a large minority group in a metropolitan area and asked how many days of unemployment they had during a certain period. The replies, in order obtained, are: 10, 15, 17, 5, 53, 62, 64, 7, 5, 12, 6, 49, 24, 13, 7, 29, 45, 20, 14, 8, 9, 34, 42, 37, 9, 18, 64, 48, 57, 31, 48, and 22. Would you conclude that this represents a random sample? Use the .01 level of significance.

Chapter 16

Bayesian Inference
and Decision Strategies

16.1   Bayes' Rule

The probabilities so far discussed refer to events which have not yet occurred or, if they did occur, the outcome is not known. An interesting and important class of problems involve *posterior probabilities*. In such problems, the event has occurred and the outcome is known; however, we take a backward look to determine the probability that a related event has also occurred. Hence, posterior probabilities are also called *backward* or *inverse* probabilities. In problems of this type there are two sets of events to be considered. One of the events in set 1 occurs first, then one of the events in set 2 occurs. We know the outcome relating to set 2. The problem is to determine the probability that a specified event in set 1 has also occurred. Let us examine such a problem.

Suppose you are having a house built and it is important that it be completed on a specified date. The interior painting must be done and the builder says that he will try to get Ed for this job since it is most likely (probability of .80) that Ed will get the job done on time. If Ed is not available, he will try to get Tim who is less likely to get the job done on time (probability of .70). Otherwise, he will have to get Joe (his assistant) who is least likely to finish the job on time (probability of .60). The probability is .30 that Ed will do the job, .50 that Tim will do it, and .20 that Joe will do it. Suppose you were away on business and return on the day the house is to be ready and find that the painting is completed. What is the probability that the job was done by Joe?

In our illustration, the first set of events relates to the painter who takes the job. There are three possible outcomes: Ed, Tim, or Joe. The second set of events relates to completion of the job. There are two possible outcomes: on time or not on time. We know the outcome for the second set: on time. Our problem is to determine the probability that the outcome from the first set was Joe. Based on the narrative, the following probabilities are known:

1. Probability that a given painter took the job:

$$P(\text{Ed}) \ \ = .30$$
$$P(\text{Tim}) = .50$$
$$P(\text{Joe}) \ = .20$$

2. Conditional probability that the job would be finished on time, given that a specified painter took the job:

$$P(\text{on time} \mid \text{Ed}) \ \ = .80$$
$$P(\text{on time} \mid \text{Tim}) = .70$$
$$P(\text{on time} \mid \text{Joe}) \ = .60$$

The sample space for this problem has 3 (painters) · 2 (painting completion possibilities) or 6 sample points. The sample space and probability distribution are presented in Table 16.1.1, as well as the marginal probabilities. The probability for the first sample point in the first column, $P(\text{Ed and on time})$, is computed by the use of Rule C as

$$P(\text{Ed and on time}) = P(\text{Ed}) \cdot P(\text{on time} \mid \text{Ed})$$
$$= (.30) \cdot (.80) = .24$$

TABLE 16.1.1

Painters	Completion of job		Total
	On time	Not on time	
Ed	(.30)(.80) = .24	.30 − .24 = .06	.30
Tim	(.50)(.70) = .35	.50 − .35 = .15	.50
Joe	(.20)(.60) = .12	.20 − .12 = .08	.20
Total	.71	.29	1.00

The other probabilities in this column are similarly computed. The marginal probability distribution for painters (in the total column of the table) is copied from the given probabilities. The second column of probabilities may be obtained by subtraction. For example, $P(\text{Ed and not on time}) = .30 - .24 =$

.06. The marginal probability distribution for completion of the job (in the total row of the table) is determined by obtaining the column totals.

The probability we are seeking is $P(\text{Joe}|\text{on time})$, the conditional probability that Joe did the job, given the knowledge (condition) that it was completed on time. We are concerned with the reduced sample space made up only of those events (sample points) which indicate that the job was done on time. Hence, we are interested only in the probabilities in the "on time" column (Table 16.1.1). Since these probabilities total .71, we determine the appropriate probability distribution for the reduced sample space by dividing the probabilities in the "on time" column by .71. The result is shown in Table 16.1.2. We can see the required probability from this distribution: $P(\text{Joe}|\text{on time}) = .17$. Consequently, the chances are 17 in 100 that the painting job was done by Joe. We may also determine from this distribution that $P(\text{Ed}|\text{on time}) = .34$ and $P(\text{Tim}|\text{on time}) = .49$. This indicates that it is most likely that Tim did the job.

TABLE 16.1.2

Painter	Job completed on time
Ed	.24/.71 = .34
Tim	.35/.71 = .49
Joe	.12/.71 = .17
Total	1.00

Let us examine the computation procedure we followed and set up an equation for computing *posterior probabilities*. We obtained the first column of probabilities in Table 16.1.1 as follows (Rule C):

$$P(\text{Ed and on time}) = P(\text{Ed}) \cdot P(\text{on time}|\text{Ed})$$
$$P(\text{Tim and on time}) = P(\text{Tim}) \cdot P(\text{on time}|\text{Tim})$$
$$P(\text{Joe and on time}) = P(\text{Joe}) \cdot P(\text{on time}|\text{Joe})$$

Then, we obtained the probabilities for the reduced sample space (Table 16.1.2) by dividing each of the above three probabilities by their sum (.71). The probability relating to Joe was computed as follows:

$$P(\text{Joe}|\text{on time}) = \frac{P(\text{Joe}) \cdot P(\text{on time}|\text{Joe})}{\sum (\text{the three probabilities})}$$

Using this result, we may express the computation of posterior probabilities in a general way, as follows: Let A_1, A_2, etc., represent the possible mutually exclusive outcomes in the first set of events and let B denote the outcome which occurred in the second set. Then, the probability that a specified possible outcome from the first set also occurred, for instance A_i, may be computed as

$$P(A_i|B) = \frac{P(A_i) \cdot P(B|A_i)}{\sum_{A_i} P(A_i) \cdot P(B|A_i)} \tag{16.1.1}$$

Equation 16.1.1 is known as *Bayes' Rule* for computing *posterior probabilities.* There has been considerable discussion pro and con concerning the use of Bayes' Rule. There is no argument with the mathematics underlying this rule; however, the probabilities used in the computations are usually unknown and must be hypothesized. Often, subjective probabilities are used. "Traditional" statisticians are reluctant to use such probabilities. "Bayesian" statisticians, on the other hand, have no such reluctance.

Let us apply Bayes' Rule to the following problem. In a city, it has been estimated that 40 percent of the residents are in the low-income class, 35 percent are in the middle-income class, and 25 percent are in the high-income class. Crime statistics show that 8 percent of the low-income class commit crimes, 2 percent of the middle-income class, and 4 percent of the high-income class. The local press reports that a crime was committed by a city resident. What is the probability that he belongs to the middle-income class? The required probability is P(middle-income|crime). We know the following probabilities:

1. Probability that a resident belongs to a given income class:

$$P(\text{low-income}) \quad = .40$$
$$P(\text{middle-income}) = .35$$
$$P(\text{high-income}) \quad = .25$$

2. Conditional probability that a resident will commit a crime:

$$P(\text{crime}|\text{low-income}) \quad = .08$$
$$P(\text{crime}|\text{middle-income}) = .02$$
$$P(\text{crime}|\text{high-income}) \quad = .04$$

Finally, applying Bayes' Rule,

$$P(\text{middle-income}|\text{crime}) = \frac{(.35)(.02)}{(.40)(.08) + (.35)(.02) + (.25)(.04)} = .14$$

16.2 Revising Probabilities

An important application of Bayes' Rule is the revision of probabilities based on new information. Suppose a buyer is considering the purchase of a large supply of a certain mechanical part. After inspecting the parts and determining that a portion of the supply is not usable, he constructs a probability distribution for the proportion of usable parts p as presented in the $P(p_i)$ column in Table 16.2.1. The probabilities in this distribution are subjective probabilities (Section 6.2), since they are based on the buyer's beliefs after his inspection of the parts. They are called *prior probabilities* because they are not based on specific information (such as sample data). According to the probability distribution, the probability that 20 percent of the supply is usable $P(p_i = .20)$ is .10, etc.

The buyer discussed the purchase of this supply with the production manager of his firm and it was agreed that purchase should be considered only if there is a high probability that at least 60 percent of the parts are usable. The buyer

TABLE 16.2.1

Proportion of usable parts in a large supply p_i: prior probability distribution and posterior probabilities.

p_i	Prior probability distribution $P(p_i)$	$P\left(\dfrac{X}{n} = .40 \middle\| p_i\right)$	$P(p_i) \cdot P\left(\dfrac{X}{n} = .40 \middle\| p_i\right)$	Posterior probability distribution $P\left(p_i \middle\| \dfrac{X}{n} = .40\right)$
.20	.10	.2048	.0205	.09
.40	.30	.3456	.1036	.45
.60	.40	.2304	.0922	.41
.80	.20	.0512	.0102	.05
Total	1.00		.2265	1.00

decides to select a sample of the parts and to use the sample results to improve his prior probabilities. He selects a sample of $n = 5$ parts and finds that $X = 2$ are usable. How can he use this new information to revise his probabilities?

Let us apply Equation 8.5.2 to determine the probability that $X = 2$ usable parts would be selected in a random sample of $n = 5$, if $p_i = .20$, as follows:

$$P(X = 2) = \binom{5}{2}(.20)^2(.80)^3 = .2048$$

Then, we may also write $P(X/n = 2/5 = .40) = .2048$. This is the *conditional probability* $P(X/n = .40 | p_i = .20)$, the probability that the sample proportion X/n will equal .40, given the condition that the population proportion p is equal to .20. Using the same approach, we compute the conditional probability that $X/n = .40$, given that p takes on each of the values in Table 16.2.1. These probabilities are shown in the $P(X/n = .40 | p_i)$ column in the table.

The revised probabilities we want are the *reverse conditional probabilities*. More specifically, compute the probability that p has a specified value, given that we obtained 2 usable parts in a sample of 5. For example, find $P(p_i = .20 | X/n = .40)$, $P(p_i = .40 | X/n = .40)$, etc. These are the *posterior probabilities* computed according to Bayes' Rule (Equation 16.1.1). In this equation $A_i = $ a specified p value and $B = $ the new information that $X/n = .40$. In terms of our illustration, we may rewrite the Bayesian probability equation as

$$P\left(p_i \middle\| \frac{X}{n} = .40\right) = \frac{P(p_i) \cdot P\left(\dfrac{X}{n} = .40 \middle\| p_i\right)}{\sum P(p_i) \cdot P\left(\dfrac{X}{n} = .40 \middle\| p_i\right)} \qquad (16.2.1)$$

Compute the *posterior probability* $P(p_i = .20 | X/n = .40)$ based on the data in Table 16.2.1, as follows: the prior probability $P(p_i = .20) = .10$, the conditional probability $P(X/n = .40 | p_i = .20) = .2048$, and the sum of the products of prior and conditional probabilities $= .2265$. Applying Equation 16.2.1, we obtain

$$P\left(p_i = .20 \middle\| \frac{X}{n} = .40\right) = \frac{(.10)(.2048)}{.2265} = .09$$

The prior probability $P(p = .20) = .10$ (the probability that 20 percent of the parts are usable) is revised to .09 based on the sample information.

The posterior probabilities (shown in the last column of the table) represent revisions of the prior probabilities. Based on the prior probabilities, the probability that at least 60 percent of the parts are usable is $.40 + .20$ or $.60$. On the other hand, the revised probabilities indicate that the probability is $.41 + .05$ or $.46$ that at least 60 percent are usable. Clearly, the sample data resulted in reducing this probability and in making it unlikely that the buyer will purchase the supply of parts.

16.3   Decision Making under Conditions of Risk

A decision situation involving alternative decisions and a set of possible outcomes identified with stated probabilities is referred to as *decision making under conditions of risk*.

A manufacturer of a certain consumer item offers a retailer a plan for buying a supply of such merchandise at $50 a unit, with the option of returning unsold units for a refund of $20 apiece if the merchandise is returned no later than the end of the fiscal year (in three months). The retailer knows from past experience that he can safely count on selling two units during the three-month period and is considering whether he should invest in a stock of up to eight units. The retailer-manufacturer agreement specifies a retail price of $90 per unit. Also, the retailer is informed that this item will become obsolete and virtually unsalable after the end of the fiscal year since a new and improved model will be put on the market. How can the retailer decide whether to invest in a stock of 3, 4, 5, 6, 7, or 8 units?

The profit for each unit sold is $90 − $50 = $40 and the loss if returned to the manufacturer is $50 − $20 = $30. Then, if 4 units are stocked and only 3 sold, the profit is $(3)(40) − 30 = 90. If 6 are stocked and 3 sold, the profit is $(3)(40) − (3)(30) = 30. If 8 are stocked and 3 sold, the profit is $(3)(40) − (5)(30) = −$30$ (a negative profit or a loss). The profit, positive or negative, is called the *payoff*. The payoffs for 3 to 8 units stocked and 3 to 8 units sold are presented in a *payoff matrix* in Table 16.3.1. Notice that the first column is headed "number wanted," not "number sold," since customers may wish to buy more than the store has to sell. Table 16.3.2 presents the probabilities that the possible outcomes (number of units wanted) will materialize based on the experience and judgment of the retailer. For example, the probability is .10 that 4 units will be wanted before the end of the fiscal year, etc.

Generally, decision making under conditions of risk is based on the decision rule: *select the decision alternative for which the expected payoff is a maximum*. The *expected payoff* for a decision alternative is the *weighted mean payoff* (Equation 4.2.4), *with the probabilities for the payoffs used as weights*. Clearly, the probability distribution for number of items wanted (Table 16.3.2) is also the probability distribution for the payoffs for a decision alternative. For example, if the probability that 3 units will be wanted is .05, then the probability

TABLE 16.3.1

Payoff matrix (dollars) for a retailer's stock (cost per item $50, selling price $90, refund if returned to the manufacturer $20).

Possible outcomes (number wanted)	Decision alternatives (number of units in stock)					
	3	4	5	6	7	8
3	$120	$ 90	$ 60	$ 30	$ 0	−$ 30
4	120	160	130	100	70	40
5	120	160	200	170	140	110
6	120	160	200	240	210	180
7	120	160	200	240	280	250
8	120	160	200	240	280	320

TABLE 16.3.2

Number of units wanted before the end of the fiscal year: probability distribution.

Number wanted	Probability
3	.05
4	.10
5	.40
6	.30
7	.10
8	.05
Total	1.00

is also .05 that the payoff is $120 if 3 units are stocked and 3 sold (Table 16.3.1) or that the payoff is − $30 if 8 are stocked and only 3 sold, etc.

Let X_i denote a payoff and $P(X_i)$ the associated probability. Applying Equation 4.2.4 and noting that $\sum P(X_i) = 1$ for a decision alternative, we may write the equation for the expected payoff $E(X)$ as

$$E(X) = \frac{\sum X_i P(X_i)}{\sum P(X_i)} = \sum X_i P(X_i) \tag{16.3.1}$$

If the decision alternative is "stock 5 units," the expected payoff is

$$E(X) = (60)(.05) + (130)(.10) + (200)(.40) + (200)(.30)$$
$$+ (200)(.10) + (200)(.05) = \$186$$

The expected payoff $E(X) = \$186$ indicates that if the same decision problem occurred again and again (theoretically) and if 5 units were stocked each time, on the average the payoff would amount to $186. Notice that the expected payoff is a mean and, like any mean, it does not necessarily equal any of the individual payoffs.

The expected payoff for the decision alternative "stock 8 units" is computed as

$$E(X) = (-30)(.05) + (40)(.10) + (110)(.40) + (180)(.30)$$
$$+ (250)(.10) + (320)(.05) = \$141.50$$

In the same way, we compute $E(X) = \$120$ if 3 units are stocked; $E(X) = \$156.50$ if 4 units are stocked; $E(X) = \$187.50$ if 6 units are stocked; and $E(X) = \$168$ if 7 units are stocked. Then, according to the decision rule, the decision should be "stock 6 units," since this leads to the maximum expected payoff.

Suppose, however, the retailer decides to stock only 4 units and evaluate the situation after a month. He will then decide whether to add to his stock. Based on his experience and judgment, he determines the probabilities that at least 2 units will be sold during the month, given each of the six possible outcomes (O_i) in Table 16.3.1. These probabilities, presented in Table 16.3.3, are conditional probabilities since they were determined on the basis that a specified outcome will materialize. For example, as shown in the table, the probability that at least 2 units will be sold during the month (event B) given that 3 will be wanted before the end of fiscal year (three months away) $P(B|O_i = 3)$ is estimated by retailer to be .05. Also, we may note from the table that $P(B|O_i = 5) = .20$, etc.

TABLE 16.3.3

Probability that at least 2 units will be sold during the month (event B), given a specified outcome (O$_i$), prior and posterior probabilities.

Number wanted (O_i)	$P(B\|O_i)$	Prior probability $P(O_i)$	Product	Posterior probability $P(O_i\|B)$
3	.05	.05	.0025	.01
4	.10	.10	.0100	.03
5	.20	.40	.0800	.28
6	.30	.30	.0900	.31
7	.60	.10	.0600	.21
8	.90	.05	.0450	.16
		Total 1.00	.2875	1.00

Suppose that 2 units were sold by the end of the month. We may use this information and the conditional probabilities from Table 16.3.3 to revise the prior probabilities presented in Table 16.3.2. Applying the procedure explained in the previous section, compute the posterior probabilities according to Bayes' Rule (Equation 16.1.1). We can rewrite Equation 16.1.1 as follows:

$$P(O_i|B) = \frac{P(O_i) \cdot P(B|O_i)}{\sum P(O_i) \cdot P(B|O_i)} \tag{16.3.2}$$

Compute the posterior probability $P(O_i = 3|B)$ based on the data in Table 16.3.3, as follows: the prior probability $P(O_i = 3) = .05$, the conditional probability $P(B|O_i = 3) = .05$, and the sum of the products of prior and conditional probabilities $= .2875$. Applying Equation 16.3.2, we obtain

$$P(O_i = 3|B) = \frac{(.05)(.05)}{.2875} = .01$$

Then, the prior probability $P(O_i = 3) = .05$ (the probability that only 3 units will be wanted within the three-month period before the fiscal year ends) is revised to .01 based on the new information. In other words, since 2 units have been sold in a month, the probability is .01 that only one more unit will be sold in the remaining two months. The other revised probabilities shown in the last column of Table 16.3.3 were computed in the same way and are similarly interpreted.

Now, applying the revised probabilities to the payoff matrix (Table 16.3.1), revised expected payoffs are computed for the decision alternatives according to Equation 16.3.1. The computations are facilitated by collecting the probability and payoff data into one table, as shown in Table 16.3.4. If 6 units are stocked, the expected payoff is

$$E(X) = (30)(.01) + (100)(.03) + (170)(.28) + (240)(.31)$$
$$+ (240)(.21) + (240)(.16) = \$214.10$$

TABLE 16.3.4

Probability and payoff data for computation of revised expected payoffs.

Possible outcomes O_i	Revised probabilities	Decision alternatives (number of units in stock)					
		3	4	5	6	7	8
3	.01	$120	$ 90	$ 60	$ 30	$ 0	$-$ 30
4	.03	120	160	130	100	70	$+$ 40
5	.28	120	160	200	170	140	110
6	.31	120	160	200	240	210	180
7	.21	120	160	200	240	280	250
8	.16	120	160	200	240	280	320

In the same way, expected payoffs are computed for the other decision alternatives. The expected payoff is $120 for 3 units stocked, $159.30 for 4 units stocked, $196.50 for 5 units stocked, $210 for 7 units stocked, and $191.20 for 8 units stocked. Clearly, using the revised probabilities, $E(X)$ is still a maximum if 6 units are stocked. Hence, the decision alternative to adopt is "stock 6 units."

In some decision situations, maximizing the expected payoff is not the best decision rule. For example, if it is necessary to keep losses as small as possible, the decision alternative with the minimum loss may be the best decision to make. Or, it may be desired to act with extreme optimism and choose the alternative which offers the maximum payoff (not the maximum *expected* payoff).

16.4   Decision Strategies

Decision making under conditions of risk, discussed in the previous section, did not involve competition among opponents; however, in such decision problems, we may look upon the situation as involving a conflict between an *opponent* and *nature*. For example, in Section 16.3 we considered the problem of a retailer who has to decide how many units to stock and with the possible six outcomes 3, 4, 5, 6, 7, or 8 units wanted by his customers. This may be looked upon as competition between the firm and nature or chance. Decision problems involving decision making under conditions of conflict, usually where intelligent (as opposed to chance) opponents are in competition with each other, are treated in a specialized field of mathematics called the *theory of games*. We will consider the simplest type of such decision-making problems.

Suppose there are only two restaurants in a town, and once a year each restaurant reviews its business operations and policies. This year restaurant I is considering whether to offer a free second cup of coffee or to offer a weekly meal ticket at a discount price. It has been definitely decided that one of these actions will be taken. Restaurant II is considering three decision strategies (alternatives): to continue its present policy of not offering a free cup of coffee or a meal ticket; to offer free a second cup of coffee; or to offer a weekly meal ticket at a discount. Let us assume that the competition between the firms is for each to try to increase its share of the market and that increasing (or decreasing) the size of the market is not a factor in this decision problem. Consequently, one firm's loss is the other firm's gain and the gains and losses for the two firms just balance out (add to zero).

The foregoing illustration is an example of a *zero-sum two-person game*. "Zero-sum" refers to the balancing out of gains and losses and "two-person" refers to the two opponents in the competitive struggle. In zero-sum two-person games each opponent has perfect information concerning the decision strategies under consideration by (available to) the competitor and the payoffs; however, each has to choose his strategy without knowing which choice was made by the other side. Table 16.4.1 presents the payoff matrix appropriate to our decision problem, where each entry in the table represents the payoff (in thousands of dollars) to restaurant I. A positive entry indicates a gain (profit) for restaurant I (and a loss for restaurant II); whereas, a negative entry indicates a loss for restaurant I (and a gain for restaurant II).

How should the opponents select their strategies? Consider restaurant I. If the free coffee strategy is selected, then the minimum gain is $0 (if restaurant II also adopts this strategy). If restaurant I chooses the meal ticket strategy, then the minimum gain is $-$2,000, a loss (if restaurant II adopts the free coffee strategy). It would certainly make sense for restaurant I to adopt the decision rule *"choose the strategy which maximizes the minimum gain,"* since there is no way to know beforehand what restaurant II will do. This is known as the *maximin criterion*. Based on this decision rule, restaurant I would select the free coffee strategy. Since this is a zero-sum decision problem, restaurant II will look

TABLE 16.4.1

*Zero-sum two-person game: payoff matrix for restaurants
I and II (in thousands of dollars payoff to restaurant I).*

| | Restaurant II strategies | | | |
Restaurant I strategies	No change	Free coffee	Meal ticket	Row minima
Free coffee	4	0	2	0 ←*maximin*
Meal ticket	2	−2	0	−2
Column maxima	4	0	2	
		↑ *minimax*		

upon the payoff matrix in Table 16.4.1 *with the signs reversed.* Hence, the minimum gain for restaurant II is − $4,000 for the no change strategy, $0 for the free coffee strategy, and − $2,000 for the meal ticket strategy. Consequently, to maximize the minimum gain, restaurant II should adopt the free coffee policy, its *maximin strategy.*

Notice that Table 16.4.1 presents a "row minima" column and a "column maxima" row. This is to assist in determining the maximin strategy for each opponent (often called each *player*). An arrow identifies the maximum of the row minima, which is the payoff of the maximin strategy for restaurant I. The payoff of the maximin strategy for restaurant II is $0; however, notice that this payoff is identified as the *minimax strategy* in Table 16.4.1, bottom row. The reason for reversing *maximin* to *minimax* is because the payoffs in Table 16.4.1 are in terms of gains (positive and negative) to restaurant I. Consequently, the minimum of the maxima for restaurant II strategies in this payoff matrix (called the *minimax*) amounts to the same thing as the *maximin* for the payoff matrix *with signs reversed.*

Notice that in the payoff matrix (Table 16.4.1), the maximum of the row minima and the minimum of the column maxima are equal, since they are both $0. In such case, the payoff $0 is called the *saddle point* of the payoff matrix. When the saddle point is 0, the game (competition) is *equitable.* If it is positive, the game favors restaurant I; if negative, the game favors restaurant II.

The players (restaurants I and II) may evaluate the payoff matrix in another way. Notice that no matter which strategy restaurant II selects, restaurant I is better off with the free coffee strategy than with the meal ticket strategy. In such a situation, for restaurant I the free coffee strategy is said to *dominate* the meal ticket strategy. Hence, restaurant I can do no better than to adopt the dominant strategy. Since each player has full knowledge of the payoff matrix, restaurant II knows this and can do no better than to minimize the loss and adopt the free coffee strategy. The selected strategies then are no different than those selected by the maximin criterion.

When a saddle point exists in a payoff matrix, the decision problem is said to be strictly determined, since neither player will want to depart from strategy moves indicated by the maximin criterion. Any other strategy moves will result in a loss to one or the other of the players compared to the payoffs from adoption of the maximin strategies.

Not every payoff matrix has a saddle point. For example, inspect the payoff matrix in Table 16.4.2 (payoffs in dollars). The maximin criterion leads to a payoff of $-\$1$ (strategy b') for player A and $\$2$ (strategy b) for player B. Since these payoffs are not equal, there is no saddle point and the game is not strictly determined. How should the players go about choosing strategies?

TABLE 16.4.2

Zero-sum two-person game: payoff matrix with no saddle point (payoffs to player A, in dollars).

Player A	Player B		Row
	a	b	minima
a'	-4	2	-4
b'	3	-1	-1 ←*maximin*
Column maxima	3	2	
		↑	
		minimax	

In discussing a decision rule for players A and B, it should be borne in mind that plays occur repeatedly at intervals, so that each player has the opportunity to observe the outcome of each play and change his strategy in the next play, if this is indicated. If A and B play their maximin strategies (as indicated in Table 16.4.2), player A will lose $\$1$; whereas, player B will find that he gains $\$1$ instead of losing $\$2$. However, if player A observes that player B followed strategy b and believes that he will continue to do so, he will be motivated to adopt strategy a' in order to gain $\$2$ instead of losing $\$1$; however, player B will certainly react and change his strategy.

This jockeying or *mixed strategy* will continue with changes in gains and losses. The important consideration in such jockeying is to keep the opponent guessing. A decision rule usually followed when there is no saddle point is to *randomize* the mixed strategy. When a *random strategy* is employed, the objective is to maximize the minimum gain by stabilizing the expected payoff, no matter which strategy is selected by the opponent. How can this be accomplished?

Let $p_1 = $ probability that player A will select strategy a' and $1 - p_1 = $ probability that he will select strategy b'. Computing the expected payoff $E(X)$ as in the previous section (Equation 16.3.1), if player B chooses strategy a, the expected payoff $E(X_1)$ for player A is

$$E(X_1) = -4p + 3(1 - p)$$
$$= -7p + 3$$

If player B chooses strategy b, the expected payoff $E(X_2)$ for player A is

$$E(X_2) = 2p - 1(1 - p)$$
$$= 3p - 1$$

Then, to stabilize the expected payoffs for player A, we equate the $E(X_i)$ so that $E(X_1) = E(X_2)$. Substituting for $E(X_i)$ according to the previous equations, we have

$$-7p + 3 = 3p - 1$$

and $p = 2/5$. Then, $1 - p = 3/5$.

Suppose player A puts 5 chips in a bowl, 2 chips marked a' (since $p = 2/5$) and 3 chips marked b' (since $1 - p = 3/5$). He may select strategy a' or b' by selecting a chip at random from the bowl. If he follows this procedure for each play, he can expect to obtain a stable expected payoff in the long run, no matter which strategy is selected by his opponent. If player B chooses strategy a, then the expected payoff to player A is

$$E(X_1) = -7\left(\frac{2}{5}\right) + 3 = -\$.20$$

If player B adopts strategy b, then the expected payoff to player A is

$$E(X_2) = 3\left(\frac{2}{5}\right) - 1 = -\$.20$$

In the long run player A may expect to lose $.20 on the average compared with a possible loss of $4 if he chooses strategy a' or a possible loss of $1 if he chooses strategy b'.

Study Problems

1. Distinguish between prior and posterior probabilities.

2. Why are posterior probabilities called backward or inverse probabilities?

3. The probability is .80 that a student will make an effort to pass a certain course, .60 that he will pass the course if an effort is made, and .10 that he will pass the course if no effort is made. If it turns out that he passed the course, what is the probability that he made an effort to pass it?

4. In Problem 3, if it turns out that he did not pass the course, what is the probability that no effort was made?

5. The probability that a man will purchase a new house is .70 if he can obtain a low-interest loan and .20 if he cannot. The probability that he can obtain a low-interest loan is .30. If you find out that he did not purchase a new house, what is the probability that he was able to obtain a low-interest loan?

6. A soft drink bottling firm buys 10 percent of its bottle closures from company A, 30 percent from company B, 20 percent from company C, and 40 percent from company D. The closures are found not to fit well 3 percent of the time for company A, 1 percent for company B, 2 percent for company C, and 1 percent for company D. If the next closure tried does not fit well, what is the probability it comes from company C?

7.   In Problem 3, determine the probability distribution for (a) the sample space pertinent to this problem and (b) the pertinent reduced sample space.

8.   In the situation described in Problem 3, what are the odds that (a) the student will pass the course and (b) the student will make an effort to pass the course?

9.   In Problem 5, determine the probability distribution relating to (a) the appropriate sample space and (b) the appropriate reduced sample space.

10.   In Problem 5, if you find out that the man did purchase a new house, what is the probability that he got a low-interest loan?

11.   Smith's campaign committee estimated the following probabilities that he would obtain the specified proportions of total votes cast in the upcoming election:

Proportion of total votes (p_i)	$P(p_i)$
.30	.05
.40	.10
.50	.35
.60	.25
.70	.15
.80	.10

A random sample of five registered voters indicated that three would vote for Smith.
      a.   Compute revised probabilities based on the sample information.
      b.   Based on the revised probabilities, what is the probability that Smith would obtain at least 60 percent of the total votes cast?

12.   A college entrance study team estimated the probabilities that various proportions of high school graduates with high grades will be admitted to an ivy league university as follows:

Proportion admitted (p_i)	$P(p_i)$
.40	.05
.50	.10
.60	.25
.70	.35
.80	.20
.90	.05

A sample of four high school graduates with high grades showed that three were admitted to an ivy league school.
      a.   Compute revised probabilities based on the sample data.
      b.   What is the probability that at most half of the high school graduates with high grades will be admitted to an ivy league school, based on the revised probabilities?

13.   A crime commission in an urban area estimated the following probabilities that certain proportions of serious crimes were committed by members of high-income families:

Proportion by high-income family members (p_i)	$P(p_i)$
.05	.15
.08	.20
.10	.20
.15	.25
.20	.10
.25	.05
.30	.05

A sample of six serious crimes showed that four were committed by members of low- or medium-income families and two by members of high-income families.

    a.   Use the new information and revise the prior probabilities presented by the crime commission.

    b.   According to the posterior probabilities, what is the probability that high-income family members commit 10 percent to 15 percent of the serious crimes?

14.   A retail firm has to decide whether to take on 5, 6, 7, or 8 salesclerks for Saturday. The policy of the firm is to employ one salesclerk for each $200 in anticipated sales. Each salesclerk is paid $20 for the day. (Ignore other factors which are involved in profit determination.) Based on past experience, the firm estimates that the probability is .20 that 5 salesclerks will be needed, .50 that 6 will be needed, .20 that 7 will be needed, and .10 that 8 will be needed.

    a.   Construct the payoff matrix for the 4 possible outcomes (5, 6, 7, or 8 salesclerks needed) and the 4 decision alternatives.

    b.   Using the estimated probability distribution for the possible outcomes, compute the expected payoff for each decision alternative.

    c.   Using the maximum expected payoff decision rule, which decision alternative should the firm select?

15.   In Problem 14, suppose the decision is delayed until Thursday night. The firm estimates based on past experience: the probability is .10 that 3 or more salesclerks will be needed on Thursday if 5 will be needed on Saturday; the probability is .30 that 3 or more will be needed on Thursday if 6 will be needed on Saturday; the probability is .60 that 3 or more will be needed on Thursday if 7 will be needed on Saturday; and the probability is .90 that 3 or more will be needed on Thursday if 8 will be needed on Saturday. It was found that 4 salesclerks were needed on Thursday.

    a.   Revise the prior probabilities relating to the possible outcomes as noted in Problem 14, using the new information.

    b.   Compute the revised expected payoffs for the decision alternatives.

    c.   Applying the maximum expected payoff decision rule, which decision alternative should the firm accept?

16.   A computer school is being organized with the intention of qualifying for government subsidy for a large part of tuition payments. The amount of government funds received depends on the enrollment plan of the school. The school will receive $5,000 for general overhead expenses and $200 per student included in the planned scope of operations. Twenty-five dollars will be deducted for each student enrolling after 35 have already enrolled and $50 per student will have to be returned to the government for the difference between actual enrollment and planned enrollment. Each student

enrolled will have to pay only $100 tuition for the training course. (In determining the school's profit for a given number of students enrolled, ignore other factors which may be involved.) It is estimated that the probability is .10 that 30 students will enroll, .30 that 40 will enroll, .40 that 50 will enroll, and .20 that 60 will enroll.

    a. Construct the payoff matrix for the possible outcomes (30, 40, 50, or 60 students enrolled) and the decision alternatives (30, 40, 50, or 60 student enrollments planned).

    b. Using the maximum expected payoff decision rule, which decision alternative should the school select?

17. In Problem 16, the computer school decided to plan for only 30 students, at first. Then, if at least 15 students enroll after a one-week registration period, a decision will be made whether to increase planned enrollment. It is estimated that the probability is .20 that at least 15 students will enroll during the first week if 30 will enroll for the semester; the probability is .40 that at least 15 will enroll during the first week if 40 will enroll for the semester; the probability is .70 that at least 15 will enroll during the first week if 50 will enroll for the semester; and the probability is .90 that at least 15 will enroll during the first week if 60 will enroll for the semester. It was determined that 20 students enrolled during the first week.

    a. Revise the prior probabilities relating to the possible outcomes presented in Problem 16, using the new information.

    b. Using the maximum expected payoff decision rule, which alternative should be adopted?

18. What is meant by decision making under conditions of risk?

19. What is meant by decision making under conditions of conflict?

20. What is a zero-sum two-person game?

21. What is the element of uncertainty in a zero-sum two-person game?

22. What is meant by the maximin strategy in a zero-sum two-person game?

23. Why does the minimax criterion for a player actually amount to the maximin strategy?

24. What is meant by saddle point?

25. What is the payoff at the saddle point for an equitable game?

26. What is meant by dominant strategy?

27. When is the decision problem strictly determined in a zero-sum two-person game?

28. What strategy is usually used when there is no saddle point?

29. There are two shopping centers (one mile apart) serving a city. Every half-year the shopping center managements review their operations and change policies and practices as appears to be required. The following payoff matrix (in thousands of dollars payoff to shopping center A) shows the decision problems facing the managements:

Shopping center A: Offer	Shopping center B: Offer trading stamps	free bus to shopping center
purchase discount	−50	100
trading stamps	0	200
free bus to shopping center	−200	0

a. Apply the maximin criterion and determine the decision moves to be made by the shopping centers.
b. Does this payoff matrix have a saddle point? If so, identify it.
c. Is this an equitable game? Why?
d. Is there a dominant strategy in this competitive struggle? Why?
e. Is this game strictly determined? Why?

30.　The following payoff matrix has been constructed for a zero-sum two-person game (in terms of payoff in dollars to player K):

Player K strategies	Player H strategies	
	D	E
A	-2	4
B	1	-6

a. Is there a saddle point? Defend your answer.
b. What strategy (be specific) should player K adopt in this game?

Part VI
Association and Prediction

Chapter 17

Regression and Correlation

17.1 Basic Concepts

An important application of statistical methods is measurement of the association or relationship between two variables. For example, is there a relationship between the amount spent by a housewife in a shopping center and the distance she must travel to get to the shopping center? How closely associated are a student's high school average and his average during the first year at college?

Table 17.1.1 presents crime rates and divorce rates for ten cities. Suppose a sociologist wishes to determine whether there is an association between crime rate and divorce rate based on these data. A *scatter diagram* (or a *dot chart*) is an effective device for displaying the relationship between two variables. Figure 17.1.1 presents the scatter diagram constructed for the data in Table 17.1.1, where X = divorce rate and Y = crime rate. This figure contains ten dots, one for each pair of X and Y observations. For city A, $X = 3.5$ and $Y = 1.8$.

Locating $X = 3.5$ on the horizontal scale, the dot for this city is placed directly above at a height corresponding to $Y = 1.8$. The other dots are plotted in the same way. The dots for cities A and B are identified in the figure.

TABLE 17.1.1

Crime rates and divorce rates per thousand population in ten cities.

City	Divorce rate X	Crime rate Y
A	3.5	1.8
B	12.1	2.4
C	16.1	3.1
D	23.7	4.8
E	24.5	6.9
F	6.0	3.0
G	18.6	4.2
H	19.5	6.2
I	15.0	4.9
J	9.1	3.5

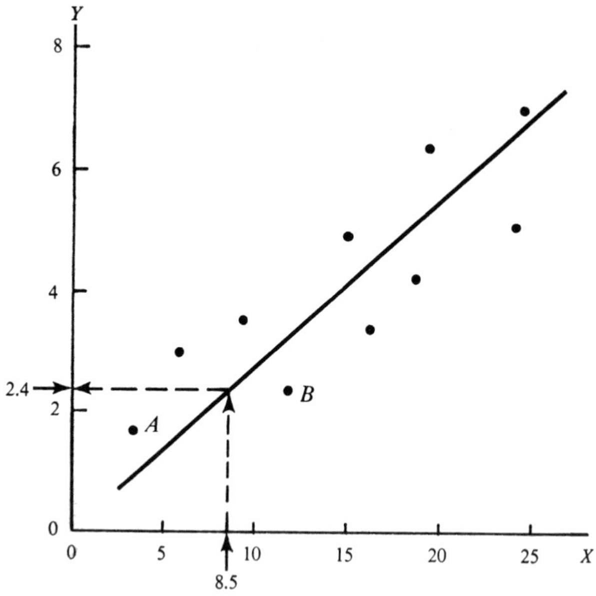

FIGURE 17.1.1

Scatter diagram and freehand regression line for divorce rate (X) and crime rate (Y) for ten cities.

Notice that the scatter of dots in Figure 17.1.1 suggests a straight line. Such a line is drawn in the figure on a freehand (judgment) basis. This line, called a *regression line*, represents the *relationship (association) between two variables*. When studying the relationship between two variables, the chief interest may be to determine whether one variable may be used to predict the other. The variable you want to predict is denoted as Y and is called the *dependent variable*. The other variable is denoted as X and is the *independent variable*. In our illustration, the sociologist wishes to predict the crime rate (Y) based on the divorce rate (X). The regression line may be used for this purpose. For example, if the divorce rate for a city is 8.5, based on the freehand regression line in Figure 17.1.1, we would predict (or estimate) the crime rate as 2.4. How this is determined is shown in the figure. Of course, a freehand regression line is subjectively determined. A more precise method for determining a regression line will be presented in the next section.

Observe that in Figure 17.1.1 low values of X are associated with low values of Y and high values of X are associated with high values of Y. When two variables are so related, they are said to be *directly* or *positively related* and the regression line slopes *upward* (from the lower left corner to the upper right corner). Such a regression line is said to be *positively sloped*. Figure 17.1.2 presents a regression line for X and Y which is *negatively sloped*. Notice that in this figure low (high) values of X are associated with high (low) values of Y, so that the X and Y variables are *inversely* or *negatively associated*.

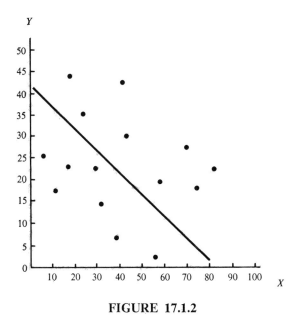

FIGURE 17.1.2

In some problems the chief interest in studying the relationship between two variables may be to determine how closely the variables are associated. In such analyses, the important measure to compute is the *coefficient of correlation*, which is a measure of the *extent* or *degree of association* between two variables. This measure will be discussed in Section 17.3. Notice that the dots in Figure

17.1.1 are more closely clustered around the regression line than in Figure 17.1.2, indicating that X and Y are more closely related or associated. In the extreme case, where all dots fall on the regression line, the relationship between X and Y is *perfect*. This is illustrated in Figure 17.1.3, part A. Clearly, a straight line could be drawn to pass through all the dots in this scatter diagram. On the other hand, part B of this figure illustrates a set of dots where X and Y are not at all related, since the pattern of dots does not suggest any regression line. The relationship between X and Y need not be expressed by a *linear* (*straight line*) regression line. For example, the dots in part C of Figure 17.1.3 suggest a *nonlinear* or *curvilinear regression line*. Such regression relationships are discussed in Chapter 18.

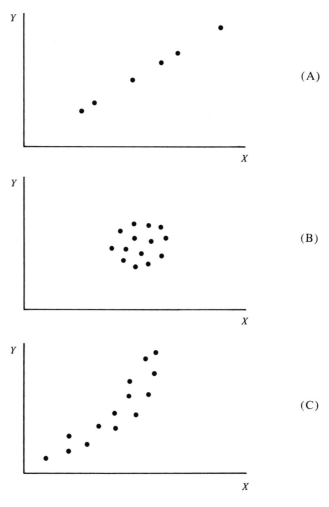

FIGURE 17.1.3

Scatter diagrams illustrating various types of relationship between X and Y.

17.2 Regression

We need an objective method for determining a "best-fitting" regression line to remove the personal judgments involved in a freehand determination. Figure 17.2.1 presents a scatter diagram for X and Y and a freehand regression line. Let Y denote the *observed value* and Y_e the corresponding *predicted* (or *estimated*) *value*. The difference between the two or the deviation between the observed and predicted values of Y for a given value of X, denoted as

$$(Y - Y_e) \tag{17.2.1}$$

indicates the error made in using Y_e as a prediction (or estimate) of Y.

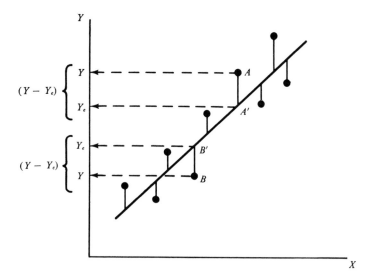

FIGURE 17.2.1

In Figure 17.2.1, the observed Y values for dots A and B are indicated by the dashed-line arrows leading from the dots to the Y-axis. The corresponding predicted Y_e values are indicated by the dashed-line arrows leading from the A' and B' points on the regression line to the Y-axis. The corresponding $(Y - Y_e)$ *deviations* are indicated in the figure. Clearly, it is desirable for a "best-fitting" line to minimize these deviations (errors). Notice in the figure that some dots deviate more from the regression line than others. However, if the freehand line is adjusted to reduce the larger deviations, it will result in increasing other deviations. Note that some dots are above the regression line (such as dot A). The deviations for such dots are positive, since Y_e is less than Y. The deviations for other dots are negative (such as for dot B), since Y_e is larger than Y.

The method generally used to obtain a best-fitting regression line is called the *method of least squares*. It is not appropriate to discuss the mathematical development of this method in an introductory text; however, this method produces a regression line which reduces the sum of the squared deviations of observed *Y* values from the regression line to a minimum. That is, if the deviation for each dot is determined according to Equation 7.2.1 and each deviation is squared, then the sum of the squared deviations will be a minimum if the regression line is determined by the method of least squares. We may express this as follows:

$$\Sigma\,(Y - Y_e)^2 \quad \text{is a minimum} \tag{17.2.2}$$

In other words, the best-fitting line of regression has the least squares property. We encountered the least squares property before when we discussed the properties of the mean (Equation 4.5.2). Another property of the least squares regression line is that *the sum of the deviations computed according to Equation 17.2.1 (not squared) is equal to zero (the positive and negative deviations just balance out)*. We may express this property as

$$\Sigma\,(Y - Y_e) = 0 \tag{17.2.3}$$

This property, too, has its counterpart in the case of the mean (Equation 4.5.1).

We may write the *general equation of a regression line* as

$$Y_e = a + bX \tag{17.2.4}$$

where *Y*, the dependent variable or the variable predicted (or estimated), is denoted as Y_e and *X* is the independent variable or the variable on which the prediction is based. This is, of course, the *general equation of a straight line*. The symbols *a* and *b* represent *constants* and are known as *regression coefficients*. When specific values are assigned to *a* and *b* a specific regression line is identified. Consider the specific regression equation

$$Y_e = 4 + 2X \tag{17.2.5a}$$

where *a* = 4 and *b* = 2. Substituting *X* = 0 into this equation, we obtain

$$Y_e = 4 + 2(0) = 4$$

Now, substitute *X* = 10 and compute

$$Y_e = 4 + 2(10) = 24$$

Hence, based on the regression equation, the predicted *Y* value is 4 if it is known that *X* = 0 and it is 24 if *X* = 10. Since two points determine a straight line, we may construct the regression line identified by Equation 17.2.5a as shown in Figure 17.2.2. This line is constructed by plotting the two points (*X* = 0, *Y* = 4) and (*X* = 10, *Y* = 24) and connecting them with a straight line.

Notice that the regression line in Figure 17.2.2 intersects the *Y*-axis at the point $Y_e = 4 = a$. This point is called the *Y-intercept*. Hence, *a determines the height of the regression line*. For example, construct the regression line

$$Y_e = 8 + 2X \tag{17.2.5b}$$

Substituting *X* = 0, we obtain $Y_e = 8$; and substituting *X* = 10, we obtain *Y* = 28. This line is also shown in Figure 17.2.2. Observe that this regression line is parallel to the first line, but at a greater height (larger *a* value).

The regression coefficient *b* represents the *slope of the line*. For example, consider the regression equation

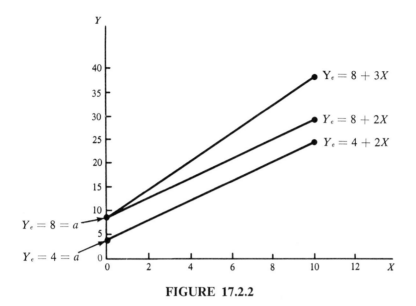

FIGURE 17.2.2

$$Y_e = 8 + 3X \qquad (17.2.5c)$$

Substituting first $X = 0$ and then $X = 10$, we obtain $Y_e = 8$ and 38, respectively. Plotting these points and connecting them with a straight line, we obtain the regression line as shown in Figure 17.2.2. Notice that this line intersects the Y-axis at $Y_e = 8$ as in the case of Equation 17.2.5b because in both equations $a = 8$; however, the slope is steeper, since the b value is higher.

A least squares linear regression equation is obtained for a set of X and Y pairs if a and b are computed as follows:

$$b = \frac{\Sigma\, xy}{\Sigma\, x^2} \qquad (17.2.6)$$

$$a = \frac{\Sigma\, Y - b \Sigma\, X}{n} \qquad (17.2.7)$$

where

$$\Sigma\, xy = \Sigma\, XY - \frac{\Sigma\, X \Sigma\, Y}{n} \qquad (17.2.8)$$

$$\Sigma\, x^2 = \Sigma\, X^2 - \frac{(\Sigma\, X)^2}{n} \qquad (17.2.9)$$

n = number of *pairs* of X and Y values

Notice that in Equation 17.2.6 for b lower case x and y are used; whereas, in the other equations upper case X and Y are used. The lower case letters denote *deviations from the mean* as follows:

$$x_i = X_i - \overline{X} \qquad (17.2.10)$$
$$y_i = Y_i - \overline{Y} \qquad (17.2.11)$$

$\overline{X}$ and $\overline{Y}$, of course, denote the means of the X's and Y's respectively. Although the subscript i is used in the foregoing equations, we will omit it generally as a matter of convenience. The following sequence of computations are suggested for determination of a least squares linear regression equation:

1. Compute $\sum X, \sum Y, \sum X^2, \sum XY$
2. Compute $\sum xy$ (Equation 17.2.8)
3. Compute $\sum x^2$ (Equation 17.2.9)
4. Compute b (Equation 17.2.6)
5. Compute a (Equation 17.2.7)

Finally, the least squares *linear* regression equation is

$$Y_e = a + bX \tag{17.2.12}$$

We will illustrate the computation of a least squares regression equation and it will become apparent that this is a simple procedure. The computational labor, however, is fairly considerable. It is very important to organize the computations carefully in order to reduce the possibility of error. (Of course, where computer facilities are available, it is easy to program the computations to determine a least squares regression equation).

Table 17.2.1 presents data collected by a metallurgist on the number of blemishes found (Y) in an area of a specified size for a metal for ten cooling periods (X) during which the metal was cooled to a specified temperature (after emergence from a production process). Let us compute the least squares linear regression equation to predict the number of blemishes based on cooling period. (Notice that the variable to be predicted is denoted as Y.) The four column totals in the table provide the information needed for step 1 in the suggested sequence of computations. (X^2, of course, denotes the square of each X value and XY, called the *cross-product*, denotes the product of the X and Y values for an X and Y pair.) Then, compute

$$\sum xy = 14{,}436 - \frac{(349)(429)}{10} = -536.1$$

$$\sum x^2 = 13{,}233 - \frac{(349)^2}{10} = 1{,}052.9$$

$$b = \frac{-536.1}{1{,}052.9} = -.5092$$

and $$a = \frac{429 - (-.5092)(349)}{10} = 60.6711$$

Finally, the *least squares linear regression equation* is

$$Y_e = 60.67 - .51X$$

Figure 17.2.3 presents the scatter diagram for the data in Table 17.2.1. Substituting $X = 20$ and $X = 50$ into the regression equation, we obtain $Y_e = 50.47$ and 35.17, respectively. Plotting the two points ($X = 20$, $Y_e = 50.47$) and ($X = 50$, $Y_e = 35.17$) on the scatter diagram, the regression line is constructed as shown in Figure 17.2.3.

TABLE 17.2.1

Number of blemishes found for various cooling periods (minutes) for a metal.

Cooling period X	Number of blemishes Y	X^2	XY
41	37	1,681	1,517
50	37	2,500	1,850
23	50	529	1,150
32	43	1,024	1,376
18	52	324	936
25	46	625	1,150
33	47	1,089	1,551
42	42	1,764	1,764
36	41	1,296	1,476
49	34	2,401	1,666
Total 349	429	13,233	14,436

Sometimes the regression equation constant *a* has a *negative value.* This indicates that the regression line intersects the *Y*-axis *below Y* = 0. The value of the regression coefficient *b* in a regression equation indicates *the increase (or decrease if b is negative) in Y per unit increase in X*. In our example, $b = -.51$ indicates that, based on the data in the analysis (Table 17.2.1), an increase of one minute in the cooling period is associated with a decrease (*b* is negative) of .51, on the average, in the number of blemishes. Or an increase of 10 minutes in the cooling period is associated with a decrease of about 5 in the number of blem-

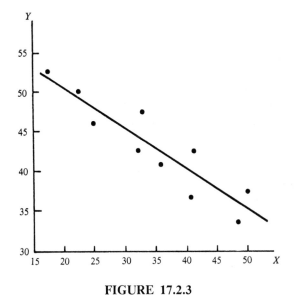

FIGURE 17.2.3

ishes, on the average. In our illustration, note that $\sum xy$ as well as b is negative. Since $\sum xy$ is in the numerator of the equation for b (Equation 17.2.6) and $\sum x^2$ (in the denominator) is always positive, b has the same sign as $\sum xy$.

Referring to Equation 17.2.7 for a, we may write

$$a = \frac{\sum Y}{n} - b \frac{\sum X}{n} \tag{17.2.13}$$

$$= \bar{Y} - b\bar{X}$$

Then, $$\bar{Y} = a + b\bar{X} \tag{17.2.14}$$

This indicates that the least squares regression line goes through the $\bar{X}$, $\bar{Y}$ point.

17.3   Correlation

Statistical analysis of the relationship between two variables X and Y distinguishes two types of problems, depending on how the observations for the independent variable X are obtained. Suppose a chemist wishes to determine the relationship between the reaction time of a chemical process and the temperature in which the chemical process takes place. Suppose he collects his data by conducting the chemical process with certain selected temperatures and observing the reaction time for each temperature. This provides data on X = temperature and Y = reaction time which permits the chemist to compute a least squares regression equation to predict reaction time based on temperature. The temperature data X are predetermined by the chemist and are not random observations; however, the reaction time data Y are random observations, since reaction time may be expected to vary even for a given temperature. In such problems, *where the observations on the independent variable X are predetermined and the observations for the dependent variable Y are random*, we have a *regression analysis*. Generally, the objective of a regression analysis is to determine a regression equation to predict Y based on X.

On the other hand, suppose a doctor wishes to determine the relationship between blood pressure and weight for middle-aged males. Suppose he selects a random sample of males from his population of interest and determines for each male in his sample the blood pressure and weight. In this problem the observations on weight and blood pressure are random. In such problems, *where the observations on both variables are random*, we have a *correlation analysis*. The objective of a correlation analysis may be to determine a regression equation to predict one of the variables (either one) based on the other, or to determine how closely the two variables are associated, or it may include both of these.

Suppose we have n X, Y pairs of random observations, so that we are concerned with a correlation problem. Using Equations 17.2.10 and 17.2.11, express each observation as a deviation from its mean. Then, we have x and y pairs instead of X and Y pairs. Some of these deviations will be positive and some negative. For X observations larger than $\bar{X}$ and Y observations larger than $\bar{Y}$, x and y are positive and for X observations smaller than $\bar{X}$ and Y observations smaller than $\bar{Y}$, x and y are negative. Figure 17.3.1 presents three scatter diagrams in terms of x and y and the four *quadrants* (quarters) of each scatter diagram (I, II, III, IV).

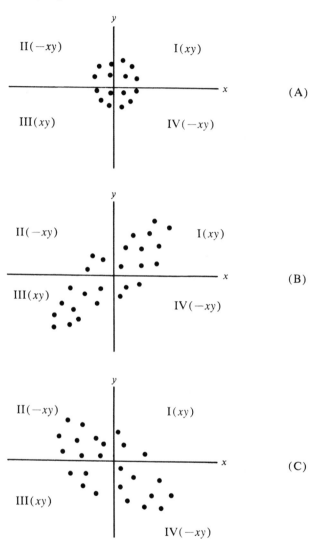

FIGURE 17.3.1

If the cross-product xy is computed for each x, y pair, some will be positive (if x and y have the same sign) and some negative (if x and y have different signs). As indicated in Figure 17.3.1, the cross-products are positive in quadrants I and III and negative in quadrants II and IV. If x and y (or X and Y) are not related (or associated) as in part A of Figure 17.3.1, the sums of the positive and negative xy products just balance out and so the sum of all the xy products ($\sum xy$)

equals zero. On the other hand, if x and y are positively related as in part B of the figure, the sum of the positive xy quantities in quadrants I and III exceed (in absolute value) the sum of the negative xy quantities so that for all the products $\sum xy$ is positive. This reflects the larger number of positive xy products as well as the fact that many of these products are larger in absolute value than the negative products. If x and y are negatively related as in part C of the figure, the opposite is true. Then, the sum of the negative xy products exceed (in absolute value) the sum of the positive products and the overall sum $\sum xy$ is negative.

The sum of the cross-products $\sum xy$ represents a measure of the association between x and y (or X and Y). If $\sum xy = 0$, X and Y are not related. If $\sum xy > 0$, X and Y are positively related and if $\sum xy < 0$, X and Y are negatively related. $\sum xy$ represents the *covariation* or *joint variation* of X and Y and may be used to construct a measure of how closely X and Y are associated. This is accomplished by converting x and y to standard units (Equation 5.4.4) and dividing the summation by $n - 1$ to obtain

$$r = \sum \frac{\frac{x}{s_X} \cdot \frac{y}{s_Y}}{n - 1} = \frac{\sum xy}{(n - 1)s_X s_Y} \qquad (17.3.1)$$

where s_X denotes the standard deviation for X (Equation 5.2.2) and s_Y denotes the standard deviation for Y computed according to the same equation but with Y substituted for X.

The reason for converting to standard units is to remove the effect on $\sum xy$ of a particular unit of measurement. For example, $\sum xy$ will be different if each variable is measured in, for example, feet instead of inches. The reason for dividing by $n - 1$ is to remove the effect on $\sum xy$ of the number of x and y pairs in the analysis. For example, $\sum xy$ would be higher (in absolute value) if n were doubled, except, of course, if there is no relationship between x and y. (Recall that $n - 1$, instead of n, was also used as the divisor for s^2 in Equation 5.2.2.) The statistic r is called the coefficient of correlation and represents a measure of how closely X and Y are associated. Substituting in Equation 17.3.1 for s_X and s_Y according to Equation 5.2.2 (in terms of X and Y, respectively) we obtain

$$r = \frac{\sum xy}{\sqrt{\sum x^2 \sum y^2}} \qquad (17.3.2)$$

This is a more convenient equation for computing r in most problems. $\sum xy$ is computed according to Equation 17.2.8, $\sum x^2$ according to Equation 17.2.9, and $\sum y^2$ as follows:

$$\sum y^2 = \sum Y^2 - \frac{(\sum Y)^2}{n} \qquad (17.3.3)$$

The coefficient of correlation r is also a measure of the goodness of fit of a regression line to a set of X and Y pairs. Figure 17.3.2 presents scatter diagrams and regression lines indicating various degrees of goodness of fit. In part A, the regression line provides a perfect fit to the scatter of dots, since all the dots fall on the line. In this case, the observed Y values are equal to the corresponding predicted Y_e values, X and Y are perfectly associated, and we are able to make perfectly correct predictions of Y based on X. In parts B and C, the dots are not on the regression line and the observed Y's deviate from the predicted Y_e's. The deviations are indicated by the dashed lines connecting the dots to the regression line and are measured as $(Y - Y_e)$, according to Equation 17.2.1. The deviations are larger in part C than in part B so that the regression line provides a poorer fit to the X and Y pairs in part C and a poorer basis for predicting Y based on X.

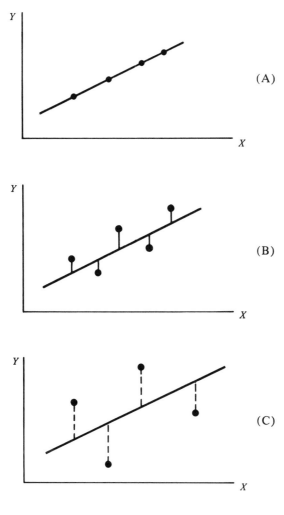

FIGURE 17.3.2

Figure 17.3.3 presents a dot A from a set of dots in a scatter diagram and the regression line. This dot is at height Y (the observed Y value) and the corresponding prediction Y_e is at point A' on the regression line. The mean $\bar{Y}$ is indicated at point A''. The deviation from A to A'' ($Y - \bar{Y}$) represents the total deviation of the observed Y from the mean $\bar{Y}$. As indicated in the figure, this total deviation is made up of two parts as follows:

$$(Y - \bar{Y}) = (Y_e - \bar{Y}) + (Y - Y_e) \qquad (17.3.4)$$

$(Y_e - \bar{Y})$ represents the deviation of the prediction Y_e from the mean $\bar{Y}$, based on the regression line. This part of the total deviation is explained by the association between Y and X and is said to be due to regression. The deviation $(Y - Y_e)$ is unexplained by the association between the two variables and represents the error in using Y_e as a prediction for Y. (This is the deviation expressed by Equation 17.2.1)

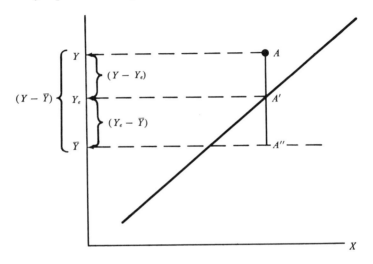

FIGURE 17.3.3

The meaning of these deviations is more easily grasped in an illustration. Suppose X denotes height and Y weight for ten-year-old girls. Then, dot A in Figure 17.3.3 represents the height X and weight Y for a particular girl. Suppose it is desired to predict the weight of a ten-year-old girl of a specified height X. Generally, weight depends upon a variety of factors, such as height, heredity, etc., however, if no information is available other than weight of the girls, the best prediction of the girl's weight is $\bar{Y}$, the mean weight for a sample of ten-year-old girls. If, however, sufficient related information is available, it is possible to produce a better prediction.

Suppose we have information on height as well as weight for a sample of ten-year-old girls. Then, the regression equation (for the regression line in Figure 17.3.3) may be computed and the predicted weight Y_e corresponding to the height X determined. The deviation $(Y_e - \bar{Y})$ represents the improvement in the use of Y_e instead of $\bar{Y}$ as the predicted weight of the girl based on the regression relationship. This deviation represents the part of the total deviation $(Y - \bar{Y})$ explained by X (height). Suppose it is determined that the weight of the ten-year-old girl of height X is Y, which differs from $\bar{Y}$ and Y_e. The deviation $(Y - Y_e)$, indicated in Figure 17.3.3, is due either to factors other than X (height) such as heredity, etc., or due to measurement errors, or a combination of factors. In any event, this deviation is unexplained by regression or by X (height).

If the total deviation $(Y - \bar{Y})$ is determined for each dot in a scatter diagram and then squared and summed for all dots, it can be shown mathematically that this total sum of squares is related to the explained and unexplained portions of the total deviation as follows:

$$\sum (Y - \bar{Y})^2 = \sum (Y_e - \bar{Y})^2 + \sum (Y - Y_e)^2 \qquad (17.3.5)$$

The total sum of squares $\sum (Y - \bar{Y})^2$ represents the total variation in Y and it will be recognized as the numerator of the equation for the variance s_Y^2 (Equation 5.2.2). The sum of squares for regression $\sum (Y_e - \bar{Y})^2$ represents the variation in Y explained by or associated with X. The sum of squares for error $\sum (Y - Y_e)^2$ represents the variation in Y unexplained or not associated with X. Equation 17.3.5 represents an important relationship in statistics (similar to Equations 14.3.2 and 14.3.3 encountered in Chapter 14). Dividing this equation through by the total variation in Y, we obtain

$$1 = \frac{\sum (Y_e - \bar{Y})^2}{\sum (Y - \bar{Y})^2} + \frac{\sum (Y - Y_e)^2}{\sum (Y - \bar{Y})^2} \qquad (17.3.6)$$

We may express Equation 17.3.6 in words as follows:

$$1 = \left(\begin{array}{c} \text{Proportion of the} \\ \text{total variation in } Y \\ \text{associated with } X \end{array} \right) + \left(\begin{array}{c} \text{Proportion of the} \\ \text{total variation in } Y \\ \text{not associated with } X \end{array} \right) \qquad (17.3.7)$$

The more closely X and Y are associated, the greater will be the proportion of the total variation in Y associated with X and the smaller will be the proportion not associated with X. In other words, the more closely X and Y are associated, the smaller will be the sums of squares for error $\sum (Y - Y_e)^2$ and the better is the fit of the regression line to the scatter of dots. Let us write

$$r^2 = \frac{\sum (Y_e - \bar{Y})^2}{\sum (Y - \bar{Y})^2} = \left(\begin{array}{c} \text{Proportion of the} \\ \text{total variation in} \\ Y \text{ associated with } X \end{array} \right) \qquad (17.3.8)$$

The proportion r^2, called the *coefficient of determination*, is a measure of how closely X and Y are related and the goodness of fit of the regression line to a set of X and Y pairs. Then, r, the coefficient of correlation, may be expressed as

$$r = \pm \sqrt{\frac{\sum (Y_e - \bar{Y})^2}{\sum (Y - \bar{Y})^2}} \qquad (17.3.9)$$

This is another formulation of the coefficient of correlation expressed in Equation 17.3.2. The sign of r ($+$ or $-$) must be chosen to agree with the sign of the regression coefficient b. Equation 17.3.2, however, is more suitable for the computation of r and provides the $+$ or $-$ sign directly.

Clearly, the lowest possible magnitude for r^2 (or any proportion) as well as for r is 0 and the highest is 1. If r is negative, it indicates a negative relationship between X and Y; and if positive, a positive relationship between X and Y. Also, b and r always have the same sign.

Let us use Equation 17.3.2 and compute the coefficient of correlation for the number of blemishes Y and cooling periods X in Table 17.2.1. We have already computed $\sum xy = -536.1$ and $\sum x^2 = 1{,}052.9$. We obtain from Table 17.2.1 $n = 10$ and $\sum Y = 429$. Squaring each Y value in the table and adding these squares, we obtain $\sum Y^2 = 18{,}717$. Apply Equation 17.3.3 to obtain

$$\sum y^2 = 18{,}717 - \frac{(429)^2}{10} = 312.9$$

Finally, applying Equation 17.3.2, we obtain

$$r = \frac{-536.1}{\sqrt{(1{,}052.9)(312.9)}} = -.93$$

A correlation coefficient of $-.93$ is, clearly, very high. How is this interpreted? We interpret it by computing the coefficient of determination $r^2 = .86$. This indicates that 86 percent of the variation in the number of blemishes Y is explained by (or associated with) the variation in the length of cooling period X. (We may also state that 86 percent of the variation in X is associated with the variation in Y.) When computing regression equations, Y always denotes the variable to be predicted and X the variable on which the prediction is based. If only the coefficient of correlation is to be computed, it does not matter which variable is denoted as Y and which is denoted as X. Either way, the same value of r will be obtained.

17.4 Optional: Computational Notes

Computation of a least squares regression equation and the coefficient of correlation are usually quite laborious. This has led to the programming of computer procedures for such computations. There are, however, various computational aids which are useful when a desk calculator is used.

In some problems the work is much easier if a constant is subtracted from each X observation and another constant is subtracted from each Y observation. For example, if the X observations are amounts between 100 and 200, it may be helpful to subtract 100 from each X value so that you may work with smaller numbers. If h is subtracted from each X and k is subtracted from each Y, we obtain the transformed variables

$$X'_i = X_i - h \tag{17.4.1}$$

$$Y'_i = Y_i - k \tag{17.4.2}$$

Then, compute

$$\sum x^2 = \sum (X')^2 - \frac{(\sum X')^2}{n} \tag{17.4.3}$$

$$\sum y^2 = \sum (Y')^2 - \frac{(\sum Y')^2}{n} \qquad (17.4.4)$$

$$\sum xy = \sum X'Y' - \frac{\sum X' \sum Y'}{n} \qquad (17.4.5)$$

The regression coefficient b is computed as before (Equation 17.2.6). The original X and Y values are used to obtain $\sum Y$ and $\sum X$ to compute a according to Equation 17.2.7. The coefficient of correlation is computed as previously (Equation 17.3.2). Sometimes subtraction of a constant results in negative quantities. This causes no problem, however, the sign should be taken into account in the computations. Sometimes only one variable is transformed. If Y is transformed by subtracting k and X is not changed, proceed as before with $Y' = Y - k$ and $X' = X$.

Sometimes, it is helpful to transform X or Y or both by dividing by a constant. Suppose each X is divided by h and each Y by k. Then

$$X' = \frac{X}{h} \qquad (17.4.6)$$

$$Y' = \frac{Y}{k} \qquad (17.4.7)$$

$$\sum (x')^2 = \sum (X')^2 - \frac{(\sum X')^2}{n} \qquad (17.4.8)$$

$$\sum (y')^2 = \sum (Y')^2 - \frac{(\sum Y')^2}{n} \qquad (17.4.9)$$

$$\sum x'y' = \sum X'Y' - \frac{\sum X' \sum Y'}{n} \qquad (17.4.10)$$

$$b = \frac{k \sum x'y'}{h \sum (x')^2} \qquad (17.4.11)$$

$$a = \frac{k \sum Y' - bh \sum X'}{n} \qquad (17.4.12)$$

and

$$r = \frac{\sum x'y'}{\sqrt{\sum (x')^2 \sum (y')^2}} \qquad (17.4.13)$$

Suppose X is transformed to X/h and Y is unchanged. Then, apply the foregoing equations with $X' = X/h$, $Y' = Y$, and $k = 1$.

Table 17.4.1 presents 83 pairs of X and Y observations. A convenient procedure for computing a least squares regression equation and the correlation coefficient for a large volume of data is to summarize the data in a two-way tally table as presented in Figure 17.4.1. The column headings indicate class intervals for X and the *stub* (on the left) indicates class intervals for Y. Equal class intervals are determined for each variable following the approach presented in Section 3.1. (Only a few classes are used in our illustration to keep it simple.) Each pair of X and Y observations in Table 17.4.1 represents one tally in the tally table.

TABLE 17.4.1

83 pairs of X and Y observations.

X	Y	X	Y	X	Y	X	Y	X	Y
101	42	111	44	102	41	114	45	106	40
109	59	99	53	105	53	94	64	118	49
115	44	104	58	104	48	90	51	99	54
106	52	89	60	108	59	98	53	89	62
107	54	99	52	93	57	100	46	90	53
104	49	88	60	103	49	108	57	109	52
95	51	112	43	91	55	104	57	87	77
90	63	106	53	90	62	105	52	81	73
103	47	91	64	100	54	90	56	97	59
98	66	91	63	90	64	90	54	87	78
91	62	98	66	94	64	98	66	95	63
94	50	92	61	114	45	90	61	109	59
95	62	100	52	108	40	102	49	85	79
85	79	91	62	95	63	91	63	110	49
89	70	93	61	81	73	87	77	103	56
98	68	88	77	85	79	87	73	89	70
93	53	87	73	93	52				

FIGURE 17.4.1

Two-way tally table.

X \ Y	80–89	90–99	100–109	110–119
40–49			⁣〣 〣	〣 //
50–59		〣 〣 〣	〣 〣 〣	
60–69	///	〣 〣 〣 〣		
70–79	〣 〣 ///			

Using the tally table, set up a two-way frequency table (also called a *correlation table* or a *scatter plot*), as shown in Figure 17.4.2. The cell frequencies are obtained from the tallies in the tally table. Column and row totals are obtained, as shown. The row totals are class frequencies for the Y values and are denoted as f_Y. The column totals are class frequencies for the X values and are denoted as f_X. The frequencies within a cell, the cell frequencies, represent pair frequencies. For example, the cell identified by the X class 100–109 and the Y class 40–49 contains the frequency 10. This indicates that 10 of the 83 pairs have X values in the 100–109 interval and Y values in the 40–49 interval.

FIGURE 17.4.2

Two-way frequency table for regression and correlation computations.

Y \ X	80–89	90–99	100–109	110–119	f_Y	V	Vf_Y	V^2f_Y	UVf
40–49			⑩ 10	⑭ 7	17	−1	−17	17	−24
50–59		⊖ 15	⊖ 15		30	0	0	0	0
60–69	⊖3 3	⊖ 20			23	1	23	23	−3
70–79	⊖13 13				13	2	26	52	−26
f_X	16	35	25	7	$n = 83$			32	92	−53
U	−1	0	1	2					
Uf_X	−16	0	25	14	23				
U^2f_X	16	0	25	28	69				

Now, transform X to U by entering 0 in the U row corresponding to the largest value of f_X (35) and entering the other values as shown in Figure 17.4.2. This follows the approach used in the shortcut procedures for computing the mean and standard deviation in Sections 4.6 and 5.6; however, now we use U instead of u. (It is not necessary to compute the class midpoints.) In the same way, transform Y to V by entering 0 in the V column corresponding to the largest value of f_Y (30) and entering the other values as shown in the figure.

The two remaining rows in Figure 17.4.2, Uf_X and U^2f_X, are easily computed. The first item in the Uf_X row (-16) is obtained as the product of the first item in the U row (-1) and the first item in the f_X row (16). The balance of the items in the Uf_X row is similarly computed. The first item in the U^2f_X row (16) is obtained as the product of the first item in the Uf_X row (-16) and the first item in the U row (-1). The balance of the U^2f_X row is determined in the same way. The columns Vf_Y and V^2f_Y are obtained in a similar manner. The first item in the Vf_Y column (-17) is the product of the first item in the V column (-1) and the first item in the f_Y column (17). The balance of the Vf_Y column is computed in a similar way. The first item in the V^2f_Y column (17) is computed as the product of the first item in the Vf_Y column (-17) and the first item in the V column (-1). The remaining items in the V^2f_Y column are similarly determined.

The last column in Figure 17.4.2, UVf, contains items which are obtained by multiplying three figures: U, V, and f (cell frequency). This is accomplished most easily in two steps. First, multiply each cell frequency by the corresponding

U value. For example, in the first row of cell frequencies, multiply the frequency 10 by the *U* value 1 and write the product 10 in a circle in the cell. The second cell frequency in this row (7) times the *U* value (2) is 14, which is written in a circle in the cell. The second step is to add all circled numbers in a row and multiply this sum by the corresponding *V* value. This product is entered in the *UVf* column.

Notice that some circles contain a dash instead of a figure. It is a good idea to place a dash in the circle for any cell where either the corresponding *U* or the corresponding *V* or both are equal to 0. Then, consider these dashes as 0 when performing the second step noted previously. This approach avoids unnecessary computations. (The cells in the column with $U = 0$ will always have a 0 in the circle. In the second step, the sum of the circled numbers in a row with $V = 0$ is multiplied by $V = 0$.)

The two-way frequency table in Figure 17.4.2 provides the data needed to compute the least squares regression equation and the correlation coefficient. The following equations are to be used:

$$\Sigma u^2 f_X = \Sigma U^2 f_X - \frac{(\Sigma U f_X)^2}{n} \tag{17.4.14}$$

$$\Sigma v^2 f_Y = \Sigma V^2 f_Y - \frac{(\Sigma V f_Y)^2}{n} \tag{17.4.15}$$

$$\Sigma uvf = \Sigma UVf - \frac{\Sigma U f_X \Sigma V f_Y}{n} \tag{17.4.16}$$

where: $u = U - \bar{U}$
$v = V - \bar{V}$
$n = \Sigma f =$ number of pairs of X and Y

$$b = \frac{c_Y \Sigma uvf}{c_X \Sigma u^2 f_X} \tag{17.4.17}$$

$$a = \frac{c_Y \Sigma V f_Y - b c_X \Sigma U f_X}{n} + Y_0 - bX_0 \tag{17.4.18}$$

where: $c_X =$ length of X class intervals
$c_Y =$ length of Y class intervals
$X_0 = X$ midpoint corresponding to $U = 0$
$Y_0 = Y$ midpoint corresponding to $V = 0$

Then, as before, the *least squares regression equation* is $Y_e = a + bX$.
The *correlation coefficient* is computed as

$$r = \frac{\Sigma uvf}{\sqrt{\Sigma u^2 f_X \Sigma v^2 f_Y}} \tag{17.4.19}$$

Applying the foregoing equations and the data in Figure 17.4.2, we obtain

$$\Sigma u^2 f_X = 69 - \frac{(23)^2}{83} = 62.6265$$

$$\Sigma v^2 f_Y = 92 - \frac{(32)^2}{83} = 79.6627$$

$$\Sigma uvf = -53 - \frac{(23)(32)}{83} = -61.8675$$

$$c_X = 10$$

$$c_Y = 10$$

$$X_0 = \frac{90 + 99}{2} = 94.5$$

$$Y_0 = \frac{50 + 59}{2} = 54.5$$

$$b = \frac{(10)(-61.8675)}{(10)(62.6265)} = -.9879$$

$$a = \frac{(10)(32) - (-.9879)(10)(23)}{83} + 54.5 - (-.9879)(94.5)$$

$$= 154.4495$$

Then,
$$Y_e = 154.45 - .99X$$

Finally,
$$r = \frac{-61.8675}{\sqrt{(62.6265)(79.6627)}} = -.88$$

Study Problems

1. Given the following sets of X and Y pairs:

(1) X	4	7	9	6	10	
Y	1	3	4	2	5	

(2) X	7	3	10	9	11	5
Y	5	13	1	3	1	10

    a. Construct the scatter diagram.
    b. Compute the least squares regression equation.
    c. Construct the computed regression line on the scatter diagram.
    d. Compute the coefficient of correlation.
    e. Compute the coefficient of determination and explain what it means.

2. The following measurements (in inches) on height and spread were obtained for a sample of 5 plants:

Height	Spread
5	4
2	4
7	5
1	1
6	7

It is desired to predict spread based on height.
    a. Construct the scatter diagram.
    b. Compute the least squares regression equation.
    c. Plot the regression equation on the scatter diagram.

d. Using the regression relationship, predict the spread of a plant with height 4 inches and 5 inches.
e. Compute the coefficient of correlation.
f. Compute the coefficient of determination and explain its meaning.

3.   The numbers of units of a product sold when the price was set at different levels (dollars per unit) were as follows:

Quantity sold	Price
3	3
6	2
3	8
4	5
4	1
2	7

It is desired to predict quantity to be sold based on price.
a. Construct the scatter diagram.
b. Compute the least squares regression equation.
c. Plot the regression line on the scatter diagram.
d. Estimate quantity to be sold when the price is $6 per unit.
e. Compute the coefficient of correlation.

4.   Given $n = 25$, $s_X = 10$, $s_Y = 15$, $\sum xy = -1,500$. Compute the coefficient of correlation for X and Y.

5.   Given $\sum x^2 = 25$, $\sum y^2 = 30$, $\sum xy = 20$. Compute the coefficient of correlation for X and Y.

6.   If 60 percent of the total variation in Y is not associated with the variation in X, compute (a) r^2 and (b) r.

7.   A random sample of eight junior high school students obtained the following grades on two tests:

Student	Test E	Test O
A	4	8
B	6	9
C	4	6
D	7	8
E	1	8
F	9	10
G	2	5
H	5	9

a. Compute a linear least squares regression equation to predict test O scores based on test E scores.
b. Applying the regression equation, what score on test O would you predict for a junior high school student who obtained 8 on test E?

  c. Compute the coefficient of correlation for tests E and O.

  d. Compute the coefficient of determination and explain its meaning.

8. What is meant by the least squares property of the best-fitting regression line for two variables X and Y?

9. How do you interpret the regression equation constant a and the regression coefficient b?

10. Indicate whether a regression analysis or a correlation analysis is involved and defend your answer:

      a. Problem 2

      b. Problem 3

      c. Problem 7

11. The following data were collected on the number of marriages for ten consecutive weeks in a city and the number of first-born children a year later:

Number of marriages	Number of first-born
10	8
20	16
14	12
17	12
12	6
23	20
17	16
15	11
19	18
13	10

      a. Determine the least squares line of regression to predict number of first-born (a year later) based on number of marriages.

      b. Using the regression equation, how many first-born (a year later) would you predict if 18 marriages were performed during a given week; if 11 marriages were performed?

      c. Compute the coefficients of correlation and determination.

12. In Problem 11:

      a. Determine the least squares regression equation to estimate the number of marriages (a year before) based on the numbers of first-born.

      b. Using the regression equation, estimate the number of marriages performed (a year before), if 14 first-born children were recorded for a week; if 17 first-born children were recorded.

      c. Compute the coefficient of correlation.

13. Compare the value of r^2 computed for Problems 11 and 12 with the product of the b coefficients obtained in the two problems. (Notice that r^2 = product of the b coefficients, except for possible rounding differences. This represents another method for computing r^2 and r.)

14. A random sample of ten firms provided the following information relating to number of employees and value of product.

Firm	Number of employees	Value of product ($000)
A	14	10
B	10	11
C	13	12
D	13	11
E	11	9
F	9	10
G	11	11
H	7	9
I	12	10
J	13	12

a.   Compute the linear regression equation to predict the value of product based on the number of employees.

b.   Compute the coefficient of correlation. How do you interpret this measure?

c.   What would you predict as the value of product for a firm with 12 employees; with 14 employees?

15.   Given $Y_e = 2.50 + 1.5X$, $r = .70$, X = score on test A, and Y = score on test B. Determine the b coefficient for the regression equation to predict test A scores based on test B scores. (Refer to Problem 13.)

16.   Given the following data on height (inches) and weight (pounds) for a sample of 11 students:

Height (X)	Weight (Y)
58	139
54	127
57	130
59	146
60	137
53	132
51	112
58	147
62	147
56	139
55	136

a.   Compute the least squares regression equation to predict weight based on height. (Use $X' = X - 50$ and $Y' = Y - 130$ for the computations.)

b.   Compute the coefficient of correlation.

17.   A sample of nine boxes of apples were selected and weighed before and after shipment:

Weight (pounds)	
X (after)	Y (before)
148	151
152	159

Weight (pounds)

X (after)	Y (before)
121	167
136	150
141	164
149	152
137	160
135	170
129	160

a. Compute the least squares linear regression equation to predict weight before shipment based on weight after shipment (Use $X' = X - 120$ and $Y' = Y - 150$ in the computations.)

b. Compute the coefficient of correlation.

18. Given scores on tests K and L for a sample of 12 subjects:

Test K scores X	Test L scores Y
90	87
100	84
120	75
150	75
170	72
180	60
200	54
200	45
220	45
220	45
220	39
230	39

a. Determine the linear regression equation to predict test L scores based on test K scores. (Use $X' = X/10$ and $Y' = Y/3$ in your computations.)

b. Determine the coefficient of correlation.

19. Given the following two-way frequency table for quantity sold (dozens of units) and price (cents per unit):

Quantity sold	Price 0–2	3–5	6–8
3– 5		1	1
6– 8	1	2	1
9–11	2		

a. Compute the least squares regression equation to predict quantity to be sold based on price.

b. Compute the coefficient of correlation.

20. Data were summarized for number of letters typed and months of experience for a sample of nine typists as follows:

Number	Months of experience			
of letters	0–4	5–9	10–14	15–19
3–5	2	1		
6–8		2		
9–11			1	2
12–14			1	

    a.   Determine the least squares regression equation to predict the number of letters typed based on the number of months of experience.

    b.   Compute the correlation coefficient.

21.   The following two-way frequency table was constructed for weekly average prices (in dollars) for a sample of common stocks and a sample of bonds for a 12-week period.

Bonds	Common stocks			
	3–5	6–8	9–11	12–14
0–2			1	
3–5		1		2
6–8		2	1	
9–11	2	1		
12–14		1		
15–17	1			

    a.   Compute the least squares regression equation to predict the weekly average bond price based on weekly average common stock price.

    b.   Compute the correlation coefficient.

22.   The following two-way frequency table presents amount of fertilizer used (pounds) and yield per plot (bushels):

Yield	Amount of fertilizer				
	50–54	55–59	60–64	65–69	70–74
100–109	1				
110–119	1	2	2		
120–129	1	1	2	1	
130–139		2	1		
140–149			2	1	1
150–159			1	2	1
160–169			2	1	

    a.   Compute the regression equation to predict yield based on amount of fertilizer used.

    b.   Compute the correlation coefficient.

Chapter 18

Further Measures of Association

18.1 Curvilinear Relationships

The general approach in computing a least squares regression equation and the coefficient of correlation when the relationship between two variables is curvilinear is the same as when it is linear; however, the application of these methods and the interpretation of the results require more discussion than is possible in this text. We will discuss nonlinear regression and correlation sufficiently to show how such problems are handled.

Table 18.1.1 presents ten pairs of X and Y observations and Figure 18.1.1 presents the scatter diagram for these data. It is clear from the figure that a straight-line regression line is not appropriate. In order to fit a least squares regression line to these data, it is necessary to transform X or Y or both so that the resulting relationship may be appropriately represented by a straight line. Let us transform X to log X.

311

TABLE 18.1.1

Ten pairs of X and Y observations.

X	Y
16	54.5
24	59.5
34	64.5
39	68.5
10	48.5
27	64.5
43	67.5
13	54.0
31	66.0
20	60.0

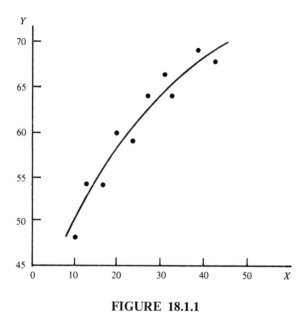

FIGURE 18.1.1

Figure 18.1.2 shows the scatter diagram for log X and Y. Clearly, this scatter diagram indicates that the relationship is linear after X is transformed to log X. Hence, we will compute the least squares linear regression equation for log X and Y and the correlation coefficient applying the equations from the previous chapter; however, in these equations, we must substitute log X for X.

First, carry out the computations in Table 18.1.2. Using the data in this table, compute

$$\sum (\log x)^2 = \sum (\log X)^2 - \frac{(\sum \log X)^2}{n}$$

$$= 19.101 - \frac{(13.676)^2}{10} = .398$$

(18.1.1)

$$\sum y^2 = \sum Y^2 - \frac{(\sum Y)^2}{n}$$

$$= 37{,}303.75 - \frac{(607.5)^2}{10} = 398.12 \qquad \textbf{(18.1.2)}$$

$$\sum (\log x)y = \sum (\log X)Y - \frac{(\sum \log X)(\sum Y)}{n}$$

$$= 843.112 - \frac{(13.676)(607.5)}{10} = 12.295 \qquad \textbf{(18.1.3)}$$

$$b = \frac{\sum (\log x)y}{\sum (\log x)^2} = \frac{12.295}{.398} = 30.892 \qquad \textbf{(18.1.4)}$$

$$a = \frac{\sum Y - b \sum \log X}{n}$$

$$= \frac{607.5 - (30.892)(13.676)}{10} = 18.500 \qquad \textbf{(18.1.5)}$$

$$r = \frac{\sum (\log x)y}{\sqrt{\sum (\log x)^2 \sum y^2}}$$

$$= \frac{12.295}{\sqrt{(.398)(398.12)}} = .98 \qquad \textbf{(18.1.6)}$$

Then, the *linear* regression equation for predicting Y based on $\log X$ is

$$Y_e = 18.500 + 30.892 \log X \qquad \textbf{(18.1.7)}$$

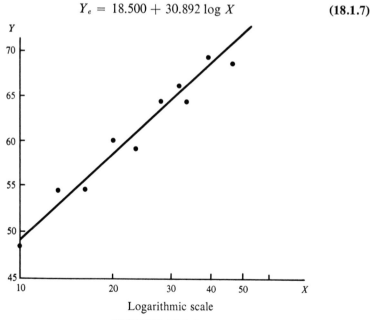

Logarithmic scale

FIGURE 18.1.2

In order to plot the regression line, we must compute the Y_e corresponding to several X values. Using $X = 10$ and $X = 45$, first transform each to logarithms to obtain $\log 10 = 1.000$ and $\log 45 = 1.653$ (Table B.9). Then, substituting into the regression equation, compute

$$Y_e = 18.500 + (30.892)(1.000) = 49.39$$
$$Y_e = 18.500 + (30.892)(1.653) = 69.56$$

TABLE 18.1.2

X, log X, and Y for the data in Table 18.1.1.

X	Log X	Y	(Log X)²	Y²	(Log X)Y
16	1.204	54.5	1.450	2,970.25	65.618
24	1.380	59.5	1.904	3,540.25	82.110
34	1.531	64.5	2.344	4,160.25	98.750
39	1.591	68.5	2.531	4,692.25	108.984
10	1.000	48.5	1.000	2,352.25	48.500
27	1.431	64.5	2.048	4,160.25	92.300
43	1.633	67.5	2.667	4,556.25	110.228
13	1.114	54.0	1.241	2,916.00	60.156
31	1.491	66.0	2.223	4,356.00	98.406
20	1.301	60.0	1.693	3,600.00	78.060
Total	13.676	607.5	19.101	37,303.75	843.112

Table 18.1.3 presents the predicted (or estimated) Y_e's corresponding to several X values, including those just computed. Using the data in this table, the linear regression line for $\log X$ and Y is constructed in Figure 18.1.2 and the curvilinear regression line for X and Y in Figure 18.1.1.

TABLE 18.1.3

Data for plotting the regression equation for the illustration in Table 18.1.1.

X	Log X	Y_e
10	1.000	49.39
15	1.176	54.83
20	1.301	58.69
25	1.398	61.69
30	1.477	64.13
35	1.544	66.20
40	1.602	67.99
45	1.653	69.56

When r is computed based on the linear relationship between X and Y, it is called the *coefficient of correlation;* however, when it is computed based on a curvilinear relationship, it is called the *index of correlation.* Furthermore, it is absolutely essential to specify the transformations used when the index of correlation is presented. For example, in our illustration, $r = .98$ for $\log X$ and Y. If X and Y were used in the computations, r would have been much lower. Or if some other type of transformation were used (in lieu of log X), r would have a different value. Consequently, r is meaningful only if the transformations used are stated.

An important problem in determining a curvilinear regression equation is to find the proper transformations to use. This is often a difficult problem, requiring a knowledge of mathematical relationships and curves. We will not consider this problem further, however, two more types of transformations will be noted which are useful in many situations.

Sometimes a linear relationship results when both X and Y are transformed to logarithms, so that the linear regression equation becomes

$$\log Y_e = a + b \log X \qquad (18.1.8)$$

Figure 18.1.3 illustrates such a regression curve fitted to a set of 13 pairs of X and Y observations. The first step in computing a log Y, log X least squares regression equation is to construct a table similar to Table 18.1.2. Such a table should contain the following seven columns: X, Y, log X, log Y, $(\log X)^2$, $(\log Y)^2$, and log X log Y. The following equations should be used:

$$\sum (\log x)^2 \qquad \text{(as shown previously)}$$

$$\sum (\log y)^2 = \sum (\log Y)^2 - \frac{(\sum \log Y)^2}{n} \qquad (18.1.9)$$

$$\sum \log x \log y = \sum \log X \log Y - \frac{\sum \log X \sum \log Y}{n} \qquad (18.1.10)$$

$$b = \frac{\sum \log x \log y}{\sum (\log x)^2} \qquad (18.1.11)$$

$$a = \frac{\sum \log Y - b \sum \log X}{n} \qquad (18.1.12)$$

$$r = \frac{\sum \log x \log y}{\sqrt{\sum (\log x)^2 \sum (\log y)^2}} \qquad (18.1.13)$$

The linear regression line for log X and log Y is plotted based on log Y_e values corresponding to selected X values. The curvilinear regression line is plotted based on $Y_e =$ antilog (log Y_e) values corresponding to the selected X values. This work is facilitated by setting up a table similar to Table 18.1.3, with the following columns: X, log X, log Y_e, and Y_e.

In some problems, a useful approach is to transform Y to $1/Y$. The linear regression equation becomes

$$\frac{1}{Y_e} = a + bX \qquad (18.1.14)$$

Figure 18.1.4 illustrates such a regression curve fitted to ten X and Y pairs of observations. The first step in computing such a least squares regression equation is to construct a table similar to Table 18.1.2 containing the seven columns X, Y, $1/Y$, $Z = k(1/Y)$, X^2, Z^2, and XZ. Sometimes, the reciprocals $1/Y$ are

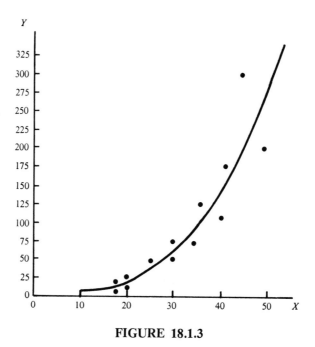

FIGURE 18.1.3

very small numbers and it is convenient to multiply them by an appropriate constant. For example, if the $1/Y$ values are quantities such as .0127, it is convenient to use 10,000 as a multiplier so that we have $10,000(.0127) = 127$. This will facilitate the computations. In general, suppose k is the constant multiplier. Then, the column heading, as noted previously, is $Z = k(1/Y)$.

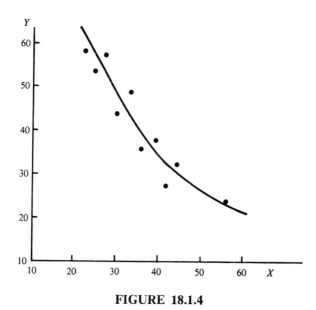

FIGURE 18.1.4

The following equations should be used:

$$\sum x^2 = \sum X^2 - \frac{(\sum X)^2}{n} \tag{18.1.15}$$

$$\sum z^2 = \sum Z^2 - \frac{(\sum Z)^2}{n} \tag{18.1.16}$$

$$\sum xz = \sum XZ - \frac{\sum X \sum Z}{n} \tag{18.1.17}$$

$$b = \frac{\sum xz}{k \sum x^2} \tag{18.1.18}$$

$$a = \frac{\sum\left(\frac{1}{Y}\right) - b \sum X}{n} \tag{18.1.19}$$

$$r = \frac{\sum xz}{\sqrt{\sum x^2 \sum z^2}} \tag{18.1.20}$$

The linear regression line for $1/Y$ and X is plotted based on $1/Y_e$ values corresponding to selected X values. The curvilinear regression line is plotted based on Y_e = reciprocal of $1/Y_e$ values corresponding to the selected X values. This work is facilitated by setting up a table similar to Table 18.1.3, with columns X, $1/Y_e$, and Y_e.

18.2   Correlation of Ranks

Suppose ten typists are ranked on experience X (1 = most experienced) and number of typing errors made Y (1 = largest number of errors) as shown in Table 18.2.1. We could compute the correlation coefficient using the approach of Chapter 17, however, a more rapid method when ranks are involved is to compute r', the *correlation of ranks*, as follows:

$$r' = 1 - \frac{6 \sum D^2}{n(n^2 - 1)} \tag{18.2.1}$$

where: D = the difference between ranks
$\quad\quad n$ = number of pairs of ranks

Applying this equation to our problem (Table 18.2.1), we obtain

$$r' = 1 - \frac{(6)(310)}{(10)(100 - 1)} = -.88$$

This is identically the same result which would be obtained using Equation 17.3.2 and indicates that the greater the experience (X), the fewer are the errors (Y).

When it is desired to determine the correlation between two variables and a rapid approximation to r is satisfactory, a typical approach is to rank the X

TABLE 18.2.1

Ten typists ranked on experience X and number of errors made Y.

Typist	X	Y	D = Y − X	D²
A	1	10	9	81
B	2	8	6	36
C	3	7	4	16
D	4	9	5	25
E	5	6	1	1
F	6	3	−3	9
G	7	5	−2	4
H	8	1	−7	49
I	9	4	−5	25
J	10	2	−8	64
			Total	310

and Y observations separately and compute the correlation of ranks r' based on the ranks. When data are ranked and ties occur, ranks should be assigned by assuming that the tied data can be ranked and then by assigning the mean of the ranks so obtained to each of the tied values. For example, suppose we have the observations 20, 25, 28, 28, 29, 31, 31, 31, and 33. Start by assigning ranks 1, 2, 3, 4, 5, 6, 7, 8, and 9. Note that rank 3 is assigned to the first 28 and 4 to the other. Then, computing the mean, $(1/2)(3 + 4) = 3.5$, we assign the rank 3.5 to each observation of 28. Also, computing the mean, $(1/3)(6 + 7 + 8) = 7$, assign the rank 7 to each observation of 31. Then, the ranks for these observations are 1, 2, 3.5, 3.5, 5, 7, 7, 7, and 9.

Study Problems

1.   Scores on two sociometric tests, D and J, were obtained for a sample of 11 subjects, as follows:

Subject	Test D X	Test J Y
A	6	9
B	19	17
C	4	9
D	6	13
E	14	13
F	16	17
G	3	5
H	18	13
I	21	15
J	12	15
K	10	11

a. Determine the least squares regression equation to predict test J score based on test D score for a subject. (Transform X to log X.)
b. Construct the scatter diagram for log X and Y and plot the regression line.
c. Construct the scatter diagram for X and Y and plot the regression curve.
d. Compute the index of correlation.

2.   A sample of 12 housewives provided the following information relating to distance from a regional shopping center and number of trips to the shopping center during a period:

Distance (miles), X	Number of trips, Y
6	15
15	5
10	6
17	4
8	12
10	9
7	10
7	18
6	17
12	4
16	3
13	6

a. Determine the least squares regression equation to predict the number of trips based on distance. (Use log Y and log X instead of Y and X.)
b. Construct the scatter diagram for log X and log Y and plot the regression line.
c. Construct the scatter diagram for X and Y and plot the regression curve.
d. Compute the index of correlation.

3.   Number of units produced and number of machine breakdowns in a factory during 12 monthly periods were as follows:

Month	Units produced Y	Number of Machine breakdowns X
1	15	2
2	12	4
3	7	10
4	13	3
5	9	9
6	6	10
7	18	2
8	9	5
9	11	6
10	10	6
11	11	4
12	8	10

   a.   Determine the least squares regression equation to predict production
   volume based on machine breakdowns. (Use $1/Y$ instead of Y.)
   b.   Construct the scatter diagram for $1/Y$ and X and plot the regression line.
   c.   Construct the scatter diagram for Y and X and plot the regression
   curve.
   d.   Compute the index of correlation.

4.   A sample of nine seniors were ranked on scholastic average and scores obtained
on test B. (Note: Highest scholastic average = 1 and highest test score = 1.) The
following table of rankings was prepared:

| | Ranks | |
| | Scholastic | Test B |
Senior	average	score
A	3	8
B	6	4
C	1	9
D	5	6
E	7	1
F	9	3
G	2	7
H	8	2
I	4	5

Compute the coefficient of correlation for this pair of rankings.

5.   A sample of 12 farms was ranked on size of farm and investment in machinery as
follows:

Size of farm	Investment
12	9
4	2
6	4
9	7
11	10
5	5
7	11
1	6
2	3
8	8
10	12
3	1

(Smallest farm = rank 1 and lowest investment = rank 1.) Compute the coefficient
of correlation.

6.   The average price of crystal chandeliers for home use and average cost of installation were determined for a sample of 15 cities, as follows:

City	Average price	Installation cost
A	$64	$34
B	76	50
C	49	27
D	66	42
E	59	32
F	68	42
G	53	34
H	70	39
I	46	31
J	56	36
K	73	47
L	56	37
M	77	44
N	62	34
O	58	24

Rank average price and average installation cost. Then compute the coefficient of rank correlation.

Chapter 19

Prediction

19.1 Relationships and Their Measurement

The methods of correlation and regression provide a useful means for measuring and evaluating the relationship between two variables and for predicting (or estimating) one based on the other. These are, however, *mechanical procedures* which may be applied to data for any pair of variables. A high correlation coefficient computed for a pair of variables does not in itself indicate a cause and effect relationship. For example, the average price of fresh vegetables and intracity commuter bus fare in a city may appear to be closely related for a period only because they both reflect the rising price effect of inflation in the economy. Or the prices of two commodities may show a close relationship, not because a price change in one causes a price change in the other, but because an ingredient common to both has risen in price. Such apparent relationships between two variables represent spurious relationships. Sometimes two variables, such as the number of accidents in a shoe factory in New York City and the amount of snowfall in Alaska, may show a close relationship for a certain period only because of chance effects. Such apparent relationships may be called *nonsense relationships*.

Correlation and regression analyses are useful to evaluate hypotheses. For example, a chemist may hypothesize that the mileage obtained from a gasoline will be increased in proportion to the amount of a chemical added to the gasoline. This hypothesis may be evaluated by determining the mileage per gallon of the gasoline for various amounts of the chemical added. Then, the coefficient of correlation may be computed for amount of the chemical added (X) and miles per gallon (Y) and the least squares regression equation determined. The correlation and regression analysis may be used to evaluate the chemist's hypothesis. Or it may be hypothesized that the number of gainfully employed in a community is a useful predictor of the number of subscriptions for a national weekly periodical in the community. This hypothesis may be evaluated by the methods of correlation and regression based on employment and subscription data.

Sometimes an exploratory type of correlation and regression analysis is useful. For example, what is the single best predictor of the demand for washing machines in a city? Is it average income earned, the number of homes purchased, the number of marriages, or the amount of accumulated savings? Such an *exploratory analysis* may be based on separate analysis, using sample data, of the relationship between the number of washing machines sold and each of the possible predictors under consideration. It goes without saying that only logically related variables should be considered in an exploratory analysis.

19.2 Regression and Correlation Models

Typically, the methods of correlation and regression are applied to sample data and inferences are made about the population. In this section, we will consider *theoretical sampling models* appropriate for formulation of statistical inferences in two-variable analyses.

Figure 19.2.1 presents a scatter diagram for two variables X and Y. The regression line shown is based on the population of X and Y pairs, however, only some of these pairs (dots) are shown. More than one Y observation is paired or associated with a given X observation, as specifically indicated by the dots which fall on the vertical lines constructed at the points $X = X_1$ and $X = X_2$. Assume that $X = $ height and $Y = $ weight for a population of adult males. Then, X_1 and X_2 represent specific heights. Of course, it is not to be expected that all males of a given height will have the same weight. Actually, for any specified height, we would expect a distribution of weights. The distribution of weights for males of a given height is called a *conditional distribution*.

Generally, we are concerned with the *conditional distribution of Y for a given value of X*. In Figure 19.2.1, the conditional distribution of Y (weight), given

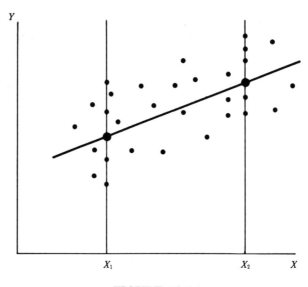

FIGURE 19.2.1

that $X = X_1$, is represented by the dots in the population which fall on the vertical line constructed at X_1. (We will speak of the distribution of all weights in the population, without regard to height, as the *unconditional distribution* of weights.) Clearly, *there is a conditional distribution of Y* for each possible value of X.

Consider the conditional distribution of Y corresponding to X_1 in Figure 19.2.1 and imagine that small coins are placed on the scatter diagram for each X_1, Y pair in the population. The coins would pile up along the vertical line at X_1, reflecting the frequency with which the different values of Y occur. The three-dimensional diagram in Figure 19.2.2 presents a possible pile-up of the coins for Y values paired with X_1 and also for Y values paired with X_2. In this figure, the largest pile-up of coins occurs in the center of each distribution. The general shape of each pile-up of coins suggest a histogram. In fact, a histogram could be constructed for the Y values in the population which are paired with a given X and it would exhibit the conditional distribution of Y. We may, of course, approximate the histogram by a smooth continuous curve.

We will denote the mean of all the Y values in the population (the unconditional distribution) as μ_Y and the mean of the conditional Y distribution as $\mu_{Y.x}$ (which indicates that this is the mean of the Y's paired or associated with a given X). We will usually refer to $\mu_{Y.x}$ as the conditional mean of Y. In Figures 19.2.1 and 19.2.2, the conditional means $\mu_{Y.X_1}$ and $\mu_{Y.X_2}$ are identified by heavy dots. We will denote the standard deviation for the population distribution of Y as σ_Y and for the conditional distribution of Y as $\sigma_{Y.x}$. We will refer to $\sigma_{Y.x}$ as the conditional standard deviation of Y.

In Section 17.3 it was noted that we distinguish between two types of two-variable analyses, regression and correlation. Two theoretical sampling models will be presented, a regression analysis model and a correlation analysis model.

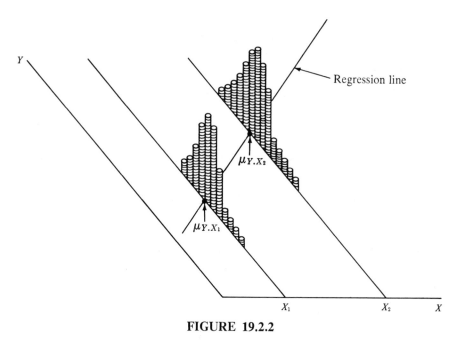

FIGURE 19.2.2

These models describe the conditional distribution of Y and provide the basis for making statistical inferences in two-variable analyses. These are called *bivariate (two-variable) models*.

Bivariate Model (Regression)

Given two variables, X independent and Y dependent. The X distribution is unspecified. The conditional distribution of Y for a given X has the following characteristics:

    1. It is normal, with mean $\mu_{Y.x}$ and standard deviation $\sigma_{Y.x}$.
    2. The conditional means $\mu_{Y.x}$ fall on a straight line and satisfy the linear regression equation $\mu_{Y.x} = \alpha + \beta X$.
    3. The conditional standard deviation $\sigma_{Y.x}$ is constant over all conditional distributions of Y (the assumption of *homoscedasticity*).

   A sample of n X, Y pairs is obtained by selecting n X values (not necessarily on a random basis) and then selecting a Y observation on a random basis from the Y values associated with each of the n X values.

Bivariate Model (Correlation)

Given a two-variable population. The variables X and Y are individually and jointly normal in distribution, so that the X, Y population represents a *bivariate normal population*. There is a Y distribution associated with a given X (the conditional distribution of Y) and an X distribution associated with a given Y

(the conditional distribution of X). The characteristics relating to the conditional distributions of Y stated in the Bivariate Model (Regression) apply to this model as well. Similar characteristics apply to the conditional distributions of X, as follows:

1. The distributions are normal, with means $\mu_{X.Y}$ and standard deviations $\sigma_{X.Y}$.
2. The conditional means $\mu_{X.Y}$ fall on a straight line and satisfy the linear regression equation $\mu_{X.Y} = (\alpha) + (\beta)Y$.
3. The conditional standard deviation $\sigma_{X.Y}$ is constant over all conditional distributions of X.

A sample of size n is obtained by selecting a random sample of n X, Y pairs from the X, Y population.

Figure 19.2.3 shows the normal conditional distributions of Y corresponding to $X = X_1$ and $X = X_2$ and the regression line passing through the conditional means $\mu_{Y.X_1}$ and $\mu_{Y.X_2}$.

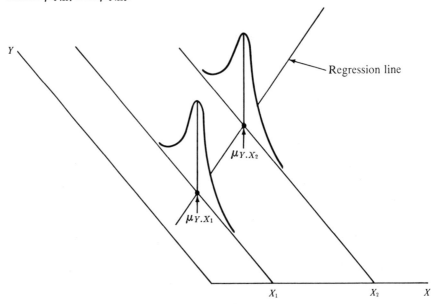

FIGURE 19.2.3

Previously, we expressed the least squares regression equation as

$$Y_e = a + bX \tag{19.2.1}$$

This denotes the regression equation computed from sample data and is actually the estimated regression equation. The *true regression equation*, as noted in the bivariate models, is

$$\mu_{Y.X} = \alpha + \beta X \tag{19.2.2}$$

for predicting (or estimating) Y based on X (called the regression of Y on X) and

$$\mu_{X.Y} = (\alpha) + (\beta)Y \qquad (19.2.3)$$

for predicting (or estimating) X based on Y (called the regression of X on Y). We use (α) and (β) with parentheses to distinguish these coefficients from the corresponding coefficients α and β for the regression of Y on X.

Notice that in Equation 19.2.2, $\mu_{Y.X}$ is shown as the predicted Y instead of Y_e as in the estimated regression equation. This is in accordance with the bivariate models which specify that the conditional means $\mu_{Y.X}$ fall on the regression line. Consequently, the predicted Y for a given X is the conditional mean of Y, if computed from the true regression equation and assuming that either of the bivariate models hold. Y_e computed from the estimated regression equation represents a point estimate of the conditional mean $\mu_{Y.X}$. The coefficients in the true regression equation α and β represent the true regression coefficients; whereas a is a *point estimate* of α and b is a *point estimate* of β.

The conditional standard deviation $\sigma_{Y.X}$ may be estimated by computing

$$s_{Y.X} = \sqrt{\frac{\sum (Y - Y_e)^2}{n - 2}} \qquad (19.2.4)$$

where $s_{Y.X}$ is called the *standard error of estimate*. Notice that $\sum (Y - Y_e)^2$, which appears in the numerator, is the sum of squares for error discussed in Section 17.3. The closer the association between X and Y and the better the fit of the regression line to the X, Y data, the smaller are the $(Y - Y_e)$ deviations and the smaller is the standard error of estimate. If X and Y are perfectly associated so that Y_e and Y are equal (all the dots in the scatter diagram fall on the regression line), the $(Y - Y_e)$ deviations equal zero and $s_{Y.X} = 0$. Hence, the smaller $s_{Y.X}$, the less is the dispersion of the X, Y pairs around the regression line, the closer is the association of X and Y, and the better is X as a predictor of Y.

A more desirable equation for the *computation* of $s_{Y.X}$ is

$$s_{Y.X} = \sqrt{\frac{1}{n - 2}\left[\sum y^2 - \frac{(\sum xy)^2}{\sum x^2}\right]} \qquad (19.2.5)$$

where $\sum y^2$, $\sum x^2$, and $\sum xy$ are computed as previously (Equations 17.3.3, 17.2.9, and 17.2.8). Let us compute the standard error of estimate for the data in Table 19.2.1, where $X =$ number of weeks of training and $Y =$ time required to perform a specified task for $n = 12$ subjects. First, compute

$$\sum x^2 = 1,620 - \frac{(120)^2}{12} = 420$$

$$\sum y^2 = 259 - \frac{(51)^2}{12} = 42.25$$

$$\sum xy = 394 - \frac{(120)(51)}{12} = -116$$

Then, applying Equation 19.2.5, we obtain

$$s_{Y.X} = \sqrt{\frac{1}{10}\left[42.25 - \frac{(-116)^2}{420}\right]} = 1.01$$

$s_{Y.X} = 1.01$ indicates that on the average the observed time required Y deviates 1.01 minutes (higher or lower) from the predicted (or estimated) time required Y_e based on the regression equation.

TABLE 19.2.1

Number of weeks of training X and time required to per-
form a certain task Y (minutes).

	X	Y	X^2	Y^2	XY
	11	3	121	9	33
	18	3	324	9	54
	2	8	4	64	16
	6	5	36	25	30
	9	6	81	36	54
	14	3	196	9	42
	8	4	64	16	32
	1	6	1	36	6
	20	1	400	1	20
	16	2	256	4	32
	4	5	16	25	20
	11	5	121	25	55
Total	120	51	1,620	259	394

Suppose the data in Table 19.2.1 were obtained as follows: The X values were predetermined by the analyst and from subjects with each specified period of training (X value), a subject was selected on a random basis, and the time required to perform the task Y determined for this subject. The bivariate model suitable for this analysis is the regression model, since the X values were predetermined, whereas the Y values are random observations. On the other hand, suppose the 12 subjects were randomly selected from the population and X and Y determined for each subject. Then, the correlation model is the suitable bivariate model, since the X and Y observations are both random.

When formulating statistical inferences, it is necessary to determine which of the two bivariate models is suitable. Inferences relating to the correlation coefficient can be made only for problems which fit the correlation model. This will be discussed in the subsequent sections of this chapter.

19.3 Interval Estimation

As already noted, the estimated regression coefficients a and b are point estimates of the true regression coefficients α and β. We may evaluate these point estimates by constructing confidence intervals, following the general procedure of Chapter 9; however, first we need to present sampling models to describe the sampling distributions of the statistics a and b.

Bivariate Model A

If samples of n pairs are selected from a population of X and Y pairs and the conditions of the regression or correlation model are satisfied, the sampling distribution of b is normal, with mean β and estimated standard error

$$s_b = \frac{s_{Y.X}}{\sqrt{\sum x^2}}$$

and the statistic

$$t = \frac{b - \beta}{s_b}$$

has the t distribution, with $n - 2$ degrees of freedom.

Bivariate Model B

If samples of n pairs are selected from a population of X and Y pairs and the conditions of the regression or correlation model are satisfied, the sampling distribution of a is normal, with mean α and estimated standard error

$$s_a = s_{Y.X} \sqrt{\frac{1}{n} + \frac{\overline{X}^2}{\sum x^2}}$$

and the statistic

$$t = \frac{a - \alpha}{s_a}$$

has the t distribution with $n - 2$ degrees of freedom.

The symbols in the foregoing models should be familiar. Following the procedure of Chapter 9, we may write the confidence intervals for β and α as follows:

$$b \pm t s_b \qquad\qquad (19.3.1)$$

$$a \pm t s_a \qquad\qquad (19.3.2)$$

We use t, as previously, to denote a t value for any specified number of df and any desired level of confidence.

Let us compute the least squares regression equation for the data in Table 19.2.1, where $n = 12$, $X =$ number of weeks of training, and $Y =$ time required to perform a task. We have already computed $\sum x^2 = 420$, $\sum y^2 = 42.25$, $\sum xy = -116$, and $s_{Y.X} = 1.01$. Then, $b = -116/420 = -.28$ and $a = [51 - (-.28)(120)]/12 = 7.05$ and the regression equation is $Y_e = 7.05 - .28X$. In order to compute a confidence interval for β, first compute (Bivariate Model A)

$$s_b = \frac{1.01}{\sqrt{420}} = .05$$

Then, we construct a .95 confidence interval for β as (Equation 19.3.1)

$$-.28 \pm (2.228)(.05) \quad \text{or} \quad -.28 \pm .11$$

where $t_{.05} = 2.228$ was determined (Table B.5) for df $= 12 - 2 = 10$. Finally, the confidence limits are

$$\text{LL} = -.28 - .11 = -.39$$
$$\text{UL} = -.28 + .11 = -.17$$

Hence, we are 95 percent confident that the interval $(-.39)$–$(-.17)$ contains the true regression coefficient β.

In order to compute a confidence interval for α, first compute (Bivariate Model B)

$$s_a = 1.01 \sqrt{\frac{1}{12} + \frac{(10)^2}{420}} = .57$$

Then, we compute a .95 confidence interval for α as (Equation 19.3.2)

$$7.05 \pm (2.228)(.57) \quad \text{or} \quad 7.05 \pm 1.27$$

The confidence limits are

$$\text{LL} = 7.05 - 1.27 = 5.78$$
$$\text{UL} = 7.05 + 1.27 = 8.32$$

Hence, we are 95 percent confident that the interval 5.78–8.32 contains the true regression coefficient α.

Important problems encountered in regression analysis relate to the question, How good is a predicted value? Clearly, we have been answering such questions by construction of confidence intervals. As discussed in Chapter 9, a confidence interval indicates the maximum error which may be made when using a statistic as a point estimate of a parameter, at a stated level of confidence. An interval constructed around a predicted value is called a *prediction interval*.

Predictions are of two types: the *average Y* for a given X or a *specific value* of Y corresponding to a given X. For example, referring to our illustration in Table 19.2.1, we may wish to predict (or estimate): (1) the average time required to perform the task (average Y) for subjects with a specified number of weeks of training X or (2) the time required for a particular subject (specific Y value) with a stated amount of training X. In each case, the predicted Y for a given X is the same; however, the maximum error which may be made at a stated level of confidence is larger for a specific prediction than for a prediction of an average. We will present a sampling model which provides a basis for constructing prediction intervals.

Bivariate Model C

If samples of n pairs are selected from a population of X and Y pairs and the conditions of the regression or correlation model are satisfied, the sampling distribution of $Y_e (= a + bX_0)$ corresponding to a specified $X = X_0$ is normal, with mean $\mu_{Y.X}$ and

1. If Y_e represents an average value, the estimated standard error is

$$s_{Y_e} = s_{Y.X} \sqrt{\frac{1}{n} + \frac{(X_0 - \overline{X})^2}{\sum x^2}}$$

2. If Y_e represents a specific value, the estimated standard error is

$$s_{Y_e} = s_{Y.X} \sqrt{1 + \frac{1}{n} + \frac{(X_0 - \overline{X})^2}{\sum x^2}}$$

3. The sampling distribution of the statistic

$$t = \frac{Y_e - \mu_{Y.X}}{s_{Y_e}}$$

has the t distribution, with $n - 2$ degrees of freedom and s_{Y_e} computed by one or the other equation, as appropriate.

When Y_e represents an average value, it is a point estimate of the conditional mean $\mu_{Y.X}$. On the other hand, when Y_e represents a specific value, it is a point estimate of the particular value predicted (or estimated). A prediction interval may be constructed as follows:

$$Y_e \pm t s_{Y_e} \qquad\qquad (19.3.3)$$

where s_{Y_e} is determined by either of the two equations, as appropriate.

Referring to our illustration in Table 19.2.1 and using the regression equation $Y_e = 7.05 - .28X$, we predict (estimate) the time required to perform the task for $X = 15$ weeks of training as

$$Y_e = 7.05 - .28(15) = 2.85 \text{ minutes}$$

Then, we may interpret $Y_e = 2.85$ minutes as the average time required for subjects with 15 weeks of training or as the predicted time required for a particular subject who has had 15 weeks of training.

Suppose $Y_e = 2.85$ minutes is interpreted as an average value. We may determine a prediction interval by first computing s_{Y_e}. Substituting $s_{Y.X} = 1.01$, $\overline{X} = 10$, $\sum x^2 = 420$, and $X_0 = 15$ (the X value used for the prediction) into the appropriate equation for s_{Y_e} in Bivariate Model C, we obtain

$$s_{Y_e} = 1.01 \sqrt{\frac{1}{12} + \frac{(15 - 10)^2}{420}} = .38$$

Then, we may construct a .95 prediction interval for the average Y_e corresponding to $X = 15$, by substituting into Equation 19.3.3 and noting that $t_{.05} = 2.228$ for $n - 2 = 10$ df. This gives

$$2.85 \pm (2.228)(.38) \qquad \text{or} \qquad 2.85 \pm .85$$

Then, the prediction limits are

$$\text{LL} = 2.85 - .85 = 2.00 \text{ minutes}$$
$$\text{UL} = 2.85 + .85 = 3.70 \text{ minutes}$$

Therefore, we are 95 percent confident that the interval 2.00–3.70 minutes includes the average time required to perform the task for subjects with 15 weeks of training.

On the other hand, if $Y_e = 2.85$ minutes is interpreted as a prediction of the time required for a particular subject who has had 15 weeks of training, we may determine a .95 prediction interval by first computing

$$s_{Y_\bullet} = 1.01 \sqrt{1 + \frac{1}{12} + \frac{(15 - 10)^2}{420}} = 1.08$$

Notice how much larger $s_{Y_\bullet}$ is now. (Compare the two equations for $s_{Y_\bullet}$.) Then the prediction interval becomes

$$2.85 \pm (2.228)(1.08) \qquad \text{or} \qquad 2.85 \pm 2.41$$

The prediction limits are

$$\text{LL} = 2.85 - 2.41 = .44 \text{ minute}$$
$$\text{UL} = 2.85 + 2.41 = 5.26 \text{ minutes}$$

Hence, we are 95 percent confident that the interval .44–5.26 minutes includes the time required for the subject who has had 15 weeks of training.

Notice that the prediction interval is wider when a specific prediction is made than for a prediction of an average value. Since the equations for $s_{Y_\bullet}$ contain the squared deviation $(X_0 - \overline{X})^2$, the estimated standard error $s_{Y_\bullet}$ is larger the further from $\overline{X}$ is X_0. Also, for $X_0 = 0$, the predicted value is $Y_e = a + b(0) = a$. When $X_0 = 0$ is substituted in $s_{Y_\bullet}$ for an average Y, $s_{Y_\bullet}$ becomes equal to s_a (Bivariate Model B). Hence, when $X = 0$, a represents the point estimate of $\mu_{Y.X}$ and the estimated standard errors $s_{Y_\bullet}$ (average value) and s_a are identical.

19.4   Tests of Significance

It is customary to test the regression and correlation coefficients obtained for *statistical significance*. That is, the null hypothesis is set up to state that the true coefficient is equal to zero. Then, the deviation of the estimated coefficient from zero in standard units is evaluated to determine how probable it is for such a deviation or a larger one to occur, if the true coefficient is actually equal to zero. This is called a *test of significance*.

We can conduct tests of significance of a and b by first computing

$$t = \frac{b - \beta}{s_b} \qquad df = n - 2 \qquad \text{(Bivariate Model A)}$$

and

$$t = \frac{a - \alpha}{s_a} \qquad df = n - 2 \qquad \text{(Bivariate Model B)}$$

Then, on the assumption that the null hypotheses $H_0: \beta = 0$ and $H_0: \alpha = 0$ are true, these equations become

$$t = \frac{b}{s_b} \tag{19.4.1}$$

and
$$t = \frac{a}{s_a} \tag{19.4.2}$$

Let us test the coefficients $b = -.28$ and $a = 7.05$ (for the illustration in Table 19.2.1) for statistical significance. We have already determined $s_b = .05$ and $s_a = .57$. Applying the foregoing equations, $t = -.28/.05 = -5.60$ for b and $t = 7.05/.57 = 12.31$ for a. Referring to Table B.5, we find that the t for df $= 12 - 2$ or 10 closest to -5.60 is $t = \pm 4.587$ for the .001 (two-tail) level of significance. The computed $t = -5.60$ for $b = -.28$ is considerably below -4.587, indicating that the b coefficient is clearly significant at better than the .001 level of significance. In other words, the probability is less than one chance in 1000 that a b of $-.28$ or one further from 0 (in either direction) would be obtained, if the true coefficient β were actually equal to 0. Thus, it is unlikely that β is equal to 0.

Typically, if the computed t is equal to or larger in magnitude than the t for the .05 level of significance, the coefficient being tested is considered to be statistically significant or significantly different from zero. If a test of significance indicates that the regression coefficient b is not statistically significant at an acceptable level of significance (for example, the .05 level), this indicates that X is worthless as a predictor of Y because there is no linear relationship between X and Y. (Of course, if there is a curvilinear relationship between X and Y, then a transformation of X or Y or both would provide a useful basis for prediction, as discussed in Chapter 18.)

Determination that $b = -.28$ is statistically significant, as just discussed, amounts to conducting a two-tail test of the null hypothesis and rejecting this hypothesis. If it is hypothesized that b has a specified sign ($+$ or $-$), then the level of significance should be evaluated as for a one-tail test. For example, in our illustration $X =$ number of weeks of training and $Y =$ time required to perform a task, it should be hypothesized that b (and β) must be negative. (Why?) Accordingly, in our previous discussion we should have taken into account that $t = -4.587$ is the critical value of t at the .0005 (lower tail) level of significance. Then $t = -5.60$ indicates that $b = -.28$ is statistically significant at better than the .0005 level of significance. In other words, the chances are less than 5 in 10,000 that a b of $-.28$ or less would be obtained if β were actually 0. Thus, it is unlikely that β is 0 or higher.

In our illustrative problem, it should be hypothesized that a (and α) is positive. Then comparing $t = 12.37$ for a with $t = 4.587$ indicates that a is statistically significant at considerably better than the .0005 (upper tail) level of significance. The chances are less than 5 in 10,000 that an a of 7.05 or higher would be obtained, if α were actually 0. Thus, it is unlikely that α is 0 or lower.

Tests of significance may be conducted for the correlation coefficient for problems which fit the correlation model. We use r to denote the coefficient of correlation computed from sample data, so that r is the estimated correlation coefficient. The true correlation coefficient, computed from population data, is usually denoted as ρ (the Greek letter rho). The following sampling model is appropriate for tests relating to the coefficient of correlation:

Bivariate Model D

If samples of n pairs are selected from a population of X and Y pairs, the conditions of the correlation model are satisfied, and the coefficient of correlation in the population is equal to 0, the sampling distribution of the statistic

$$t = \sqrt{\frac{(n-2)r^2}{1-r^2}}$$

has the t distribution with $n - 2$ degrees of freedom.

Suppose in a correlation analysis $n = 14$, $r = .60$, and $r^2 = .36$. Then,

$$t = \sqrt{\frac{(14-2)(.36)}{1-.36}} = 2.60$$

Referring to Table B.5, for 12 df $t = 2.179$ for the .05 (two-tail) level of significance and $t = 2.681$ for the .02 (two-tail) level of significance. (The one-tail level of significance is not permitted in this test.) The computed $t = 2.60$ is closer to $t = 2.681$, so that $r = .60$ is statistically significant at close to the .02 level of significance. In other words, the chances are a little higher than 2 in 100 that an r of .60 or one further from 0 (either way) could be obtained if $\rho = 0$. Thus, it is unlikely that $\rho = 0$ and it may be concluded that X and Y are linearly related.

Testing r for statistical significance (when the *correlation model* is applicable) is equivalent to testing b for statistical significance. These two tests will always give the same results.

Study Problems

1. For each problem noted, indicate which Bivariate Model (regression model or correlation model) is appropriate. The problems noted are those which follow Chapter 17. Defend your choice.
        a. Problem 2   b. Problem 3
        c. Problem 7   d. Problem 14

2. For each problem noted (problems which follow Chapter 17) compute (a) $s_{Y.X}$, (b) s_b, (c) s_a, (d) .95 confidence interval for β, and (e) .95 confidence interval for α.
        (1) Problem 3   (2) Problem 7   (3) Problem 14

3. For each problem noted (those which follow Chapter 17), compute (a) Y_e corresponding to the specified X_0 value, (b) s_{Y_e}, assuming Y_e represents an average value, (c) .95 prediction interval for Y_e, (d) s_{Y_e}, assuming Y_e represents a specific value, and (e) .95 prediction interval for Y_e.

     (1) Problem 3, $X_0 = 6$
     (2) Problem 7, $X_0 = 8$
     (3) Problem 14, $X_0 = 12$

4.   For each problem noted (those which follow Chapter 17) test the regression coefficient b for statistical significance (use the results obtained in Problem 2 of this chapter):

     a. Problem 3. (According to economic theory, quantity sold and price are typically negatively related, so that β is expected to be negative.)
     b. Problem 7. (Two-tail test.)
     c. Problem 14.

5.   In Problem 7, Chapter 17, test the regression constant a for statistical significance (two-tail test). Use the results obtained in Problem 2 of this chapter.

6.   A random sample of 20 X and Y pairs has a coefficient of correlation of .83. Test this r for statistical significance.

7.   Given a correlation analysis, with $n = 15$ and $r = .70$. Test r for statistical significance.

Chapter 20

Multiple Regression and Correlation

20.1   General Concepts

The methods of regression and correlation presented so far have limited application since they are suitable only for two-variable problems. In this chapter we will consider the methods of *multiple regression and correlation* which are applicable to problems involving *three or more variables*. For example, is the volume of automobile sales related to the three variables income per capita, average price of automobiles, and the size of the suburban population? Could we estimate the amount of a chemical contained in a product by its boiling point and weight? Could we predict how long it will take a teen-ager to perform a task based on his test scores on spatial perception, mechanical ability, and finger dexterity? Each of these problems involve a dependent variable (automobile sales, amount of a chemical, how long it will take a teen-ager to perform a task) and two or more independent variables.

Conceptually, multiple correlation and regression is similar to simple (two-variable) correlation and regression; however, the statistical theory, the computations required, and the interpretation are more complicated. Although we will concern ourselves primarily, in this chapter, with three-variable problems, various comments will be made concerning problems involving more than three variables. This will provide a good basic understanding of multiple correlation and regression concepts and procedures.

It is convenient to adopt a somewhat different notation for multiple regression and correlation. Using this notation, we may write the *two-variable regression equation* as

$$X_{1e} = a_{1.2} + b_{12}X_2 \qquad (20.1.1)$$

We use X_1 instead of Y for the dependent variable and X_{1e} replaces Y_e for the predicted value. Then, $a_{1.2}$ replaces a and b_{12} replaces b. Notice that the subscripts 1.2 for a and 12 for b identify the dependent variable as X_1 and the independent variable as X_2.

A *three-variable regression equation* is written as

$$X_{1e} = a_{1.23} + b_{12.3}X_2 + b_{13.2}X_3 \qquad (20.1.2)$$

where $a_{1.23}$ is the *regression equation constant* for the dependent variable X_1 and independent variables X_2 and X_3. There is a regression coefficient associated with each independent variable. The coefficient for X_2 is $b_{12.3}$ and for X_3 is $b_{13.2}$. Similarly, we may write the *regression equation for four variables* as

$$X_{1e} = a_{1.234} + b_{12.34}X_2 + b_{13.24}X_3 + b_{14.23}X_4 \qquad (20.1.3)$$

What is the interpretation of a regression equation such as Equation 20.1.2 for three variables? What is the interpretation of the constant $a_{1.23}$ and coefficients such as $b_{12.3}$? A regression equation for two variables (Equation 20.1.1) denotes a straight line in two-dimensional space determined by X_1 and X_2 and represents the linear relationship between X_1 and X_2. Equation 20.1.2 for three variables denotes a plane in three-dimensional space and represents the linear relationship between X_1 and X_2 and also between X_1 and X_3. In Figure 20.1.1 plane A represents the regression plane in the X_1, X_2, X_3 space. Plane B is constructed perpendicular to the X_3 axis, so that X_3 has a constant value over the entire plane B. Notice that plane B intersects plane A in a straight line, B'B''. This line represents the relationship between X_1 and X_2, with X_3 held constant and the slope of this line is $b_{12.3}$, a regression coefficient in Equation 20.1.2. Consequently, $b_{12.3}$ is the slope of the linear relationship between X_1 and X_2 for a constant value of X_3.

If X_3 is held constant at any stated value, the entire plane B will shift along the X_3 axis and will be perpendicular to this axis at the corresponding point on the X_3 scale of values. The slope of the line of intersection with plane A (B'B''), $b_{12.3}$, will not change. Clearly, $b_{12.3}$ as well as b_{12} represents the slope of a regression line; however, $b_{12.3}$ is a "more refined" measure than b_{12} since it represents the expected change in X_1 corresponding to a unit change in X_2 when X_3 is held

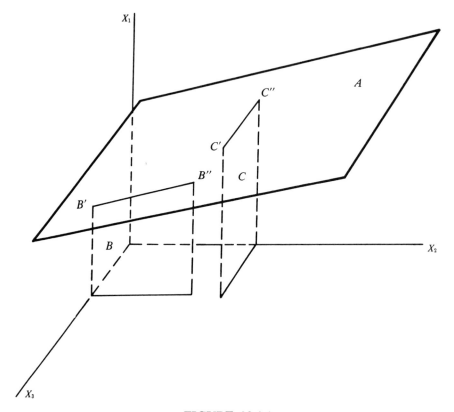

FIGURE 20.1.1

constant. The slope b_{12} represents the expected change in X_1 for a unit change in X_2 without taking into consideration the effect of X_3. Generally, $b_{12.3}$ and b_{12} will not be equal.

Plane C in Figure 20.1.1 is constructed perpendicular to the X_2 axis, so that X_2 has a constant value over this entire plane. Notice that plane C intersects plane A in the straight line C'C'' which has the slope $b_{13.2}$, a regression coefficient in Equation 20.1.2. Consequently, $b_{13.2}$ indicates the expected change in X_1 corresponding to a unit change in X_3, with X_2 held constant, and C'C'' represents the linear regression relationship between X_1 and X_3, with X_2 held constant. The regression equation constant $a_{1.23}$ indicates the point on the X_1 axis where it is intersected by regression plane A, so that it represents the height of the regression plane. This is similar to the interpretation of $a_{1.2}$ in a two-variable regression equation.

A multiple regression equation accomplishes statistically the control of relevant variables which are controlled directly in a laboratory. For example, a chemist may wish to determine the relationship of the reaction time of a chemical process to the quantity of a catalyst introduced when the temperature is held constant. The chemist would perform his experiments in a constant temperature environment. Such control of relevant variables is not always possible. For example, the business analyst who wishes to study the relationship of department store sales volume to personal income when the price level is held constant cannot (certainly in a free economy) arbitrarily control the price level. Multiple correlation and regression analysis provides such control for him; however, not in the same way as in a laboratory. In a multiple regression equation, such as Equation 20.1.2, the effect on sales volume (X_1) of a unit change in personal income (X_2) is determined with the effect of the price level (X_3) on both X_1 and X_2 removed. This *net result* is the regression coefficient $b_{12.3}$. Similarly, $b_{13.2}$ indicates the *net effect* on X_1 of a unit change in X_3 with the effect of X_2 on both X_1 and X_3 eliminated.

A multiple regression equation may define a *curved surface*, instead of a plane, as indicated in Figure 20.1.2, surface A. Plane B is constructed perpendicular to the X_3 axis and its intersection with the curved surface A is curvilinear, indicating that the regression relationship between X_1 and X_2, with the effect of X_3 eliminated, is curvilinear. Plane C in this figure is perpendicular to the X_2 axis and its intersection with surface A is linear, indicating that the regression relationship between X_1 and X_3, with the effect of X_2 removed, is linear. We will consider only linear relationships in the subsequent discussion in this chapter.

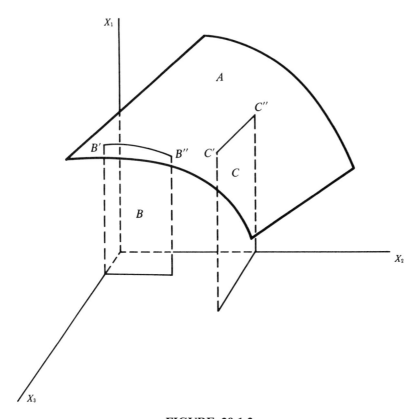

FIGURE 20.1.2

In a four-variable regression equation (Equation 20.1.3), as well as in equations with more variables, the coefficients have a similar interpretation. For example, the constant $a_{1.234}$ indicates the height of the regression surface and $b_{12.34}$ measures the expected change in X_1 corresponding to a unit change in X_2, after the effects of X_3 and X_4 have been eliminated. Note that $b_{12.34}$ is "more refined" or "more net" than $b_{12.3}$ which takes into account only the effect of X_3 or b_{12} which does not take into account the effect of any relevant variable.

When four or more variables are included in a regression equation, the equation represents a configuration in *hyperspace*. For example, with four variables, the configuration is in four-dimensional space, and with five variables, the configuration is in five-dimensional space. Hence, we cannot present the regression equation graphically when more than three variables are involved.

The regression coefficient b_{12} in a two-variable regression equation is called the *total regression coefficient* or, more often, the *simple* or *zero-order regression coefficient*. "Zero-order" means that no variables are held constant. Then, $b_{12.3}$ is the *first-order partial regression coefficient* (one variable, X_3, held constant) and $b_{12.34}$ is the *second-order partial regression coefficient* (two variables, X_3 and X_4, held constant).

We may designate the coefficient of correlation for two variables X_1 and X_2 as r_{12} (instead of just r). In a multiple correlation analysis, we compute correlation coefficients to measure the closeness of relationship between the dependent variable X_1 and each of the independent variables (X_2, X_3, etc.) separately and also between X_1 and the independent variables as a group. The correlation coefficient r_{12} is called the *total correlation coefficient* and, more often, the *simple* or *zero-order correlation coefficient*, where "zero-order" has the same meaning as before (no variables held constant).

In a multiple correlation analysis involving three variables X_1 (dependent) and X_2 and X_3 (independent), we compute the *partial correlation coefficients* $r_{12.3}$ and $r_{13.2}$. The coefficient $r_{12.3}$ measures the closeness of the relationship between X_1 and X_2, with X_3 held constant (the effect of X_3 removed from X_1 and X_2). This is called a *first-order partial correlation coefficient* (one variable held constant). The coefficient $r_{13.2}$ has a similar interpretation. We also compute $R_{1.23}$, the *coefficient of multiple correlation*, to measure the degree of association between X_1 and the independent variables X_2 and X_3 together.

In a four-variable (X_1, X_2, X_3, and X_4) analysis, we compute the *second-order partial correlation coefficients* (two variables held constant) $r_{12.34}$, $r_{13.24}$, and $r_{14.23}$ and the *multiple correlation coefficient* $R_{1.234}$.

20.2   Regression

In Chapter 17 it was shown that the scatter diagram is an effective device for displaying the relationship between two variables. When three variables are involved, the scatter diagram is three-dimensional, as shown in Figure 20.2.1. In fitting a *regression plane* to this scatter diagram, a "best fit" is obtained by the method of least squares, as in the case of two-variable problems. As before,

X_1 denotes the independent variable and X_{1e}, its predicted or estimated value. The deviations $(X_1 - X_{1e})$ are shown as vertical lines connecting the dots to the regression plane in Figure 20.2.1. (The hollow dots are below the plane.) Following Equation 17.2.2, the *least squares regression plane* is determined so that

$$\sum (X_1 - X_{1e})^2 \quad \text{is a minimum} \tag{20.2.1}$$

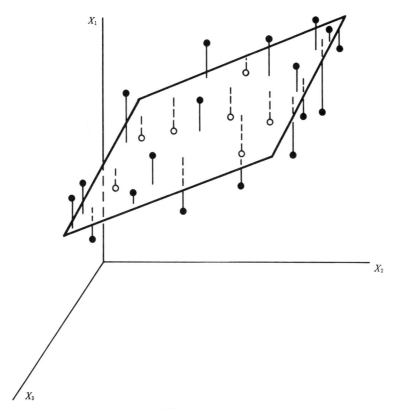

FIGURE 20.2.1

Computation of a *least squares multiple regression equation* for three variables requires computation of $a_{1.23}$, $b_{12.3}$, and $b_{13.2}$ so that the foregoing equation holds. The computational procedure involved is not difficult, but it is long and laborious. Consequently, the computations are usually computerized where computer facilities are available. Desk calculator computations are feasible, but the work should be performed carefully and systematically. The computational procedure will be presented in terms of an illustration, as a help to the reader. Since the computations are so involved, it is advisable to carry out the computations using a minimum of 4 or 5 decimals and 6 would be better. (The computations used in our illustration were obtained from a computer and the results were displayed generally with a fewer number of decimals. Therefore, we will be able to show only three decimals.)

Table 20.2.1 presents the average amount spent for commodity A (X_1), average price of a competing commodity (X_2), and average price of commodity A (X_3) for a random sample of $n = 14$ weeks. Let us compute a *linear multiple regression equation* to predict (or estimate) the average amount spent for commodity A based on the price of the competing commodity and the price of commodity A.

TABLE 20.2.1

Average amount spent for commodity A (X_1, $), average price of a competing commodity (X_2, ¢), average price of commodity A (X_3, ¢) for a random sample of 14 weeks.

X_1	X_2	X_3	$X_1{}^2$	$X_2{}^2$	$X_3{}^2$	X_1X_2	X_1X_3	X_2X_3
2	4	8	4	16	64	6	16	32
19	9	6	361	81	36	171	114	54
3	3	12	9	9	144	9	36	36
17	9	5	289	81	25	153	85	45
15	5	3	225	25	9	75	45	15
6	2	7	36	4	49	12	42	14
8	8	9	64	64	81	64	72	72
17	9	4	289	81	16	153	68	36
14	5	2	196	25	4	70	28	10
9	4	7	81	16	49	36	63	28
5	6	10	25	36	100	30	50	60
2	5	14	4	25	196	10	28	70
7	8	11	49	64	121	56	77	88
8	7	9	64	49	81	56	72	63
Total 132	84	107	1,696	576	975	903	796	623

In Table 20.2.1, the basic data (X_1, X_2, and X_3) are presented in the first three columns. The second three columns present the squares of the observations ($X_1{}^2$, $X_2{}^2$, and $X_3{}^2$) and the last three columns present the cross-products (X_1X_2, X_1X_3, and X_2X_3). The nine column totals are the basic input into the computations. These column totals are the sums ($\sum X_1$, $\sum X_2$, and $\sum X_3$), the sums of squares ($\sum X_1{}^2$, $\sum X_2{}^2$, and $\sum X_3{}^2$), and the sums of cross-products ($\sum X_1X_2$, $\sum X_1X_3$, and $\sum X_2X_3$). Using these totals, compute the sums of squares and sums of cross-products in terms of deviations from the mean (also called the *corrected sums*), as follows:

$$\sum x_1{}^2 = \sum X_1{}^2 - \frac{(\sum X_1)^2}{n} \tag{20.2.2}$$

$$\sum x_2{}^2 = \sum X_2{}^2 - \frac{(\sum X_2)^2}{n} \tag{20.2.3}$$

$$\sum x_3{}^2 = \sum X_3{}^2 - \frac{(\sum X_3)^2}{n} \tag{20.2.4}$$

$$\sum x_1x_2 = \sum X_1X_2 - \frac{\sum X_1 \sum X_2}{n} \tag{20.2.5}$$

$$\Sigma\ x_1x_3 = \Sigma\ X_1X_3 - \frac{\Sigma\ X_1\ \Sigma\ X_3}{n} \tag{20.2.6}$$

$$\Sigma\ x_2x_3 = \Sigma\ X_2X_3 - \frac{\Sigma\ X_2\ \Sigma\ X_3}{n} \tag{20.2.7}$$

where
$$x_{1i} = X_{1i} - \overline{X}_1 \tag{20.2.8}$$
$$x_{2i} = X_{2i} - \overline{X}_2 \tag{20.2.9}$$
$$x_{3i} = X_{3i} - \overline{X}_3 \tag{20.2.10}$$

n = number of sets of X_1; X_2, and X_3 observations

After taking into account the new notation, it will be apparent that similar equations were presented in Chapter 17 for two-variable problems. Noting that in our illustration (Table 20.2.1) $n = 14$, apply the preceding equations and the column totals as required to obtain

$$\Sigma\ x_1{}^2 = 1{,}696 - \frac{(132)^2}{14} = 451.429$$

$$\Sigma\ x_2{}^2 = 576 - \frac{(84)^2}{14} = 72.000$$

In the same manner, we obtain $\Sigma\ x_3{}^2 = 157.214$, $\Sigma\ x_1x_2 = 111.000$, $\Sigma\ x_1x_3 = -212.857$, and $\Sigma\ x_2x_3 = -19.000$.

Using the foregoing results compute the *simple regression coefficients:*

$$b_{12} = \frac{\Sigma\ x_1x_2}{\Sigma\ x_2{}^2} \tag{20.2.11}$$

$$b_{21} = \frac{\Sigma\ x_1x_2}{\Sigma\ x_1{}^2} \tag{20.2.12}$$

$$b_{13} = \frac{\Sigma\ x_1x_3}{\Sigma\ x_3{}^2} \tag{20.2.13}$$

$$b_{31} = \frac{\Sigma\ x_1x_3}{\Sigma\ x_1{}^2} \tag{20.2.14}$$

$$b_{23} = \frac{\Sigma\ x_2x_3}{\Sigma\ x_3{}^2} \tag{20.2.15}$$

$$b_{32} = \frac{\Sigma\ x_2x_3}{\Sigma\ x_2{}^2} \tag{20.2.16}$$

It should be noted that b_{12} and b_{21} are different coefficients (also b_{13} and b_{31}, b_{23} and b_{32}). As previously discussed, b_{12} indicates that X_1 is the dependent variable and X_2 is the independent variable. The reverse is true for b_{21} (where the regression equation is $X_{2e} = a_{2.1} + b_{21}X_1$). Applying the preceding equations to our problem, we obtain

$$b_{12} = \frac{111.000}{72.000} = 1.542 \qquad b_{21} = \frac{111.000}{451.429} = .246$$

In the same manner, we obtain $b_{13} = -1.354$, $b_{31} = -.472$, $b_{23} = -.121$, and

$b_{32} = -.264$. (Notice the similarity of the foregoing equations to Equation 17.2.6 for b_{12}.)

As a step toward computation of the partial regression coefficients, compute the *simple correlation coefficients* as follows:

$$r_{12} = \sqrt{b_{12}b_{21}} \qquad \text{(sign of } r_{12} \text{ same as for } b_{12}) \qquad (20.2.17)$$

$$r_{13} = \sqrt{b_{13}b_{31}} \qquad \text{(sign of } r_{13} \text{ same as for } b_{13}) \qquad (20.2.18)$$

$$r_{23} = \sqrt{b_{23}b_{32}} \qquad \text{(sign of } r_{23} \text{ same as for } b_{23}) \qquad (20.2.19)$$

In Equation 20.2.17, if we substitute for b_{12} and b_{21} in accordance with Equations 20.2.11 and 20.2.12, it will be seen immediately that Equation 20.2.17 for r_{12} is equivalent to Equation 17.3.2. Applying these equations to our illustration, we obtain

$$r_{12}^2 = .379 \qquad r_{12} = .62$$
$$r_{13}^2 = .638 \qquad r_{13} = -.80$$
$$r_{23}^2 = .032 \qquad r_{23} = .18$$

Using the preceding results, compute the *partial regression coefficients* as follows:

$$b_{12.3} = \frac{b_{12} - b_{13}b_{32}}{1 - r_{23}^2} \qquad (20.2.20)$$

$$b_{13.2} = \frac{b_{13} - b_{12}b_{23}}{1 - r_{23}^2} \qquad (20.2.21)$$

Applying these equations, we obtain

$$b_{12.3} = \frac{1.542 - (-1.354)(-.264)}{1 - .032} = 1.223$$

$$b_{13.2} = \frac{-1.354 - (1.542)(-.121)}{1 - .032} = -1.206$$

Notice that $b_{12} \neq b_{12.3}$ and $b_{13} \neq b_{13.2}$, as expected. Noting that $b_{12} = 1.542$, we may state that, on the average, an increase of one cent in the price of the competing commodity (X_2) was associated with an increase of about \$1.54 in the average amount spent on commodity A (X_1). It is to be expected that the price of commodity A (X_3) had an important effect on the average amount spent on the commodity, so that it should be taken into account in the analysis. Then $b_{12.3} = 1.223$ indicates that, after removing the effect of the price of commodity A, a one cent increase in the price of the competing commodity was associated, on the average, with an increase of about \$1.22 in the average amount spent on commodity A.

Similarly, $b_{13} = -1.354$ indicates that on the average, a one cent increase in the price of commodity A (X_3) was associated with a decrease of about \$1.35 in the average amount spent on the commodity. However, $b_{13.2} = -1.206$ indicates that on the average, after taking into account the effect of the price of the competing commodity (X_2), a one cent rise in the price of commodity A was associated with a \$1.21 drop in the average amount spent on the commodity.

Notice that in Equations 20.2.20 and 20.2.21 for the partial regression coefficients the denominator is $1 - r_{23}^2$, indicating that the higher r_{22}^2, the smaller is the denominator and the higher are the coefficients. If X_2 and X_3 are closely associated, r_{23}^2 will be close to unity and the partial regression coefficients will be very large. If $r_{23}^2 = 1$, these coefficients cannot be determined. Consequently, *if X_2 and X_3 are closely associated, then one of these variables should be discarded from the analysis or replaced by another logical variable.*

An important question in multiple regression analysis relates to the relative importance of the independent variables in explaining the variation in X_1. Is X_2 or X_3 more important in its effect on X_1? It is not usually possible to compare $b_{12.3}$ and $b_{13.2}$ directly for this purpose. If the regression equation is expressed in terms of standard units for the variables involved, we would obtain the partial regression coefficients in terms of these units, called the *B coefficients*, as follows:

$$B_{12.3} = b_{12.3} \frac{s_2}{s_1} = b_{12.3} \sqrt{\frac{\sum x_2^2}{\sum x_1^2}} \qquad (20.2.22)$$

$$B_{13.2} = b_{13.2} \frac{s_3}{s_1} = b_{13.2} \sqrt{\frac{\sum x_3^2}{\sum x_1^2}} \qquad (20.2.23)$$

Applying these equations to our illustration, we obtain $B_{12.3} = .49$ and $B_{13.2} = -.71$. $B_{12.3}$ indicates that an increase of one standard unit in the price of the competing commodity (X_2) has resulted in an increase of .49 standard units in the average amount spent on commodity A (X_1); whereas, an increase of one standard unit in the price of commodity A (X_3), according to $B_{13.2}$, has resulted in a decrease of .71 standard unit in the average amount spent (X_1). Clearly, changes in X_3 have had a greater effect on X_1 than changes in X_2, as well as an effect in the opposite direction.

Finally, the *regression equation constant* is computed as follows:

$$a_{1.23} = \frac{\sum X_1 - b_{12.3} \sum X_2 - b_{13.2} \sum X_3}{n}$$

$$= \frac{132 - (1.223)(84) - (-1.206)(107)}{14} = 11.306 \qquad (20.2.24)$$

Notice the similarity between the foregoing equation and Equation 17.2.7 for $a_{1.2}$. The *multiple regression equation* just determined is

$$X_{1e} = 11.31 + 1.22X_2 - 1.21X_3$$

20.3 Correlation

Given X_1 dependent and X_2 and X_3 independent, the *first-order partial correlation coefficients* are computed as

$$r_{12.3} = \frac{r_{12} - r_{13}r_{23}}{\sqrt{(1 - r_{13}^2)(1 - r_{23}^2)}} \qquad (20.3.1)$$

$$r_{13.2} = \frac{r_{13} - r_{12}r_{23}}{\sqrt{(1 - r_{12}^2)(1 - r_{23}^2)}} \qquad (20.3.2)$$

Notice in these equations that if X_2 and X_3 are closely related, the partial correlation coefficients will be very high and if $r_{23} = 1$, indeterminate. This also shows that X_2 and X_3 should not both be included in a multiple correlation analysis if they are highly correlated. Applying these equations to our problem (from the previous section), we obtain $r_{12.3} = .80$ and $r_{13.2} = -.89$.

How are the partial coefficients of correlation interpreted? Recall that in Chapter 17 it was shown that a simple correlation coefficient r_{12} is interpreted in terms of its square r_{12}^2, the coefficient of determination. For example, $r_{12} = .70$ indicates that 49 percent of the variation in X_1 is associated with or explained by the variation in X_2. The square of a partial correlation coefficient is called the *coefficient of partial determination* and provides a basis for interpreting the partial correlation coefficient. The coefficient of partial determination represents the proportion of the residual variation in X_1 which is explained by a given independent variable, where the residual variation in X_1 means the variation in this variable unexplained by the other independent variable (or variables) in the analysis.

The partial coefficient of correlation $r_{12.3} = .80$ is interpreted by computing $r_{12.3}^2 = .64$ which indicates that 64 percent of the variation in the average amount spent for commodity A (X_1) is explained by the variation in the price of the competing commodity (X_2), after eliminating the variation in X_1 explained by the variation in the price of commodity A (X_3). The coefficient $r_{13.2} = -.89$ indicates that 79 percent of the variation in X_1 is explained by the variation in X_3, after eliminating the variation in X_1 explained by X_2.

Comparing $r_{12}^2 = .38$ with $r_{12.3}^2 = .64$ reveals that X_1 and X_2 are more closely associated than appears to be the case before the masking effect of X_3 is removed from X_1 and X_2. Comparing $r_{13}^2 = .64$ with $r_{13.2}^2 = .79$ reveals a similar masking effect of X_2 on the relationship of X_1 and X_3.

It is instructive to examine the meaning of the coefficient of partial determination in a little more detail. In Section 17.3 we introduced the concept of the *total variation in the dependent variable* (Y or X_1). We may express this as

$$\sum (X_1 - \bar{X}_1)^2 = \sum x_1^2 = \text{total variation in } X_1 \qquad (20.3.3)$$

Now, let us partition this total variation, using the data in our illustration:

1. Total variation in $X_1 = \sum x_1^2$ 451.429
2. Variation in X_1 explained by $X_3 = r_{13}^2 \sum x_1^2$ 288.012
3. Difference: Variation in X_1 not explained by $X_3 = (1 - r_{13}^2) \cdot \sum x_1^2$. *This is the residual variation in X_1*   163.417
4. Residual variation in X_1 explained by $X_2 = r_{12.3}^2(1 - r_{13}^2) \cdot \sum x_1^2$ 104.587
5. Difference: Variation in X_1 *not explained by regression* (X_2 and X_3 together) $= (1 - r_{12.3}^2)(1 - r_{13}^2) \sum x_1^2$ 58.830
6. Variation in X_1 *explained by regression* (X_2 and X_3 together) = Row 2 + Row 4 = $[r_{13}^2 + r_{12.3}^2(1 - r_{13}^2)] \sum x_1^2$ 392.599

Notice that, in the preceding analysis, the residual variation in X_1 is shown in Row 3. Then $r_{12.3}^2$ indicates that 64 percent of this residual variation is explained by X_2 (Row 4). The sum of the variation in X_1 explained by X_3 (Row 2) and the residual variation in X_1 explained by X_2 (Row 4) represents the portion of the total variation in X_1 explained by the multiple correlation analysis. This is shown in Row 6 and represents the combined effect of X_2 and X_3 in explaining the variation in X_1. This portion of the total variation in X_1 provides the basis for computing a measure of correlation to indicate the effectiveness of X_2 and X_3 together in explaining the variation in X_1.

Notice that according to Equation 17.3.8 the coefficient of determination represents the proportion that the variation in X_1 explained by X_2 is of the total variation in X_1. Following this approach, we may compute the *coefficient of multiple determination* $R_{1.23}^2$ by dividing the variation in X_1 explained by X_2 and X_3 together (Row 6) by the total variation in X_1 (Row 1). This gives

$$R_{1.23}^2 = \frac{\text{Row } 6}{\text{Row } 1} = \frac{[r_{13}^2 + r_{12.3}^2(1 - r_{13}^2)] \sum x_1^2}{\sum x_1^2} \tag{20.3.4}$$

$$= r_{13}^2 + r_{12.3}^2(1 - r_{13}^2)$$

Then, the square root of $R_{1.23}^2$ or $R_{1.23}$ is called the *coefficient of multiple correlation*. This coefficient is always positive.

A more convenient equation for *computing* $R_{1.23}$ is

$$R_{1.23} = \sqrt{\frac{b_{12.3} \sum x_1 x_2 + b_{13.2} \sum x_1 x_3}{\sum x_1^2}} \tag{20.3.5}$$

Applying this equation to our illustration we obtain $R_{1.23}^2 = .87$ and $R_{1.23} = .93$. Hence, 87 percent of the variation in the average amount spent on commodity A is explained jointly by the variation in the price of the competing commodity and the price of commodity A.

Often the multiple correlation coefficient $R_{1.23}$ is adjusted for degrees of freedom to correct for an upward bias when $R_{1.23}$ computed from sample data is used as an estimate of the correlation in the population. The *adjusted* or *corrected multiple correlation coefficient*, denoted as $R'_{1.23}$, is computed as follows:

$$R'_{1.23} = \sqrt{1 - (1 - R_{1.23}^2)\frac{n-1}{n-3}} \tag{20.3.6}$$

In our illustration, $R'_{1.23} = .92$.

20.4 Tests of Significance

We will present two theoretical sampling models, similar to the two bivariate sampling models presented in Section 19.2, to be used in formulating statistical inferences in three-variable regression and correlation analyses.

Multivariate Model (Regression)

Given three variables, X_1 dependent, X_2 and X_3 independent. The X_2 and X_3 distributions are unspecified. The distribution of X_1 values associated with a given pair of X_2 and X_3 values, called the conditional distribution of X_1, has the following characteristics:

1. It is normal with mean $\mu_{1.23}$ and standard deviation $\sigma_{1.23}$.
2. The conditional means $\mu_{1.23}$ satisfy the true linear multiple regression relationship $\mu_{1.23} = \alpha_{1.23} + \beta_{12.3}X_2 + \beta_{13.2}X_3$ for the X_1, X_2, X_3 population.
3. The conditional standard deviation $\sigma_{1.23}$ is constant over all conditional distributions of X_1.

A sample of n sets of X_1, X_2, and X_3 values is obtained by selecting, first, n pairs of X_2, X_3 values (not necessarily on a random basis). Then an X_1 observation is selected on a random basis from the X_1 values associated with each of the n pairs of X_2, X_3 values.

Multivariate Model (Correlation)

In the X_1, X_2, X_3 population, the variables are individually and jointly normal, making up a *multivariate normal population*. There is a distribution for each variable associated with a given pair of values specified for the other two, called the conditional distribution of the variable; so that there are conditional distributions for X_1, X_2, and X_3. The characteristics of the conditional distributions are similar to those stated in the Multivariate Model (Regression).

A sample of size n is obtained by selecting a random sample of n sets of X_1, X_2, X_3 values from the X_1, X_2, X_3 population.

As indicated in the multivariate models, the *true* linear multiple regression equation

$$\mu_{1.23} = \alpha_{1.23} + \beta_{12.3}X_2 + \beta_{13.2}X_3 \qquad (20.4.1)$$

corresponds to the *estimated* linear multiple regression equation based on sample data

$$X_{1e} = a_{1.23} + b_{12.3}X_2 + b_{13.2}X_3 \qquad (20.4.2)$$

If we formulate the null hypothesis H_0: $\beta_{12.3} = 0$ and if either multivariate model holds, we can conduct a test of significance of the regression coefficient $b_{12.3}$ by computing

$$t = \frac{b_{12.3} - \beta_{12.3}}{s_{b_{12.3}}} \qquad df = n - 3 \qquad (20.4.3)$$

which, after taking H_0 into account, becomes

$$t = \frac{b_{12.3}}{s_{b_{12.3}}} \qquad df = n - 3 \qquad (20.4.4)$$

The *standard error of* $b_{12.3}$, $s_{b_{12.3}}$, is computed as

$$s_{b_{12.3}} = \frac{s_{1.23}}{\sqrt{\sum x_2^2(1 - r_{23}^2)}} \qquad (20.4.5)$$

where $s_{1.23}$ is the *standard error of estimate* and obtained as follows:

$$s_{1.23} = \sqrt{\frac{\sum x_1^2 - b_{12.3} \sum x_1 x_2 - b_{13.2} \sum x_1 x_3}{n - 3}} \qquad (20.4.6)$$

The standard error of estimate $s_{1.23}$ represents an estimate of the conditional standard deviation $\sigma_{1.23}$.

Let us conduct a test of significance for $b_{12.3} = 1.223$ computed for the data in Table 20.2.1, where X_1 = average amount spent for commodity A, X_2 = average price of a competing commodity, and X_3 = average price of commodity A. Applying Equation 20.4.6 (with $\sum x_1^2 = 451.429$, $\sum x_1 x_2 = 111.000$, $\sum x_1 x_3 = -212.857$, $b_{12.3} = 1.223$, $b_{13.2} = -1.206$, and $n = 14$), we obtain $s_{1.23} = 2.315$. Applying Equation 20.4.5 (with $\sum x_2^2 = 72.000$, $r_{23}^2 = .032$), we obtain $s_{b_{12.3}} = .277$. Finally, employing Equation 20.4.4, we obtain $t = 4.413$ for $b_{12.3}$ (df = $14 - 3 = 11$). Referring to Table B.5 for df = 11, we find $t = 4.437$ for the .001 two-tail level of significance. The computed t (= 4.413) is clearly significant at close to the .001 two-tail level.

Similarly, to test $b_{13.2}$ for statistical significance, formulate the hypothesis H_0: $\beta_{13.2} = 0$. Then compute the *standard error* for this statistic as

$$s_{b_{13.2}} = \frac{s_{1.23}}{\sqrt{\sum x_3^2 (1 - r_{23}^2)}} \qquad (20.4.7)$$

Then, compute

$$t = \frac{b_{13.2} - \beta_{13.2}}{s_{b_{13.2}}} \qquad \text{df} = n - 3$$

$$= \frac{b_{13.2}}{s_{b_{13.2}}} \qquad \text{df} = n - 3 \qquad (20.4.8)$$

Referring to our illustrative problem, where $b_{13.2} = -1.206$, apply Equation 20.4.7 (with $\sum x_3^2 = 157.214$ and the previously noted results) to obtain $s_{b_{13.2}} = .188$. Then applying Equation 20.4.8, we obtain $t = -6.429$ (df = 11). Noting from Table B.5 df = 11 that $t = 4.437$ for the .001 two-tail level of significance, we conclude that $b_{13.2} = -1.206$ is statistically significant at better than the .001 two-tail level.

The partial and multiple coefficients of correlation may be tested for statistical significance only if the Multivariate Model (Correlation) holds. We will use $\rho_{12.3}$, $\rho_{13.2}$, and $\rho_{1.23}$ to denote the population parameters corresponding to $r_{12.3}$, $r_{13.2}$, and $R_{1.23}$, respectively. If we formulate the null hypotheses H_0: $\rho_{12.3} = 0$ and H_0: $\rho_{13.2} = 0$ and if the correlation model holds, we can conduct tests of significance for the *partial correlation coefficients* by computing

$$t = \sqrt{\frac{(n - 3)r_{12.3}^2}{1 - r_{12.3}^2}} \qquad \text{df} = n - 3 \qquad (20.4.9)$$

and

$$t = \sqrt{\frac{(n - 3)r_{13.2}^2}{1 - r_{13.2}^2}} \qquad \text{df} = n - 3 \qquad (20.4.10)$$

Referring to our illustrative problem (where $r_{12.3} = .80$ and $r_{13.2} = -.89$), we compute $t = 4.414$ for $r_{12.3}$ (df = 11) using Equation 20.4.9 and $t = -6.430$ for $r_{13.2}$ (df = 11), using Equation 20.4.10. Noting from Table B.5 for df = 11 that

$t = 4.437$ for the .001 two-tail level of significance (as before) we conclude that both of these partial regression coefficients are statistically significant at close to or better than the .001 two-tail level.

The multiple correlation coefficient may be tested for statistical significance by an F test, if the correlation model holds. Formulate the null hypothesis H_0: $\rho_{1.23} = 0$ and compute

$$F = \frac{(n-3)R_{1.23}{}^2}{2(1-R_{1.23}{}^2)} \qquad df = 2, n-3 \qquad (20.4.11)$$

Since the multiple correlation coefficient ($\rho_{1.23}$ and R) can only be positive, a one-tail test must be used. Applying Equation 20.4.11 to test $R_{1.23} = .93$ for our problem, we obtain $F = 36.646$. Referring to Table B.8, we see $F = 7.21$ (df $= 2,11$) for the .01 level of significance. Clearly, $F = 36.646$ indicates that $R_{1.23} = .93$ is statistically significant at better than the .01 level.

20.5 Interval Estimation

Confidence and prediction intervals may be constructed for multiple regression problems which fit either of the multivariate models. The regression coefficients $b_{12.3}$ and $b_{13.2}$ are *point estimates* of the parameters $\beta_{12.3}$ and $\beta_{13.2}$, respectively. Confidence intervals may be computed as follows:

$$b_{12.3} \pm ts_{b_{12.3}} \qquad df = n-3 \qquad (20.5.1)$$

and

$$b_{13.2} \pm ts_{b_{13.2}} \qquad df = n-3 \qquad (20.5.2)$$

with the *standard errors* $s_{b_{12.3}}$ and $s_{b_{13.2}}$ computed according to Equations 20.4.5 and 20.4.7. Referring to our illustrative problem (Table 20.2.1), using the results previously obtained, and $t = 2.201$ for 11 df (Table B.5), we compute .95 confidence intervals as

$b_{12.3} = 1.223$: $\qquad 1.223 \pm (2.201)(.277) \qquad$ or $\qquad 1.223 \pm .610$
$\qquad\qquad\qquad\qquad$ LL $= .613 \qquad$ UL $= 1.833$

$b_{13.2} = -1.206$: $\qquad -1.206 \pm (2.201)(.188) \qquad$ or $\qquad -1.206 \pm .414$
$\qquad\qquad\qquad\qquad$ LL $= -1.620 \qquad$ UL $= -.792$

The coefficient $a_{1.23}$ is a point estimate of $\alpha_{1.23}$. We may construct a confidence interval for $a_{1.23}$ as follows:

$$a_{1.23} \pm ts_{a_{1.23}} \qquad df = n-3 \qquad (20.5.3)$$

where the *standard error* of $a_{1.23}$ is computed as

$$s_{a_{1.23}} = \sqrt{\frac{s_{1.23}{}^2}{n} + s_{b_{12.3}}{}^2\,\overline{X}_2{}^2 + s_{b_{13.2}}{}^2\,\overline{X}_3{}^2 - 2s_{b_{12.3}}s_{b_{13.2}}r_{23}\,\overline{X}_2\overline{X}_3} \qquad (20.5.4)$$

In our illustration, $a_{1.23} = 11.306$. Let us compute a .95 confidence interval. Using the data $s_{1.23} = 2.315$, $s_{b_{12.3}} = .277$, $s_{b_{13.2}} = .188$, $n = 14$, $r_{23} = .18$, $\overline{X}_2 = 6.000$, and $\overline{X}_3 = 7.643$, we compute $s_{a_{1.23}} = 2.090$ (Equation 20.5.4). The .95 confidence interval is computed as follows:

$$11.306 \pm (2.201)(2.090) \quad \text{or} \quad 11.306 \pm 4.600$$
$$\text{LL} = 6.706 \quad \text{UL} = 15.906$$

If we assign values to the independent variables, for example let $X_2 = X_{20}$ and $X_3 = X_{30}$, the corresponding predicted (or estimated) value X_{1e} represents a point estimate of the conditional mean $\mu_{1.23}$ for the X_1 distribution corresponding to X_{20} and X_{30}. Let $X_2 = 5$ and $X_3 = 6$. Entering these values into the regression equation $X_{1e} = 11.31 + 1.22X_2 - 1.21X_3$ gives $X_{1e} = 10.20$. In other words, if the average price of the competing commodity is $X_2 = 5\cancel{c}$ and the average price of commodity A is $X_3 = 6\cancel{c}$, the average amount that will be spent on commodity A is predicted as $10.20.

Prediction intervals may be computed as

$$X_{1e} \pm t s_{X_{1e}} \qquad df = n - 3 \qquad (20.5.5)$$

where the *standard error of a prediction* $s_{X_{1e}}$ is computed as follows if X_{1e} represents an average prediction of X_1 when $X_2 = X_{20}$ and $X_3 = X_{30}$:

$$s_{X_{1e}} = \sqrt{\frac{s_{1.23}{}^2}{n} + Q^2} \qquad (20.5.6)$$

where $Q^2 = s_{b_{12.3}}{}^2 (X_{20} - \overline{X}_2)^2 + s_{b_{13.2}}{}^2 (X_{30} - \overline{X}_3)^2$
$$- 2 s_{b_{12.3}} s_{b_{13.2}} r_{23}(X_{20} - \overline{X}_2)(X_{30} - \overline{X}_3) \quad (20.5.7)$$

and where $s_{X_{1e}}$ is computed as follows if X_{1e} represents a specific prediction of X_1 when $X_2 = X_{20}$ and $X_3 = X_{30}$:

$$s_{X_{1e}} = \sqrt{s_{1.23}{}^2 + \frac{s_{1.23}{}^2}{n} + Q^2} \qquad (20.5.8)$$

Referring to our illustration, $s_{1.23} = 2.315$, $s_{b_{12.3}} = .277$, $s_{b_{13.2}} = .188$, $r_{23} = .18$, $\overline{X}_2 = 6.000$, $\overline{X}_3 = 7.643$, $X_{20} = 5$, $X_{30} = 6$, and $n = 14$. Then applying Equation 20.5.6, $s_{X_{1e}}$ for an *average prediction* is .720 and for a *specific prediction*, applying Equation 20.5.8, it is 2.430. We may construct a .95 prediction interval applying these data and noting that $t = 2.201$ for 11 df (Table B.5). The .95 prediction interval for the *average prediction* $X_{1e} = 10.20$ is

$$10.20 \pm (2.201)(.720) \quad \text{or} \quad 10.20 \pm 1.58$$
$$\text{LL} = \$8.62 \quad \text{UL} = \$11.78$$

and for the *specific prediction* $X_{1e} = 10.20$, it is

$$10.20 \pm (2.201)(2.430) \quad \text{or} \quad 10.20 \pm 5.35$$
$$\text{LL} = \$4.85 \quad \text{UL} = \$15.55$$

Study Problems

1.   A correlation and regression analysis is to be run with $X_1 =$ number of crimes committed, $X_2 =$ number of dropouts from school, and $X_3 =$ average family income. The data are to be obtained from a sample of cities in a specified region.

a. Express the regression equation appropriate to this problem (in symbols).

    b. Explain the meaning of the regression equation constant and the regression coefficients.

    c. List the partial correlation coefficients pertinent to this problem and explain their meaning.

    d. Interpret the coefficient of multiple correlation.

2.   Explain the meaning of b_{12}, $b_{12.3}$, $b_{12.34}$, and $b_{12.345}$.

3.   Explain the meaning of $a_{1.2}$ and $a_{1.23}$.

4.   What is the meaning of the

    a. Total regression coefficient

    b. Simple regression coefficient

    c. Zero-order regression coefficient

    d. First-order regression coefficient

5.   Explain the meaning of r_{12}, $r_{12.3}$, $r_{13.2}$, $R_{1.23}$, and $r_{12.34}$.

6.   What is a second-order partial correlation coefficient?

7.   Two tests (P and K) were administered to a sample of 15 subjects. These subjects were also rated on leadership by a panel of experts after their records were carefully studied. The following sums, sums of squares, and sums of cross-products were computed for X_1 = leadership rating, X_2 = test P score, and X_3 = test K score:

$$\sum X_1 = 202 \qquad \sum X_1^2 = 3{,}258 \qquad \sum X_1X_2 = 3{,}629$$
$$\sum X_2 = 248 \qquad \sum X_2^2 = 4{,}372 \qquad \sum X_1X_3 = 2{,}507$$
$$\sum X_3 = 162 \qquad \sum X_3^2 = 2{,}078 \qquad \sum X_2X_3 = 2{,}873$$

Compute the

    a. Corrected sums

    b. Six simple regression coefficients

    c. Three simple coefficients of determination and the three simple correlation coefficients

    d. Partial (first-order) regression coefficients

    e. B coefficients

    f. Regression equation constant and show the linear multiple regression equation

    g. Partial (first-order) correlation coefficients

    h. Multiple correlation coefficient

    i. Adjusted (or corrected) multiple correlation coefficient

8.   The following data were collected for a sample of ten workers in a large manufacturing firm:

Worker	X_1	X_2	X_3
A	8	3	8
B	15	2	10
C	1	5	7
D	16	1	12
E	8	4	10
F	9	3	9
G	6	2	4
H	3	5	8
I	6	4	10
J	9	1	5

Where: X_1 = number of defective units produced
$\qquad$ X_2 = years of experience
$\qquad$ X_3 = total units produced (in dozens)
Compute the
$\qquad$ a.$\quad$ Simple regression coefficients (six)
$\qquad$ b.$\quad$ Simple correlation coefficients (six)
$\qquad$ c.$\quad$ Least squares linear multiple regression equation
$\qquad$ d.$\quad$ *B* coefficients
$\qquad$ e.$\quad$ Partial correlation coefficients
$\qquad$ f.$\quad$ Multiple coefficients of correlation and determination
$\qquad$ g.$\quad$ Adjusted coefficient of multiple correlation
9.$\quad$ The following data were collected for a sample of 12 cities

X_1	X_2	X_3
14	6	12
8	3	14
12	5	10
5	2	8
15	8	16
10	10	4
11	4	10
11	7	12
13	6	16
7	3	10
3	1	8
8	8	6

Where: X_1 = number of live births per 1,000 population
$\qquad$ X_2 = number of marriages per 1,000 population
$\qquad$ X_3 = average family income (in $000)
Compute (a) through (g) as for Problem 8.

10.$\quad$ In Problem 7:
$\qquad$ a.$\quad$ Compute the
$\qquad\qquad$ (1)$\quad$ Standard error of estimate ($s_{1.23}$)
$\qquad\qquad$ (2)$\quad$ Standard errors of the partial regression coefficients
$\qquad\qquad$ (3)$\quad$ Standard error of $a_{1.23}$ ($s_{a_{1.23}}$)
$\qquad\qquad$ (4)$\quad$ Standard error of a predicted average value ($X_2 = 14$, $X_3 = 9$)
$\qquad\qquad$ (5)$\quad$ Standard error of a predicted specific value ($X_2 = 14$, $X_3 = 9$)
$\qquad$ b.$\quad$ Test for statistical significance at the .05 level of significance:
$\qquad\qquad$ (1)$\quad$ Partial regression coefficients
$\qquad\qquad$ (2)$\quad$ Partial correlation coefficients
$\qquad\qquad$ (3)$\quad$ Multiple correlation coefficient
$\qquad$ c.$\quad$ Compute confidence intervals at the .95 level of confidence for the
$\qquad\qquad$ (1)$\quad$ Partial regression coefficients
$\qquad\qquad$ (2)$\quad$ Multiple regression equation constant
$\qquad$ d.$\quad$ Compute the predicted value X_{1e} for $X_2 = 14$ and $X_3 = 9$.
$\qquad$ e.$\quad$ Compute prediction intervals for the X_{1e} value obtained in (d) when
$\qquad\qquad$ (1)$\quad$ X_{1e} is an average prediction
$\qquad\qquad$ (2)$\quad$ X_{1e} is a specific prediction

11.   In Problem 8, (a) through (e), as for Problem 10. For part (d): $X_2 = 3$ and $X_3 = 6$.

12.   In Problem 9, (a) through (e), as for Problem 10. For part (d): $X_2 = 6$ and $X_3 = 10$.

Appendix A
Additional Topics

A.1 Standard Test Scores

A test score standing by itself is not meaningful. It must be evaluated with reference to the average level of scores on the same test taken by others, as well as the score variation. Clearly, if ninth grade students obtained a mean of 60 on a science aptitude test with a standard deviation of 10 points, a score of 69 would look quite good, since it is nearly one standard deviation above the mean. On the other hand, if these students obtained a mean score of 120 on a language aptitude test with a standard deviation of 5 points, a score of 105 would not look so good, since it is three standard deviations below the mean.

Transformation of a score (the *raw score*) to *standard units* z_i (Equation 5.4.3) is a widely used procedure for evaluating and comparing test scores. The z_i scores are called *standard scores*. When transformed to z_i (as discussed in Section 5.4), the standard score distribution will have a mean of zero and a standard deviation of unity. When constructing standard scores for a set of raw scores, it is customary to predetermine the standard score distribution mean and standard deviation so that they will have values other than 0 and 1, respectively. In general, a population of raw scores X_i with mean μ and standard deviation σ may be transformed to standard z'_i scores, with mean μ' and standard deviation σ' as follows:

$$z'_i = \frac{\sigma'}{\sigma}(X_i - \mu) + \mu'$$

Suppose a population of raw scores has $\mu = 70$ and $\sigma = 8$ and it is desired to transform these scores to standard scores with $\mu' = 100$ and $\sigma' = 20$. Substituting into the foregoing equation, we have

$$z'_i = \frac{20}{8}(X_i - 70) + 100$$

Then for raw scores $X_i = 40$ and 87, we obtain the standardized scores $z_i = 25$ and 143, respectively. Standardized scores are usually positive, whole numbers.

A.2 Time Series Trends

Typically, long-term trend analysis for a *time series* is based on annual data. In some analyses a freehand trend line is adequate, however, as a rule, a more objective least squares trend line (or curve) is preferred.

A time series is made up of a set of measurements corresponding to a time unit sequence. For example, the dollar volume of TV sales per year recorded for a period of years is a time series, or the average weight of pigs at a fixed time each week recorded for a period of weeks is a time series. The methods for computing regression equations discussed in Part VI of this text are applicable to time series trend determination; however, in many time series analyses the computations may be considerably simplified. This is true for the trend equations presented in this section.

In a trend analysis, we are concerned with two variables. One variable is the *time variable* (the independent variable). The other variable is the data under study (the dependent variable). For example, in the time series "number of foreign-born members admitted to a fraternal society per year" the independent variable X is year and the dependent variable Y is the number of foreign-born members admitted per year.

It is often convenient to transform the time variable by coding. For an odd number of years, assign the code 0 to the middle year, assign -1, -2, etc., to successively earlier years, and 1, 2, etc., to successively later years. Then for a nine-year period, the coded independent variable X will have the values -4, -3, -2, -1, 0, 1, 2, 3, 4 corresponding to year 1, year 2, etc. For an even number of years, assign $-.5$ to the earlier of the two middle years and -1.5, -2.5, etc., to successively earlier years. Then assign .5 to the later of the two middle years and 1.5, 2.5, etc., to successively later years. The coded X variable for an eight-year period will have the values -3.5, -2.5, -1.5, $-.5$, .5, 1.5, 2.5, 3.5 corresponding to year 1, year 2, etc. After coding, $\sum X = 0$. This helps simplify the computations.

A *least squares linear trend equation* $Y_e = a + bX$ (similar to a *linear regression equation*) may be computed based on Y and the coded X variable as follows:

$$b = \frac{\sum XY}{\sum X^2} \qquad a = \frac{\sum Y}{n}$$

The coefficient b represents the *trend increment* if it is positive (the year-to-year trend increase) and the *trend decrement*, if negative. The constant a represents the trend value for the origin year (the year coded 0).

If time series data are plotted on semilog paper (log Y on the vertical axis) and a linear trend is indicated, then the least squares semilog trend equation log $Y_e = a + bX$ is appropriate. This is computed, based on Y and the coded X variable, as follows:

$$b = \frac{\sum X \log Y}{\sum X^2} \qquad a = \frac{\sum \log Y}{n}$$

A useful curvilinear trend equation is the *parabola* $Y_e = a + bX + cX^2$, where X is the coded time variable. A *parabolic least squares trend equation* is computed as follows:

$$b = \frac{\sum XY}{\sum X^2}$$

$$c = \frac{n \sum X^2Y - \sum X^2 \sum Y}{n \sum X^4 - (\sum X^2)^2}$$

$$a = \frac{\sum Y - c \sum X^2}{n}$$

The parabola is a versatile trend form which fits various types of dish-bowl-shaped trends. The larger the value of c, the deeper is the dish-bowl shape of the trend. If c is negative, the computed trend has an inverted dish-bowl appearance.

A.3 Seasonal Variation

Identification and measurement of seasonal variation can be accomplished in various ways. Today, however, the method nearly always used is the *ratio-to-moving average method*.

Table A.3.1 presents a table layout for the seasonal variation computations for Y = monthly number of first-born in a city. First compute the 12-month moving totals, as shown in the table. The first total shown is 488, which is the sum of the first 12 monthly Y's (the total count for the first year, 1964). Notice that 488 is placed between June and July, the middle of the 12-month period. The second total is 522, the sum of the 12 monthly counts from February 1964 through January 1965, and is placed between July and August, the middle of this 12-month period.

TABLE A.3.1

Monthly number of first-born children in a city (Y) for 1964–73.

Month and year		Y	12-month moving total	Moving total centered	12-month moving average centered	SI
1964	Jan	25				
	Feb	20				
	Mar	30				
	Apr	32				
	May	40				
	June	33				
			488			
	July	32		1,010	42.08	.7605
			522			
	Aug	50		1,082	45.08	1.1091
			560			
	Sept	60	.	1,169	48.71	1.2318
	.	.	.	.	.	.
	.	.	.	.	.	.
	.	.	.	.	.	.
	.	.	.	.	.	.
	.	.	.	.	.	.
	.	.	.	.	.	.

The 12-month period for which the total is obtained moves down the column of months, one month at a time, dropping the earliest month and adding the next later month. The last total obtained is the sum of the last 12 months of counts and is placed between June and July for the last year of data, the middle of this 12-month period.

We then center these totals so that a total may be associated with a specific month. This is accomplished by computing 2-month moving totals of the 12-month moving totals and are called *moving totals, centered,* as shown in the table. The first centered moving total in the column is 1,010, obtained by adding the first two 12-month moving totals (488 and 522). Notice that 1,010 is placed between 488 and 522 so that it becomes the moving total, centered for July 1964. The other centered moving totals are obtained in the same manner. There is a centered moving total associated with each month, except for the first six months of the first year and the last six months of the last year.

Each centered moving total is the sum of 24 monthly counts. Hence, each is divided by 24 to give the centered 12-month moving averages shown in the table. These moving averages represent the combined trend-cycle movement in the data. The next step is to divide each monthly Y by the corresponding moving average to obtain the ratios of observed Y's to the moving averages. (This step gives the method its name.) These ratios are denoted as SI for seasonal and irregular ratios for the time series. The first SI ratio in the table is $(Y = 32)/(\text{moving average} = 42.08) = .7605$ for July 1964.

The SI ratios are averaged month by month. That is, the average is computed for the January ratios, for the February ratios, etc. Often the arithmetic mean is the average computed for each month; however, if the SI ratios for each month show extreme values, the median or a modified arithmetic mean is preferred. A *modified arithmetic mean* is computed for a month by first discarding the SI ratios which are extremely small or extremely large and computing the arithmetic mean for the remaining ratios.

Finally, the 12 SI ratio averages are added and an *adjustment factor* is computed as 12.000/total. Each SI ratio average is multiplied by this adjustment factor to insure that these *adjusted* or *forced* SI ratio averages add to 12.000, except for rounding. These forced monthly averages are the *seasonal factors.* The set of 12 monthly seasonal factors computed for the data in Table A.3.1 may look something like .703, .751, .832, 1.250, 1.306, .815, .708, 1.305, 1.208, 1.187, 1.004, and .931 for January, February, March, etc.

The *seasonal factor* for a month indicates the average relationship of the monthly data (counts) to the annual monthly mean for the month. For example, the January factor .703 indicates that typically the January count is 70.3 percent of the yearly average for a month. Similarly, the April seasonal factor (or *seasonal index*) 1.250 indicates that typically the April count is 125.0 percent of the yearly average for a month, etc. The seasonal factors we have just computed are called *stable seasonal factors,* compared to *moving seasonal factors* which change from year to year.

A.4   Index Numbers

An index number expresses the value of a time series variable at a given time as a percentage of its value at a reference period in time. Suppose the price of a commodity currently is 50¢ per dozen compared with 40¢ per dozen a year ago. Computing $(50/40) \cdot 100 = 125$ percent, we obtain the *price index* 125 (without the percent sign). This indicates that the current price or the given price is 25 percent above the year ago price. The year ago price is the *reference period* price and is usually called the *comparison period* price or *base period* price.

Index numbers may be classified, based on the content of the index, into simple index numbers and composite index numbers. *A simple index number presents the relative (percentage) movements relating to a single commodity.* For example, a price index for a specified make and grade of men's socks is a simple price index. There are also simple quantity indexes and simple value indexes.

Simple index numbers are easy to compute and understand. We may express a *simple price index* I_p, with P_i = the given year price and P_0 = the base year price, as

$$I_p = \frac{P_i}{P_0} \cdot 100$$

The base year price P_0 is often a *base period* price which represents an average yearly price for a period of several years. When comparing the movement in an index number between two years, relative, not absolute, changes should be considered. If a price index is 113.4 for 1967 and 124.0 for 1969, it is more meaningful to speak of a 9.3 percent rise in the index than a rise of 10.4 points, especially when comparing two index number series constructed on different base periods.

Simple quantity index numbers may be constructed by a similar equation, but with Q (for quantity) substituted for P. Also, *simple value index numbers* may be constructed by substituting $V = PQ$ for P into the equation, where V denotes value. Of course, monthly index numbers, as well as yearly index numbers, may be constructed. In a monthly price index, P_i denotes the given (current) monthly price and P_0, the average monthly price for the base period.

A composite index number presents the relative movements relating to a group of commodities. Based on the method of construction, composite index numbers may be classified into simple composite index numbers and weighted composite index numbers. There are simple composite price indexes, quantity indexes, and value indexes; however, there are only price and quantity weighted composite index numbers. The more important index numbers fall in this category. The *Consumer Price Index* computed by the United States Bureau of Labor Statistics is a weighted composite price index. The *Index of Industrial Production* computed by the United States Federal Reserve Board is a weighted composite quantity index.

The *simple composite value index* I_v, with $V_i = P_i Q_i$ denoting the given (current) value and $V_0 = P_0 Q_0$ denoting the base period value, computed as

$$I_v = \frac{\Sigma V_i}{\Sigma V_0} \cdot 100 = \frac{\Sigma P_i Q_i}{\Sigma P_0 Q_0} \cdot 100$$

provides the basis for constructing weighted composite indexes. The numerator of this equation represents the sum of the given year values for a set of commodities (for example, the various foods purchased by an urban family) and the denominator, the sum of the base year values. Such an index is called a *simple aggregative value index*.

If in the foregoing equation the base year quantities Q_0 are substituted for the current year quantities Q_i, we obtain the *weighted aggregative price index* (since only price differs from the base period to the given period) as follows:

$$I_p = \frac{\Sigma P_i Q_0}{\Sigma P_0 Q_0} \cdot 100$$

This is usually called the *Laspeyres Index* (named for the man who first devised it). Some analysts object to this type of index because they feel that the base year quantities become outdated after a time and are not representative of the current distribution of quantities. In order to cope with this problem, some analysts use the given year quantities Q_i instead and compute

$$I_p = \frac{\Sigma P_i Q_i}{\Sigma P_0 Q_i} \cdot 100$$

This is called a *Paache Index* (named for the man who devised it).

Although the *Paache Index* is preferred to the *Laspeyres Index*, the *Laspeyres Index* is used more often for two reasons. First, base year quantities are easier to use because they do not change from year to year. Secondly, this index is comparable from year to year because the quantities do not change.

The base period for an index should be a recent period and it should be selected to represent a period of stable economic conditions. If several index number series are constructed, it is desirable that they all have the same base period.

A.5 Significant Figures and Rounding

Data obtained by *counting* (*enumeration data*) are always exact and all the digits in such data are significant. (In this section, we assume mistakes are not made.) On the other hand, data obtained by *measurement* are never exact. *Measurement data* (such as 18.06 inches or 15 gallons) are always approximate. The non-zero digits in such data are always significant. The zeros are significant only under certain conditions.

Consider the measurement data (inches): 348, 34.8, 3.48, .348, .0348, and .00348. Each of these numbers contains three significant digits (3, 4, 8). The position of the decimal point has no bearing on the number of significant digits. The zeros after the decimal point and before the first non-zero digit are not counted as significant figures. Zeros such as these are used only to place the decimal point.

A recorded measurement such as 348 inches belongs to the sequence 347, 348, 349, represents a length closer to 348 than to 347 or 349, and is a *rounded number* (rounded to the nearest inch). On the other hand, 348.2 inches belongs to the sequence 348.1, 348.2, 348.3, represents a length closer to 348.2 than to 348.1 or 348.3, and is a number rounded to the nearest tenth of an inch. This number has four significant figures.

Consider the following recorded data (pounds): 75, 75.0, and 75.00, where the zeros are called "trailing zeros." The weight 75 lbs has two significant figures and belongs in the sequence 74, 75, 76 since it is rounded to the nearest pound. The weight 75.0 lbs contains three significant figures (7, 5, 0) and belongs to the sequence 74.9, 75.0, 75.1, since it is rounded to the nearest tenth of a pound. The weight 75.00 lbs contains four significant figures (7, 5, 0, 0) and belongs to the sequence 74.99, 75.00, 75.01, since it is rounded to the nearest hundredth of a pound. Clearly, "trailing zeros" in such numbers are significant figures.

In the same way, the recorded measurement 6.3 has two significant figures, 6.30 has three significant figures, and 6.300 has four significant figures. Also, .3 and 0.3 have one significant figure each, .30 has two significant figures and 0.30000 has five significant figures. As indicated, "trailing zeros" in a decimal are significant figures. "Inside zeros" are always significant (8.01, 6,004).

The number of significant figures in a number is a measure of its accuracy. The number 19 tons has two significant figures and is said to have *two-digit accuracy*. The number 19.0 has more digit accuracy than 19 since it contains three significant digits and so possesses three-digit accuracy.

Recorded data with trailing zeros, such as 2,000 tons, contain an indefinite number of significant figures. If 2,000 tons is correct to the nearest ton, then it has four significant figures and belongs in the sequence 1,999, 2,000, 2,001. If it is rounded to the nearest ten tons, it contains only three significant figures (2, 0, 0) and belongs to the sequence 1,990, 2,000, 2,010. If it is rounded to the nearest hundred tons, it has only two significant figures (2, 0) and belongs to the sequence 1,900, 2,000, 2,100; etc. In such situations, it is a good idea to identify the number of significant digits by placing a dot or a bar just above the last significant figure. For example, 50,000 indicates a number with two significant figures and 1,000̄ is a number with four significant figures. If the number of significant figures is not indicated in some way, then it is safest to assume that none of the trailing zeros are significant.

Statistical analysts in all fields face the problem of rounding. For example, if a school has 7,642 pupils, for certain purposes it is better to report this as 7,640̄ or 8̄,000 pupils. When count data are rounded, it is necessary to indicate how many significant figures are included.

Three simple rounding rules are generally followed. If 684,328 is to be rounded to thousands, then 684 is retained and 328 dropped and the number becomes 684 thousand or 684,000 (684̄,000 is more informative). Notice that the first dropped digit is 3. *Whenever the first dropped digit is less than 5, do not change the retained figures. If the first dropped digit is greater than 5 or if it is 5 followed by at least one non-zero number, add one unit to the retained portion.* Then, 831,782 is rounded to 831,800 (831,8̇00 is better) and 413,500,001 is rounded to 414,000,000 (41$\overset{.}{4}$,000,000 is better). If the first digit to be dropped is 5 either standing alone or followed only by zeros, we find ourselves exactly in the middle. In such cases, it is good practice to "randomize the rounding." This is accomplished by raising the retained portion if it is odd and leaving it unchanged if it is even. The rounded number in such instances will always be even.

The foregoing rounding procedure is *fixed decimal point rounding* since rounding is to a *specified number of places. Floating decimal point rounding* is often more desirable, especially where precision is important as in scientific or engineering computations. Such rounding is to a *specified level of digit accuracy.*

When rounded numbers are used in computations, how should the results be reported? If two rounded numbers are to be multiplied, determine the number of significant figures in the one which has the least number of significant figures. This is the number of significant figures to show in the product. If a rounded number is to be divided into another rounded number, determine the number of significant figures in the one with the least number of significant figures. This is the number of significant figures to show in the quotient. The square root of a number should be rounded to the same number of significant figures as in the number itself.

When adding or subtracting rounded numbers the answer should be rounded so that no figure is retained which is in a position to the right of the last significant figure in any of the rounded numbers.

A.6   Statistical Table Construction

Statistical tables are prepared to provide information. Hence, they should be easy to read and understand. There are only a few rules to follow, such as a table should exhibit clarity in presentation and should be able to "stand on its own" without requiring reference to a text for interpretation.

The discussion in this section is based to a large extent on the *Bureau of the Census Manual of Tabular Presentation* by Bruce L. Jenkinson. Figures A.6.1 and A.6.2 present table-forms taken from this publication, with modifications, which indicate certain basic features of statistical tables. As indicated in the figures, the *heading* appears first in a table and usually includes the *table number, title*, and sometimes a *headnote*. Of course, a table number is not needed if a report contains only one table. Usually the title should be a brief statement describing the content of the table and should answer the questions What? How classified? Where? and When?

The *stub* is the portion of the table which describes the content of the lines or rows of the table. The stub usually appears at the left end of the table and has a *stubhead* at the head which describes the content of the stub column. The stub also contains *center head captions* to describe sets of rows and also *line* or *row captions*. As indicated in Figure A.6.2, a center head caption and its related rows make up a *block*.

The *boxhead*, identified in Figure A.6.1, is made up of *column heads*, or *captions, spanner heads* or *spanners*, and *panels* (Figure A.6.2). A spanner

HEADING→ TABLE 6.—AGE OF ALL PERSONS AND OF CITIZENS BY SEX, FOR THE UNITED STATES, URBAN AND RURAL: 1970

[Age classification based on completed years]

Area and age	All persons			Citizens[1]		
	Total	Male	Female	Total	Male	Female
UNITED STATES						
All ages......				769		
Under 5 years...				26		
5 to 14 years.....				115		
15 to 24 years....				139		
25 to 34 years....				178		
35 to 44 years....				205		
45 and over.....				106		
21 and over...	988	475	513	567	302	265
URBAN						
All ages......				453		
Under 5 years...				15		
5 to 14 years.....				73		
15 to 24 years....				86		
25 to 34 years....				104		
35 to 44 years....				116		
45 and over.....				59		
21 and over.....				328		
RURAL						
All ages......				316		
Under 5 years...				11		
5 to 14 years.....				42		
15 to 24 years....				53		
25 to 34 years....				74		
35 to 44 years....				89		
45 and over.....				47		
21 and over.....				239		

←BOXHEAD

STUB→

←FIELD

[1] Includes both native and naturalized.

SOURCE: Department of Commerce, Bureau of the Census, Nineteenth Census Reports, *Population*, Vol. I, p. 385.

SOURCE: Bruce L. Jenkinson, *Bureau of the Census Manual of Tabular Presentation*, p. 10 (with modifications).

FIGURE A.6.1

Data are for illustrative purposes only.

Table No.—Title of Table

[Headnote]

PANEL ─ ─ ─ ─ ─ ─ ─ ─ ─ ─ ─ ─

Stubhead	Spanner head			Spanner head[1]		
	Column head	Column head	Column head	Column head	Column head	Column head
⌐CENTER HEAD						
Total line caption				Cell		
Line caption......				Cell		
BLOCK→ Line caption......				Cell		
Line caption......				Cell		
Line caption......				Cell		
Line caption......				Cell		
Line caption......				Cell		
LINE──→ Line caption......	Cell	Cell	Cell	Cell	Cell	Cell
CENTER HEAD						
Total line caption				Cell		
Line caption......				Cell		
Line caption......				Cell		
Line caption......				Cell		
Line caption......				Cell		
Line caption......				Cell		
Line caption......				Cell		
Line caption......				Cell		
CENTER HEAD						
Total line caption				Cell		
Line caption......				Cell		
Line caption......				Cell		
Line caption......				Cell		
Line caption......				Cell		
Line caption......				Cell		
Line caption......				Cell		
Line caption......				Cell		

[1] Footnote.
Source note.

Source: same as for Figure A.6.1., p. 11 (with modifications).

FIGURE A.6.2

spreads over two or more related columns and provides a more general description of the data in the columns than the column heads. A panel consists of a group of column heads with the associated spanner.

The *field* (Figure A.6.1) is that portion of the table which contains the data provided by the table and is made up of cells, rows or lines, and columns. The location of an individual item of data is called a *cell*. *Footnotes* (Figure A.6.1) often appear at the bottom of a statistical table to explain or qualify some aspect of the table. A *reference symbol* is used to identify a footnote with the part of the table to which it refers. A careful statement of the source of the data usually appears in a *source note* at the bottom of the table.

Construction of a statistical table requires careful planning. Every effort should be made to keep it simple. A well-constructed table could be lengthy and still be easily read and interpreted. A problem to be resolved in table construction is how to classify the data and which classifications should be placed in the stub and which in the boxhead. The solution to this problem depends on the purpose to be served by the table and on how many classifications are needed. Generally speaking, it is better to make a table longer than wide, so that usually the stub will contain more categories than the boxhead. This contributes to readability of the table. It is easier to compare data which are side-by-side in columns than data which are in adjacent rows. Hence, the most important comparisons to be shown by a table should be placed in the columns, if at all possible. Of course, if many classifications are involved, then it may be necessary to place them in the stub.

Important parts of a table are the *summary* or *total rows* and *columns*. Reading and interpreting a table is helped considerably by carefully placed vertical and horizontal rulings and spacing, as indicated in the figures.

A.7 Graphic and Pictorial Presentation

The general objective of a chart is to provide relatively quick understanding of a relationship. A chart should, therefore, be simple in content (contain only one or at most two relationships) and in design (use rounded numbers, sensibly short titles, appropriate abbreviations). It should not be necessary to refer to a text or a table in order to understand a chart. A chart is meant to provide only approximate information; whereas a statistical table provides precise information.

Figure A.7.1 presents various components of a chart. Chart *titles* should be brief yet as complete as possible, but should not repeat the information provided elsewhere in the chart. *Subtitles* are sometimes used to explain the chart contents. Generally, make a chart wider than tall. As a guide, a chart should be 65 percent to 75 percent as tall as it is wide. *Grid lines*, to be most helpful, should not be too few or too numerous and should be only enough to aid the eye in making comparisons or in estimating values.

The *base line* is the principal line of reference in a chart and should be made heavier (bolder) than the grid rulings. Figure A.7.2 illustrates a *line graph*. Notice the qualifying note "Seasonally Adjusted Annual Rates" (which provides important information about the data) and the use of abbreviations ($Thous., Dec., etc.). Observe that the vertical scale *values* (0, 300, 600), as well as the *differences* $(300 - 0 = 300$ and $600 - 300 = 300)$, are round figures. This makes the chart more readable.

Various problems and pointers in chart construction are illustrated in

CHART NO.

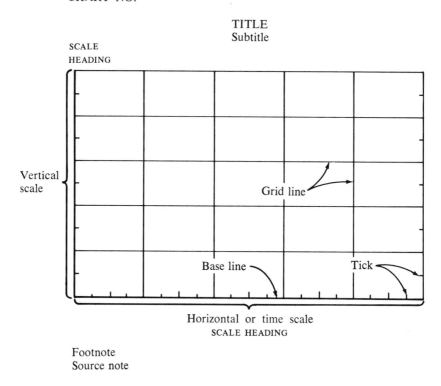

FIGURE A.7.1

Components of a chart.

Figure A.7.3. Chart (A) in this figure illustrates a chart with acceptable proportions and a *zero base line*. Chart (B) presents the same chart, without a zero base line. Notice how it gives the false impression that two values are equal to zero. A zero line should always be included. If it must be omitted, the user should be alerted to this by some device such as a jagged line or "lightening mark" as in Chart (C) or "tear lines" as in Chart (F).

Chart (D) in Figure A.7.3 shows a horizontal scale which is too stretched. Chart (E) shows the same line graph with the vertical scale too stretched. These charts are based on the same data as the line graph in Chart (A). Notice how the stretched scales distort the visual impression. Chart (F) shows that a vertical scale should not be made to accommodate extreme values and how such values may be indicated. Chart (G) illustrates an index number line graph, with the 100 base line drawn in as a heavy line. Such a chart does not show a zero line.

Bar charts represent a favorite chart form. A vertical bar chart is illustrated in Figure A.7.4. Figure A.7.5 presents a grouped or multi-unit horizontal bar chart. Figure A.7.6 presents pie or circle charts, where each segment size is determined by the appropriate proportion of 360° for the category.

CHART 5

RETAIL SALES IN JEROME CITY, KANSAS
(INCLUDING SUSAN TOWNSHIP)

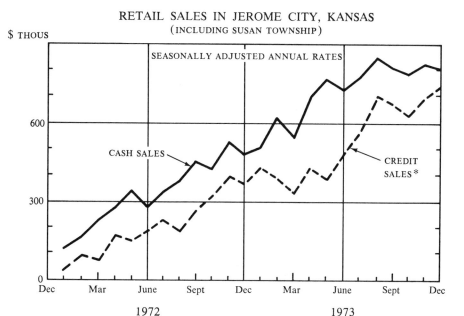

* Includes 30-day shopping plates.
SOURCE: Honig County Commerce Department.

FIGURE A.7.2

Line graph.

A.8   Semilog Line Graphs

In an *arithmetic line graph*, such as in Figure A.7.2, both vertical and horizontal scales are *arithmetic scales* (*equal scale distances represent equal amounts*). Sometimes line graphs are constructed with a *logarithmic scale* (also called a *ratio scale*) for the vertical axis and an arithmetic scale for the horizontal axis. Such graphs are called *semilog* or *ratio graphs. In a logarithmic scale, equal scale distances represent equal percent changes.*

Figure A.8.1 presents a logarithmic or ratio scale and four sets of scale values. The first set, called a *basic scale*, has two similar portions or two cycles, a lower cycle and an upper cycle. Each cycle goes from 1 through 9 to the next 1. The basic scale is usually printed on semilog graph paper and identifies the major grid lines.

When semilog graph paper is used, the basic scale values are not appropriate. Suitable scale values must be substituted. Scale 1 values in the figure illustrate a

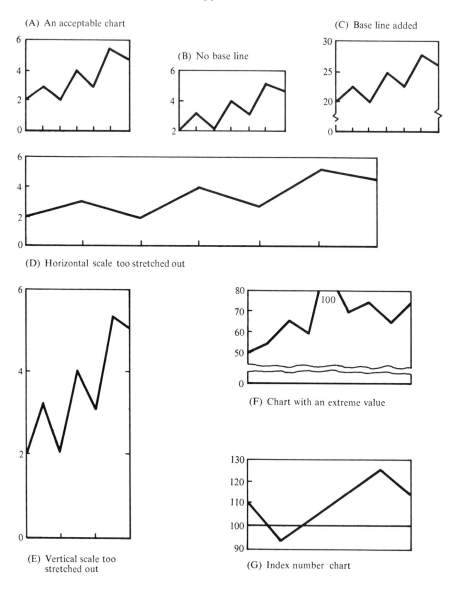

FIGURE A.7.3

Some problems and pointers in chart construction.

possible set of two-cycle ratio scale values. Notice that the vertical distance between scale values 10 and 20 is larger than between 20 and 30. As you move to higher scale values, the vertical distance becomes smaller, until you reach the second cycle of scale values. Then the same sequence of declining vertical distances occurs again. This reflects the declining percent changes within a cycle. For example, 20 is 100 percent more than 10, 30 is 50 percent more than 20, 40

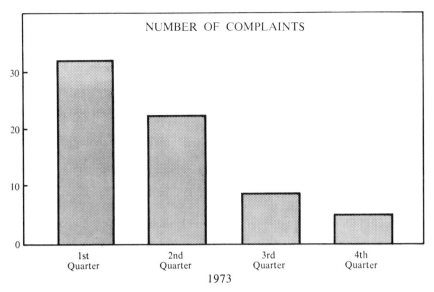

FIGURE A.7.4

Vertical bar chart.

VACANCY RATES IN MULTI-FAMILY
RESIDENTIAL STRUCTURES
FEBRUARY 1973

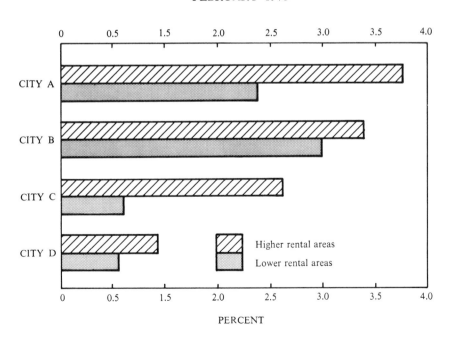

FIGURE A.7.5

Grouped or multi-unit horizontal bar chart.

371

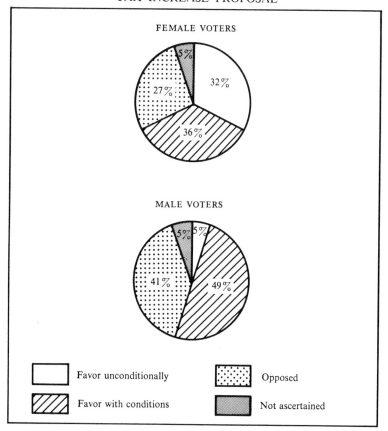

VOTER OPINIONS ON A
TAX INCREASE PROPOSAL

FIGURE A.7.6

Pie or circle chart.

is 33 percent more than 30, etc. In the second cycle, 200 is 100 percent more than 100, 300 is 50 percent more than 200, etc.

When constructing a semilog graph, the general scale in Figure A.8.1 shows how to construct a suitable scale of values for the ratio scale. A starting value is selected (call it Y). This value must not be zero or negative. Successive major grid values are integral multiples of this starting value ($2Y$, $3Y$, etc.). The top value for the first cycle is ten times the starting value ($10Y$). The second cycle of values is also determined as integral multiples of its starting value which is $10Y$ ($2 \cdot 10Y = 20Y$, $3 \cdot 10Y = 30Y$, etc.), as indicated in the general scale in Figure A.8.1.

Examine the set of data to be charted and determine the lowest and highest values. Select a starting scale value of 10 or an integral multiple of 10 which is equal to or less than the lowest value. Also, select an integral multiple of 10 equal

Basic scale		Scale 1	Scale 2	General scale	
1		1,000	100,000	$100\,Y$	
	9	900	90,000	$90\,Y$	
	8	800	80,000	$80\,Y$	
	7	700	70,000	$70\,Y$	
	6	600	60,000	$60\,Y$	
	5	500	50,000	$50\,Y$	
one cycle	4	400	40,000	$40\,Y$	one cycle
	3	300	30,000	$30\,Y$	
	2	200	20,000	$20\,Y$	
	1	100	10,000	$10\,Y$	
	9	90	9,000	$9\,Y$	
	8	80	8,000	$8\,Y$	
	7	70	7,000	$7\,Y$	
	6	60	6,000	$6\,Y$	
	5	50	5,000	$5\,Y$	
one cycle	4	40	4,000	$4\,Y$	one cycle
	3	30	3,000	$3\,Y$	
	2	20	2,000	$2\,Y$	
	1	10	1,000	Y	

FIGURE A.8.1

Logarithmic or ratio scales.

to or higher than the highest value which will represent the highest required scale value. If the highest scale value is no more than ten times as large as the starting scale value, one-cycle semilog graph paper is needed. If it is more than ten times as large, but no more than 100 times as large, two-cycle graph paper is needed. This may be seen by referring to the general scale in Figure A.8.1. If the highest scale value is more than 100 times the starting value, then three-or-more-cycle graph paper is needed. Scale 2 illustrates a two-cycle scale starting with 1,000.

Study Problems

1. A student obtained a score of 145 on a mathematics achievement test. The population of test scores has a mean of 120 and a standard deviation of 22. Transform this student's test score to a standardized score from a population with mean 50 and standard deviation 10.

2. The student in Problem 1 obtained a score of 290 on a history test, where the population mean test score is 250 and the standard deviation is 40. Transform this test score to a standardized score from a population with mean 50 and standard deviation 10.

3. In Problems 1 and 2, which test score (145 on mathematics achievement or 290 on the history test) indicates better performance? Why?

4. A job applicant obtained a score of 150 on a clerical aptitude test and 35 on a mechanical ability test. The population mean score is 220 for the clerical aptitude test (with standard deviation 30) and 50 for the mechanical ability test (with standard deviation 12).
    a. Transform each test score to a standard score from a population with mean 100 and standard deviation 20.
    b. Does this job applicant show more clerical aptitude or more mechanical ability? Why?

5. The annual number of power failures in a factory during 1963–1972 was 6, 8, 7, 8, 10, 9, 10, 11, 10, and 12.
    a. Construct a line graph of the annual number of power failures.
    b. Compute a least squares linear trend equation. (Use coded X for years.)
    c. Compute the trend values for 1971 and 1974.
    d. Construct the trend line on the graph.

6. The number of customer complaints per month in a department store during a ten-month period was 15, 13, 12, 13, 10, 8, 9, 8, 7, and 8.
    a. Compute a least squares trend equation for log Y (number of complaints) and coded X values for months.
    b. Plot a line graph of the observed data.
    c. Plot the computed trend equation.
    d. Compute the trend value for the month following the ten-month period.

7. The numbers of one-family home foreclosures in a community for 1962–1972 were as follows: 4, 3, 5, 7, 8, 10, 8, 11, 9, 8, and 9.
    a. Compute a least squares parabolic trend equation. (Use coded X for years.)
    b. Plot the line graph of the observed data.

c. Construct the parabolic trend on the graph.
d. Compute the 1973 trend value.

8. The quantity of a commodity produced each year in a 12-year period (1961–1972) was as follows (in tons): 3, 9, 7, 8, 11, 12, 11, 18, 17, 21, 53, and 63.
a. Compute a four-year moving average, centered.
b. Plot the observed yearly data and the four-year moving averages.

9. The ratio-to-moving average method was applied to the dollar volume of women's apparel sales in a region and the following SI ratios were computed:

	1966	1967	1968	1969	1970	1971	1972
Jan		.701	.698	.751	.712	.694	.745
Feb		.600	.662	.613	.611	.590	.616
Mar		1.003	1.064	1.062	1.010	1.029	1.018
Apr		1.132	1.049	1.206	1.200	1.099	1.100
May		1.025	1.013	1.004	1.030	1.010	1.020
June		.982	1.011	1.009	.984	.990	1.103
July	1.000	1.010	.996	.992	1.030	1.004	
Aug	.980	.999	.976	.974	.982	.979	
Sept	1.200	1.115	1.246	1.204	1.196	1.111	
Oct	.889	.905	.884	.879	.986	.890	
Nov	.901	.909	.900	.920	.915	.896	
Dec	1.304	1.308	1.294	1.300	1.292	1.308	

Compute the final monthly seasonal factors using the following averages of the SI ratios:
a. Arithmetic mean
b. Median
c. Modified arithmetic mean (Omit the highest and lowest SI ratio for each month.)

10. Given the following monthly data on dollar volume of order backlogs (in millions of dollars) for 1968–1972:

	1968	1969	1970	1971	1972
Jan	3	9	10	15	18
Feb	3	7	10	11	16
Mar	4	11	14	17	20
Apr	4	13	16	24	26
May	5	20	22	31	38
June	13	31	36	49	59
July	16	31	42	50	65
Aug	15	26	42	48	63
Sept	13	13	25	26	39
Oct	9	6	14	17	22
Nov	3	5	11	15	19
Dec	14	14	27	30	38

a. Using the method of ratio-to-moving average, compute the seasonal factors. (Use the mean of the SI ratios.)

b. Compute the seasonally adjusted dollar volume of backlog orders for 1972 (actual dollar volume for a month divided by seasonal factor for the month).

c. Construct line graphs of the original data and the moving averages.

d. For 1972, construct a line graph of the original data and the seasonally adjusted data.

11. What is meant by the base period price in a price index?

12. What is meant by (a) simple price index and (b) composite price index?

13. Compare the relative advantages and disadvantages of the Laspeyres and Paache price indexes.

14. Given the following data on price and quantity sold for a product:

Year	Quantity sold (1,000 dozen)	Price (dollars)
1967	20	4
1968	25	5
1969	30	6
1970	25	5
1971	35	10
1972	30	7

Compute:

a. Simple price index with 1967 as the base year

b. Simple quantity index with 1969 as the base year

c. Simple price index with 1967–1969 as the base period

d. Simple aggregative value index with 1972 as the base year

15. Given the following data relating to the sales of a large firm:

Category	Unit	Quantity sold 1970	Quantity sold 1972	Average price ($) 1970	Average price ($) 1972
Ribbons	Yard	500	800	.75	1.50
Yard goods	Yard	2,000	2,500	1.50	3.00
Paints	Gallon	10,000	15,200	5.90	7.50
Fertilizer	Pound	3,000	4,000	1.25	1.80

Compute the following weighted composite price index numbers for 1972, with 1970 as the base year:

a. Laspeyres Index   b. Paache Index

16. Which of the following recorded data are exact and which are approximate:

52 feet	32 shirts	74 tons
14 employees	186 dresses	19 weeks
.32 cms	12 ozs	25 lbs

17. How many significant figures in each of the following:

218 suits	.00020040 min	128,000 cms
8.700 cms	6,542 coats	0.1600 ozs
23.04 ft	98,00$\bar{0}$ lbs	.00803 mi

18. How many significant figures in each of the following:

8,400	67,000	4,900
30	52,000,000	2,700

19. What is the digit accuracy of each of the following:

47,000	53.07	73,000
24	34.000	.080003
0.042300	.0068	3,487,001

20. Round to the nearest whole number:

74.0001	19.5	978.500
.500001	42.75	29.5
436.0	503.50	322.5

21. Round to hundredths:

0.112	1.0278	368.16001
.00675	323.205001	29.050

22. Round to three significant figures:

.0006945	397,000	.0064001
24.0007	6.0800	.4655

23. Express the following computational results to the proper number of significant digits:

$835 \cdot .01$	$46 \cdot .1 \cdot 220$
$2,222 \div 1.1$	$.18 - .009 - .06$
$1.8 + 2.03 + 16$	$2.641 \div .03$

Appendix B
Tables

TABLE B.1

Approximate number of classes to use when constructing a frequency distribution.

Number of items of data to be summarized	Approximate number of classes to use
15–29	5
30–59	6
60–99	7
100–199	8
200–499	9
500–999	10
1,000–1,999	11
2,000–3,999	12
4,000–7,999	13
8,000–14,999	14
15,000–34,999	15
35,000–69,999	16
70,000–149,999	17
150,000–299,999	18
300,000–499,999	19
500,000 and over	20

SOURCE: Table 3.2.5 of Sidney J. Armore, *Introduction to Statistical Analysis and Inference for Psychology and Education*, © 1966, by John Wiley & Sons, Inc., New York, N.Y. Generally based on the results obtained following the procedure in H. A. Sturges, *The Choice of a Class Interval, Journal of the American Statistical Association*, Vol. 21, 1926.

TABLE B.2

Chebyshev's inequality: proportion of a set of quantities which is within a distance of k σ's from the mean (p' or more) and the proportion which is beyond k σ's from the mean (p or less).

k	$p = \dfrac{1}{k^2}$	$p' = 1 - \dfrac{1}{k^2}$
1.00	1.00	0
1.05	.90	.10
1.15	.75	.25
1.41	.50	.50
1.50	.44	.56
2.00	.25	.75
2.50	.16	.84
3.00	.11	.89
3.16	.10	.90
3.50	.08	.92
4.00	.06	.94
4.50	.05	.95
5.00	.04	.96
6.00	.03	.97
7.00	.02	.98
10.00	.01	.99

TABLE B.3

Table of random digits.

Line \ Col.	(1)	(2)	(3)	(4)	(5)	(6)
1	10480	15011	01536	02011	81647	91646
2	22368	46573	25595	85393	30995	89198
3	24130	48360	22527	97265	76393	64809
4	42167	93093	06243	61680	07856	16376
5	37570	39975	81837	16656	06121	91782
6	77921	06907	11008	42751	27756	53498
7	99562	72905	56420	69994	98872	31016
8	96301	91977	05463	07972	18876	20922
9	89579	14342	63661	10281	17453	18103
10	85475	36857	53342	53988	53060	59533
11	28918	69578	88231	33276	70997	79936
12	63553	40961	48235	03427	49626	69445
13	09429	93969	52636	92737	88974	33488
14	10365	61129	87529	85689	48237	52267
15	07119	97336	71048	08178	77233	13916
16	51085	12765	51821	51259	77452	16308
17	02368	21382	52404	60268	89368	19885
18	01011	54092	33362	94904	31273	04146
19	52162	53916	46369	58586	23216	14513
20	07056	97628	33787	09998	42698	06691
21	48663	91245	85828	14346	09172	30168
22	54164	58492	22421	74103	47070	25306
23	32639	32363	05597	24200	13363	38005
24	29334	27001	87637	87308	58731	00256
25	02488	33062	28834	07351	19731	92420
26	81525	72295	04839	96423	24878	82651
27	29676	20591	68086	26432	46901	20849
28	00742	57392	39064	66432	84673	40027
29	05366	04213	25669	26422	44407	44048
30	91921	26418	64117	94305	26766	25940
31	00582	04711	87917	77341	42206	35126
32	00725	69884	62797	56170	86324	88072
33	69011	65795	95876	55293	18988	27354
34	25976	57948	29888	88604	67917	48708
35	09763	83473	73577	12908	30883	18317
36	91567	42595	27958	30134	04024	86385
37	17955	56349	90999	49127	20044	59931
38	46503	18584	18845	49618	02304	51038
39	92157	89634	94824	78171	84610	82834
40	14577	62765	35605	81263	39667	47358
41	98427	07523	33362	64270	01638	92477
42	34914	63976	88720	82765	34476	17032
43	70060	28277	39475	46473	23219	53416
44	53976	54914	06990	67245	68350	82948
45	76072	29515	40980	07391	58745	25774
46	90725	52210	83974	29992	65831	38857
47	64364	67412	33339	31926	14883	24413
48	08962	00358	31662	25388	61642	34072
49	95012	68379	93526	70765	10592	04542
50	15664	10493	20492	38391	91132	21999

SOURCE: Page 1 of *Table of 105,000 Random Decimal Digits*, Statement No. 4914, May, 1949, File No. 261-A-1, Interstate Commerce Commission, Washington, D.C.

(7)	(8)	(9)	(10)	(11)	(12)	(13)	(14)
69179	14194	62590	36207	20969	99570	91291	90700
27982	53402	93965	34095	52666	19174	39615	99505
15179	24830	49340	32081	30680	19655	63348	58629
39440	53537	71341	57004	00849	74917	97758	16379
60468	81305	49684	60672	14110	06927	01263	54613
18602	70659	90655	15053	21916	81825	44394	42880
71194	18738	44013	48840	63213	21069	10634	12952
94595	56869	69014	60045	18425	84903	42508	32307
57740	84378	25331	12566	58678	44947	05585	56941
38867	62300	08158	17983	16439	11458	18593	64952
56865	05859	90106	31595	01547	85590	91610	78188
18663	72695	52180	20847	12234	90511	33703	90322
36320	17617	30015	08272	84115	27156	30613	74952
67689	93394	01511	26358	85104	20285	29975	89868
47564	81056	97735	85977	29372	74461	28551	90707
60756	92144	49442	53900	70960	63990	75601	40719
55322	44819	01188	65255	64835	44919	05944	55157
18594	29852	71585	85030	51132	01915	92747	64951
83149	98736	23495	64350	94738	17752	35156	35749
76988	13602	51851	46104	88916	19509	25625	58104
90229	04734	59193	22178	30421	61666	99904	32812
76468	26384	58151	06646	21524	15227	96909	44592
94342	28728	35806	06912	17012	64161	18296	22851
45834	15398	46557	41135	10367	07684	36188	18510
60952	61280	50001	67658	32586	86679	50720	94953
66566	14778	76797	14780	13300	87074	79666	95725
89768	81536	86645	12659	92259	57102	80428	25280
32832	61362	98947	96067	64760	64584	96096	98253
37937	63904	45766	66134	75470	66530	34693	90449
39972	22209	71500	64568	91402	42416	07844	69618
74087	99547	81817	42607	43808	76655	62028	76630
76222	36086	84637	93161	76038	65855	77919	88006
26575	08625	40801	59920	29841	80150	12777	48501
18912	82271	65424	69774	33611	54262	85963	03547
28290	35797	05998	41688	34952	37888	38917	88050
29880	99730	55536	84855	29080	09250	79656	73211
06115	20542	18059	02008	73708	83517	36103	42791
20655	58727	28168	15475	56942	53389	20562	87338
09922	25417	44137	48413	25555	21246	35509	20468
56873	56307	61607	49518	89656	20103	77490	18062
66969	98420	04880	45585	46565	04102	46880	45709
87589	40836	32427	70002	70663	88863	77775	69348
94970	25832	69975	94884	19661	72828	00102	66794
11398	42878	80287	88267	47363	46634	06541	97809
22987	80059	39911	96189	41151	14222	60697	59583
50490	83765	55657	14361	31720	57375	56228	41546
59744	92351	97473	89286	35931	04110	23726	51900
81249	35648	56891	69352	48373	45578	78547	81788
76463	54328	02349	17247	28865	14777	62730	92277
59516	81652	27195	48223	46751	22923	32261	85653

TABLE B.4

Table of areas under the normal curve.

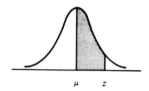

μ   z

Area from the mean to a specified value of z (the shaded area).

z	.00	.01	.02	.03	.04	.05	.06	.07	.08	.09
0.0	.0000	.0040	.0080	.0120	.0160	.0199	.0239	.0279	.0319	.0359
0.1	.0398	.0438	.0478	.0517	.0557	.0596	.0636	.0675	.0714	.0753
0.2	.0793	.0832	.0871	.0910	.0948	.0987	.0026	.1064	.1103	.1141
0.3	.1179	.1217	.1255	.1293	.1331	.1368	.1406	.1443	.1480	.1517
0.4	.1554	.1591	.1628	.1664	.1700	.1736	.1772	.1808	.1844	.1879
0.5	.1915	.1950	.1985	.2019	.2054	.2088	.2123	.2157	.2190	.2224
0.6	.2257	.2291	.2324	.2357	.2389	.2422	.2454	.2486	.2517	.2549
0.7	.2580	.2611	.2642	.2673	.2704	.2734	.2764	.2794	.2823	.2852
0.8	.2881	.2910	.2939	.2967	.2995	.3023	.3051	.3078	.3106	.3133
0.9	.3159	.3186	.3212	.3238	.3264	.3289	.3315	.3340	.3365	.3389
1.0	.3413	.3438	.3461	.3485	.3508	.3531	.3554	.3577	.3599	.3621
1.1	.3643	.3665	.3686	.3708	.3729	.3749	.3770	.3790	.3810	.3830
1.2	.3849	.3869	.3888	.3907	.3925	.3944	.3962	.3980	.3997	.4015
1.3	.4032	.4049	.4066	.4082	.4099	.4115	.4131	.4147	.4162	.4177
1.4	.4192	.4207	.4222	.4236	.4251	.4265	.4279	.4292	.4306	.4319
1.5	.4332	.4345	.4357	.4370	.4382	.4394	.4406	.4418	.4429	.4441
1.6	.4452	.4463	.4474	.4484	.4495	.4505	.4515	.4525	.4535	.4545
1.7	.4554	.4564	.4573	.4582	.4591	.4599	.4608	.4616	.4625	.4633
1.8	.4641	.4649	.4656	.4664	.4671	.4678	.4686	.4693	.4699	.4706
1.9	.4713	.4719	.4726	.4732	.4738	.4744	.4750	.4756	.4761	.4767
2.0	.4772	.4778	.4783	.4788	.4793	.4798	.4803	.4808	.4812	.4817
2.1	.4821	.4826	.4830	.4834	.4838	.4842	.4846	.4850	.4854	.4857
2.2	.4861	.4864	.4868	.4871	.4875	.4878	.4881	.4884	.4887	.4890
2.3	.4893	.4896	.4898	.4901	.4904	.4906	.4909	.4911	.4913	.4916
2.4	.4918	.4920	.4922	.4925	.4927	.4929	.4931	.4932	.4934	.4936
2.5	.4938	.4940	.4941	.4943	.4945	.4946	.4948	.4949	.4951	.4952
2.6	.4953	.4955	.4956	.4957	.4959	.4960	.4961	.4962	.4963	.4964
2.7	.4965	.4966	.4967	.4968	.4969	.4970	.4971	.4972	.4973	.4974
2.8	.4974	.4975	.4976	.4977	.4977	.4978	.4979	.4979	.4980	.4981
2.9	.4981	.4982	.4982	.4983	.4984	.4984	.4985	.4985	.4986	.4986

Source: Derived from *Tables of Normal Probability Functions, Applied Mathematics Series 23*, National Bureau of Standards, Washington, D.C.

z	.00	.01	.02	.03	.04	.05	.06	.07	.08	.09
3.0	.4987	.4987	.4987	.4988	.4988	.4989	.4989	.4989	.4990	.4990
3.1	.4990	.4991	.4991	.4991	.4992	.4992	.4992	.4992	.4993	.4993
3.2	.4993	.4993	.4994	.4994	.4994	.4994	.4994	.4995	.4995	.4995
3.3	.4995	.4995	.4995	.4996	.4996	.4996	.4996	.4996	.4996	.4997
3.4	.4997	.4997	.4997	.4997	.4997	.4997	.4997	.4997	.4997	.4998

TABLE B.5

t distribution: table of areas in both tails of the distribution.

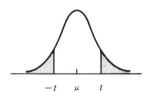

df	Probability (Area in both tails combined)								
	.5	.4	.3	.2	.1	.05	.02	.01	.001
1....	1.000	1.376	1.963	3.078	6.314	12.706	31.821	63.657	636.619
2....	.816	1.061	1.386	1.886	2.920	4.303	6.965	9.925	31.598
3....	.765	.978	1.250	1.638	2.353	3.182	4.541	5.841	12.941
4....	.741	.941	1.190	1.533	2.132	2.776	3.747	4.604	8.610
5....	.727	.920	1.156	1.476	2.015	2.571	3.365	4.032	6.859
6....	.718	.906	1.134	1.440	1.943	2.447	3.143	3.707	5.959
7....	.711	.896	1.119	1.415	1.895	2.365	2.998	3.499	5.405
8....	.706	.889	1.108	1.397	1.860	2.306	2.896	3.355	5.041
9....	.703	.883	1.100	1.383	1.833	2.262	2.821	3.250	4.781
10....	.700	.879	1.093	1.372	1.812	2.228	2.764	3.169	4.587
11....	.697	.876	1.088	1.363	1.796	2.201	2.718	3.106	4.437
12....	.695	.873	1.083	1.356	1.782	2.179	2.681	3.055	4.318
13....	.694	.870	1.079	1.350	1.771	2.160	2.650	3.012	4.221
14....	.692	.868	1.076	1.345	1.761	2.145	2.624	2.977	4.140
15....	.691	.866	1.074	1.341	1.753	2.131	2.602	2.947	4.073
16....	.690	.865	1.071	1.337	1.746	2.120	2.583	2.921	4.015
17....	.689	.863	1.069	1.333	1.740	2.110	2.567	2.898	3.965
18....	.688	.862	1.067	1.330	1.734	2.101	2.552	2.878	3.922
19....	.688	.861	1.066	1.328	1.729	2.093	2.539	2.861	3.883
20....	.687	.860	1.064	1.325	1.725	2.086	2.528	2.845	3.850
21....	.686	.859	1.063	1.323	1.721	2.080	2.518	2.831	3.819
22....	.686	.858	1.061	1.321	1.717	2.074	2.508	2.819	3.792
23....	.685	.858	1.060	1.319	1.714	2.069	2.500	2.807	3.767
24....	.685	.857	1.059	1.318	1.711	2.064	2.492	2.797	3.745
25....	.684	.856	1.058	1.316	1.708	2.060	2.485	2.787	3.725
26....	.684	.856	1.058	1.315	1.706	2.056	2.479	2.779	3.707
27....	.684	.855	1.057	1.314	1.703	2.052	2.473	2.771	3.690
28....	.683	.855	1.056	1.313	1.701	2.048	2.467	2.763	3.674
29....	.683	.854	1.055	1.311	1.699	2.045	2.462	2.756	3.659
30....	.683	.854	1.055	1.310	1.697	2.042	2.457	2.750	3.646
40....	.681	.851	1.050	1.303	1.684	2.021	2.423	2.704	3.551
60....	.679	.848	1.046	1.296	1.671	2.000	2.390	2.660	3.460
120....	.677	.845	1.041	1.289	1.658	1.980	2.358	2.617	3.373
∞	.674	.842	1.036	1.282	1.645	1.960	2.326	2.576	3.291

SOURCE: Table B.5 is taken from Table III of Fisher and Yates: *Statistical Tables for Biological, Agricultural and Medical Research*, published by Oliver & Boyd, Edinburgh, and by permission of the authors and publishers.

TABLE B.6

Chi-square distribution: critical values.

Area in the upper tail (the shaded area).

d.f.	.995	.99	.975	.95	.05	.025	.01	.005
1	.0000393	.000157	.000982	.00393	3.841	5.024	6.635	7.879
2	.0100	.0201	.0506	.103	5.991	7.378	9.210	10.597
3	.0717	.115	.216	.352	7.815	9.348	11.345	12.838
4	.207	.297	.484	.711	9.488	11.143	13.277	14.860
5	.412	.554	.831	1.145	11.070	12.832	15.086	16.750
6	.676	.872	1.237	1.635	12.592	14.449	16.812	18.548
7	.989	1.239	1.690	2.167	14.067	16.013	18.475	20.278
8	1.344	1.646	2.180	2.733	15.507	17.535	20.090	21.955
9	1.735	2.088	2.700	3.325	16.919	19.023	21.666	23.589
10	2.156	2.558	3.247	3.940	18.307	20.483	23.209	25.188
11	2.603	3.053	3.816	4.575	19.675	21.920	24.725	26.757
12	3.074	3.571	4.404	5.226	21.026	23.337	26.217	28.300
13	3.565	4.107	5.009	5.892	22.362	24.736	27.688	29.819
14	4.075	4.660	5.629	6.571	23.685	26.119	29.141	31.319
15	4.601	5.229	6.262	7.261	24.996	27.488	30.578	32.801
16	5.142	5.812	6.908	7.962	26.296	28.845	32.000	34.267
17	5.697	6.408	7.564	8.672	27.587	30.191	33.409	35.718
18	6.265	7.015	8.231	9.390	28.869	31.526	34.805	37.156
19	6.844	7.633	8.907	10.117	30.144	32.852	36.191	38.582
20	7.434	8.260	9.591	10.851	31.410	34.170	37.566	39.997
21	8.034	8.897	10.283	11.591	32.671	35.479	38.932	41.401
22	8.643	9.542	10.982	12.338	33.924	36.781	40.289	42.796
23	9.260	10.196	11.689	13.091	35.172	38.076	41.638	44.181
24	9.886	10.856	12.401	13.848	36.415	39.364	42.980	45.558
25	10.520	11.524	13.120	14.611	37.652	40.646	44.314	46.928
26	11.160	12.198	13.844	15.379	38.885	41.923	45.642	48.290
27	11.808	12.879	14.573	16.151	40.113	43.194	46.963	49.645
28	12.461	13.565	15.308	16.928	41.337	44.461	48.278	50.993
29	13.121	14.256	16.047	17.708	42.557	45.722	49.588	52.336
30	13.787	14.953	16.791	18.493	43.773	46.979	50.892	53.672

Probability of exceeding the critical value

Source: Table 8 of *Biometrika Tables for Statisticians*, Volume I, by permission of the *Biometrika* trustees.

TABLE B.7

F distribution: critical values which mark off, in the upper tail, .05 of the total area under the distribution curve.

denom \ numer	1	2	3	4	5	6	7	8	9	10	12	15	20	24	30	40	60	120	∞
1	161	200	216	225	230	234	237	239	241	242	244	246	248	249	250	251	252	253	254
2	18.5	19.0	19.2	19.2	19.3	19.3	19.4	19.4	19.4	19.4	19.4	19.4	19.4	19.5	19.5	19.5	19.5	19.5	19.5
3	10.1	9.55	9.28	9.12	9.01	8.94	8.89	8.85	8.81	8.79	8.74	8.70	8.66	8.64	8.62	8.59	8.57	8.55	8.53
4	7.71	6.94	6.59	6.39	6.26	6.16	6.09	6.04	6.00	5.96	5.91	5.86	5.80	5.77	5.75	5.72	5.69	5.66	5.63
5	6.61	5.79	5.41	5.19	5.05	4.95	4.88	4.82	4.77	4.74	4.68	4.62	4.56	4.53	4.50	4.46	4.43	4.40	4.37
6	5.99	5.14	4.76	4.53	4.39	4.28	4.21	4.15	4.10	4.06	4.00	3.94	3.87	3.84	3.81	3.77	3.74	3.70	3.67
7	5.59	4.74	4.35	4.12	3.97	3.87	3.79	3.73	3.68	3.64	3.57	3.51	3.44	3.41	3.38	3.34	3.30	3.27	3.23
8	5.32	4.46	4.07	3.84	3.69	3.58	3.50	3.44	3.39	3.35	3.28	3.22	3.15	3.12	3.08	3.04	3.01	2.97	2.93
9	5.12	4.26	3.86	3.63	3.48	3.37	3.29	3.23	3.18	3.14	3.07	3.01	2.94	2.90	2.86	2.83	2.79	2.75	2.71
10	4.96	4.10	3.71	3.48	3.33	3.22	3.14	3.07	3.02	2.98	2.91	2.85	2.77	2.74	2.70	2.66	2.62	2.58	2.54
11	4.84	3.98	3.59	3.36	3.20	3.09	3.01	2.95	2.90	2.85	2.79	2.72	2.65	2.61	2.57	2.53	2.49	2.45	2.40
12	4.75	3.89	3.49	3.26	3.11	3.00	2.91	2.85	2.80	2.75	2.69	2.62	2.54	2.51	2.47	2.43	2.38	2.34	2.30
13	4.67	3.81	3.41	3.18	3.03	2.92	2.83	2.77	2.71	2.67	2.60	2.53	2.46	2.42	2.38	2.34	2.30	2.25	2.21
14	4.60	3.74	3.34	3.11	2.96	2.85	2.76	2.70	2.65	2.60	2.53	2.46	2.39	2.35	2.31	2.27	2.22	2.18	2.13
15	4.54	3.68	3.29	3.06	2.90	2.79	2.71	2.64	2.59	2.54	2.48	2.40	2.33	2.29	2.25	2.20	2.16	2.11	2.07
16	4.49	3.63	3.24	3.01	2.85	2.74	2.66	2.59	2.54	2.49	2.42	2.35	2.28	2.24	2.19	2.15	2.11	2.06	2.01
17	4.45	3.59	3.20	2.96	2.81	2.70	2.61	2.55	2.49	2.45	2.38	2.31	2.23	2.19	2.15	2.10	2.06	2.01	1.96
18	4.41	3.55	3.16	2.93	2.77	2.66	2.58	2.51	2.46	2.41	2.34	2.27	2.19	2.15	2.11	2.06	2.02	1.97	1.92
19	4.38	3.52	3.13	2.90	2.74	2.63	2.54	2.48	2.42	2.38	2.31	2.23	2.16	2.11	2.07	2.03	1.98	1.93	1.88
20	4.35	3.49	3.10	2.87	2.71	2.60	2.51	2.45	2.39	2.35	2.28	2.20	2.12	2.08	2.04	1.99	1.95	1.90	1.84
21	4.32	3.47	3.07	2.84	2.68	2.57	2.49	2.42	2.37	2.32	2.25	2.18	2.10	2.05	2.01	1.96	1.92	1.87	1.81
22	4.30	3.44	3.05	2.82	2.66	2.55	2.46	2.40	2.34	2.30	2.23	2.15	2.07	2.03	1.98	1.94	1.89	1.84	1.78
23	4.28	3.42	3.03	2.80	2.64	2.53	2.44	2.37	2.32	2.27	2.20	2.13	2.05	2.01	1.96	1.91	1.86	1.81	1.76
24	4.26	3.40	3.01	2.78	2.62	2.51	2.42	2.36	2.30	2.25	2.18	2.11	2.03	1.98	1.94	1.89	1.84	1.79	1.73
25	4.24	3.39	2.99	2.76	2.60	2.49	2.40	2.34	2.28	2.24	2.16	2.09	2.01	1.96	1.92	1.87	1.82	1.77	1.71
30	4.17	3.32	2.92	2.69	2.53	2.42	2.33	2.27	2.21	2.16	2.09	2.01	1.93	1.89	1.84	1.79	1.74	1.68	1.62
40	4.08	3.23	2.84	2.61	2.45	2.34	2.25	2.18	2.12	2.08	2.00	1.92	1.84	1.79	1.74	1.69	1.64	1.58	1.51
60	4.00	3.15	2.76	2.53	2.37	2.25	2.17	2.10	2.04	1.99	1.92	1.84	1.75	1.70	1.65	1.59	1.53	1.47	1.39
120	3.92	3.07	2.68	2.45	2.29	2.18	2.09	2.02	1.96	1.91	1.83	1.75	1.66	1.61	1.55	1.50	1.43	1.35	1.25
∞	3.84	3.00	2.60	2.37	2.21	2.10	2.01	1.94	1.88	1.83	1.75	1.67	1.57	1.52	1.46	1.39	1.32	1.22	1.00

Degrees of freedom for numerator

Degrees of freedom for denominator

SOURCE: M. Merrington and C. M. Thompson, "Tables of percentage points of the inverted beta (F) distribution," Biometrika, Vol. 33 (1943), by permission of the Biometrika trustees.

388

TABLE B.8

F distribution: critical values which mark off, in the upper tail, .01 of the total area under the distribution curve.

Degrees of freedom for denominator (rows) × Degrees of freedom for numerator (columns)

df (denom)	1	2	3	4	5	6	7	8	9	10	12	15	20	24	30	40	60	120	∞
1	4,052	5,000	5,403	5,625	5,764	5,859	5,928	5,982	6,023	6,056	6,106	6,157	6,209	6,235	6,261	6,287	6,313	6,339	6,366
2	98.5	99.0	99.2	99.2	99.3	99.3	99.4	99.4	99.4	99.4	99.4	99.4	99.4	99.5	99.5	99.5	99.5	99.5	99.5
3	34.1	30.8	29.5	28.7	28.2	27.9	27.7	27.5	27.3	27.2	27.1	26.9	26.7	26.6	26.5	26.4	26.3	26.2	26.1
4	21.2	18.0	16.7	16.0	15.5	15.2	15.0	14.8	14.7	14.5	14.4	14.2	14.0	13.9	13.8	13.7	13.7	13.6	13.5
5	16.3	13.3	12.1	11.4	11.0	10.7	10.5	10.3	10.2	10.1	9.89	9.72	9.55	9.47	9.38	9.29	9.20	9.11	9.02
6	13.7	10.9	9.78	9.15	8.75	8.47	8.26	8.10	7.98	7.87	7.72	7.56	7.40	7.31	7.23	7.14	7.06	6.97	6.88
7	12.2	9.55	8.45	7.85	7.46	7.19	6.99	6.84	6.72	6.62	6.47	6.31	6.16	6.07	5.99	5.91	5.82	5.74	5.65
8	11.3	8.65	7.59	7.01	6.63	6.37	6.18	6.03	5.91	5.81	5.67	5.52	5.36	5.28	5.20	5.12	5.03	4.95	4.86
9	10.6	8.02	6.99	6.42	6.06	5.80	5.61	5.47	5.35	5.26	5.11	4.96	4.81	4.73	4.65	4.57	4.48	4.40	4.31
10	10.0	7.56	6.55	5.99	5.64	5.39	5.20	5.06	4.94	4.85	4.71	4.56	4.41	4.33	4.25	4.17	4.08	4.00	3.91
11	9.65	7.21	6.22	5.67	5.32	5.07	4.89	4.74	4.63	4.54	4.40	4.25	4.10	4.02	3.94	3.86	3.78	3.69	3.60
12	9.33	6.93	5.95	5.41	5.06	4.82	4.64	4.50	4.39	4.30	4.16	4.01	3.86	3.78	3.70	3.62	3.54	3.45	3.36
13	9.07	6.70	5.74	5.21	4.86	4.62	4.44	4.30	4.19	4.10	3.96	3.82	3.66	3.59	3.51	3.43	3.34	3.25	3.17
14	8.86	6.51	5.56	5.04	4.70	4.46	4.28	4.14	4.03	3.94	3.80	3.66	3.51	3.43	3.35	3.27	3.18	3.09	3.00
15	8.68	6.36	5.42	4.89	4.56	4.32	4.14	4.00	3.89	3.80	3.67	3.52	3.37	3.29	3.21	3.13	3.05	2.96	2.87
16	8.53	6.23	5.29	4.77	4.44	4.20	4.03	3.89	3.78	3.69	3.55	3.41	3.26	3.18	3.10	3.02	2.93	2.84	2.75
17	8.40	6.11	5.19	4.67	4.34	4.10	3.93	3.79	3.68	3.59	3.46	3.31	3.16	3.08	3.00	2.92	2.83	2.75	2.65
18	8.29	6.01	5.09	4.58	4.25	4.01	3.84	3.71	3.60	3.51	3.37	3.23	3.08	3.00	2.92	2.84	2.75	2.66	2.57
19	8.19	5.93	5.01	4.50	4.17	3.94	3.77	3.63	3.52	3.43	3.30	3.15	3.00	2.92	2.84	2.76	2.67	2.58	2.49
20	8.10	5.85	4.94	4.43	4.10	3.87	3.70	3.56	3.46	3.37	3.23	3.09	2.94	2.86	2.78	2.69	2.61	2.52	2.42
21	8.02	5.78	4.87	4.37	4.04	3.81	3.64	3.51	3.40	3.31	3.17	3.03	2.88	2.80	2.72	2.64	2.55	2.46	2.36
22	7.95	5.72	4.82	4.31	3.99	3.76	3.59	3.45	3.35	3.26	3.12	2.98	2.83	2.75	2.67	2.58	2.50	2.40	2.31
23	7.88	5.66	4.76	4.26	3.94	3.71	3.54	3.41	3.30	3.21	3.07	2.93	2.78	2.70	2.62	2.54	2.45	2.35	2.26
24	7.82	5.61	4.72	4.22	3.90	3.67	3.50	3.36	3.26	3.17	3.03	2.89	2.74	2.66	2.58	2.49	2.40	2.31	2.21
25	7.77	5.57	4.68	4.18	3.86	3.63	3.46	3.32	3.22	3.13	2.99	2.85	2.70	2.62	2.53	2.45	2.36	2.27	2.17
30	7.56	5.39	4.51	4.02	3.70	3.47	3.30	3.17	3.07	2.98	2.84	2.70	2.55	2.47	2.39	2.30	2.21	2.11	2.01
40	7.31	5.18	4.31	3.83	3.51	3.29	3.12	2.99	2.89	2.80	2.66	2.52	2.37	2.29	2.20	2.11	2.02	1.92	1.80
60	7.08	4.98	4.13	3.65	3.34	3.12	2.95	2.82	2.72	2.63	2.50	2.35	2.20	2.12	2.03	1.94	1.84	1.73	1.60
120	6.85	4.79	3.95	3.48	3.17	2.96	2.79	2.66	2.56	2.47	2.34	2.19	2.03	1.95	1.86	1.76	1.66	1.53	1.38
∞	6.63	4.61	3.78	3.32	3.02	2.80	2.64	2.51	2.41	2.32	2.18	2.04	1.88	1.79	1.70	1.59	1.47	1.32	1.00

SOURCE: M. Merrington and C. M. Thompson, "Tables of percentage points of the inverted beta (F) distribution," Biometrika, Vol. 33 (1943), by permission of the Biometrika trustees.

389

TABLE B.9

Common logarithms.

N	0	1	2	3	4	5	6	7	8	9
10	0000	0043	0086	0128	0170	0212	0253	0294	0334	0374
11	0414	0453	0492	0531	0569	0607	0645	0682	0719	0755
12	0792	0828	0864	0899	0934	0969	1004	1038	1072	1106
13	1139	1173	1206	1239	1271	1303	1335	1367	1399	1430
14	1461	1492	1523	1553	1584	1614	1644	1673	1703	1732
15	1761	1790	1818	1847	1875	1903	1931	1959	1987	2014
16	2041	2068	2095	2122	2148	2175	2201	2227	2253	2279
17	2304	2330	2355	2380	2405	2430	2455	2480	2504	2529
18	2553	2577	2601	2625	2648	2672	2695	2718	2742	2765
19	2788	2810	2833	2856	2878	2900	2923	2945	2967	2989
20	3010	3032	3054	3075	3096	3118	3139	3160	3181	3201
21	3222	3243	3263	3284	3304	3324	3345	3365	3385	3404
22	3424	3444	3464	3483	3502	3522	3541	3560	3579	3598
23	3617	3636	3655	3674	3692	3711	3729	3747	3766	3784
24	3802	3820	3838	3856	3874	3892	3909	3927	3945	3962
25	3979	3997	4014	4031	4048	4065	4082	4099	4116	4133
26	4150	4166	4183	4200	4216	4232	4249	4265	4281	4298
27	4314	4330	4346	4362	4378	4393	4409	4425	4440	4456
28	4472	4487	4502	4518	4533	4548	4564	4579	4594	4609
29	4624	4639	4654	4669	4683	4698	4713	4728	4742	4757
30	4771	4786	4800	4814	4829	4843	4857	4871	4886	4900
31	4914	4928	4942	4955	4969	4983	4997	5011	5024	5038
32	5051	5065	5079	5092	5105	5119	5132	5145	5159	5172
33	5185	5198	5211	5224	5237	5250	5263	5276	5289	5302
34	5315	5328	5340	5353	5366	5378	5391	5403	5416	5428
35	5441	5453	5465	5478	5490	5502	5514	5527	5539	5551
36	5563	5575	5587	5599	5611	5623	5635	5647	5658	5670
37	5682	5694	5705	5717	5729	5740	5752	5763	5775	5786
38	5798	5809	5821	5832	5843	5855	5866	5877	5888	5899
39	5911	5922	5933	5944	5955	5966	5977	5988	5999	6010
40	6021	6031	6042	6053	6064	6075	6085	6096	6107	6117
41	6128	6138	6149	6160	6170	6180	6191	6201	6212	6222
42	6232	6243	6253	6263	6274	6284	6294	6304	6314	6325
43	6335	6345	6355	6365	6375	6385	6395	6405	6415	6425
44	6435	6444	6454	6464	6474	6484	6493	6503	6513	6522
45	6532	6542	6551	6561	6571	6580	6590	6599	6609	6618
46	6628	6637	6646	6656	6665	6675	6684	6693	6702	6712
47	6721	6730	6739	6749	6758	6767	6776	6785	6794	6803
48	6812	6821	6830	6839	6848	6857	6866	6875	6884	6893
49	6902	6911	6920	6928	6937	6946	6955	6964	6972	6981
50	6990	6998	7007	7016	7024	7033	7042	7050	7059	7067
51	7076	7084	7093	7101	7110	7118	7126	7135	7143	7152
52	7160	7168	7177	7185	7193	7202	7210	7218	7226	7235
53	7243	7251	7259	7267	7275	7284	7292	7300	7308	7316
54	7324	7332	7340	7348	7356	7364	7372	7380	7388	7396

N	0	1	2	3	4	5	6	7	8	9
55	7404	7412	7419	7427	7435	7443	7451	7459	7466	7474
56	7482	7490	7497	7505	7513	7520	7528	7536	7543	7551
57	7559	7566	7574	7582	7589	7597	7604	7612	7619	7627
58	7634	7642	7649	7657	7664	7672	7679	7686	7694	7701
59	7709	7716	7723	7731	7738	7745	7752	7760	7767	7774
60	7782	7789	7796	7803	7810	7818	7825	7832	7839	7846
61	7853	7860	7868	7875	7882	7889	7896	7903	7910	7917
62	7924	7931	7938	7945	7952	7959	7966	7973	7980	7987
63	7993	8000	8007	8014	8021	8028	8035	8041	8048	8055
64	8062	8069	8075	8082	8089	8096	8102	8109	8116	8122
65	8129	8136	8142	8149	8156	8162	8169	8176	8182	8189
66	8195	8202	8209	8215	8222	8228	8235	8241	8248	8254
67	8261	8267	8274	8280	8287	8293	8299	8306	8312	8319
68	8325	8331	8338	8344	8351	8357	8363	8370	8376	8382
69	8388	8395	8401	8407	8414	8420	8426	8432	8439	8445
70	8451	8457	8463	8470	8476	8482	8488	8494	8500	8506
71	8513	8519	8525	8531	8537	8543	8549	8555	8561	8567
72	8573	8579	8585	8591	8597	8603	8609	8615	8621	8627
73	8633	8639	8645	8651	8657	8663	8669	8675	8681	8686
74	8692	8698	8704	8710	8716	8722	8727	8733	8739	8745
75	8751	8756	8762	8768	8774	8779	8785	8791	8797	8802
76	8808	8814	8820	8825	8831	8837	8842	8848	8854	8859
77	8865	8871	8876	8882	8887	8893	8899	8904	8910	8915
78	8921	8927	8932	8938	8943	8949	8954	8960	8965	8971
79	8976	8982	8987	8993	8998	9004	9009	9015	9020	9025
80	9031	9036	9042	9047	9053	9058	9063	9069	9074	9079
81	9085	9090	9096	9101	9106	9112	9117	9122	9128	9133
82	9138	9143	9149	9154	9159	9165	9170	9175	9180	9186
83	9191	9196	9201	9206	9212	9217	9222	9227	9232	9238
84	9243	9248	9253	9258	9263	9269	9274	9279	9284	9289
85	9294	9299	9304	9309	9315	9320	9325	9330	9335	9340
86	9345	9350	9355	9360	9365	9370	9375	9380	9385	9390
87	9395	9400	9405	9410	9415	9420	9425	9430	9435	9440
88	9445	9450	9455	9460	9465	9469	9474	9479	9484	9489
89	9494	9499	9504	9509	9513	9518	9523	9528	9533	9538
90	9542	9547	9552	9557	9562	9566	9571	9576	9581	9586
91	9590	9595	9600	9605	9609	9614	9619	9624	9628	9633
92	9638	9643	9647	9652	9657	9661	9666	9671	9675	9680
93	9685	9689	9694	9699	9703	9708	9713	9717	9722	9727
94	9731	9736	9741	9745	9750	9754	9759	9763	9768	9773
95	9777	9782	9786	9791	9795	9800	9805	9809	9814	9818
96	9823	9827	9832	9836	9841	9845	9850	9854	9859	9863
97	9868	9872	9877	9881	9886	9890	9894	9899	9903	9908
98	9912	9917	9921	9926	9930	9934	9939	9943	9948	9952
99	9956	9961	9965	9969	9974	9978	9983	9987	9991	9996

TABLE B.10

Table of squares and square roots.

Part A: Table of N^2; $N = 0.0$ through 99.9.

N	.0	.1	.2	.3	.4	.5	.6	.7	.8	.9
0	0.00	0.01	0.04	0.09	0.16	0.25	0.36	0.49	0.64	0.81
1	1.00	1.21	1.44	1.69	1.96	2.25	2.56	2.89	3.24	3.61
2	4.00	4.41	4.84	5.29	5.76	6.25	6.76	7.29	7.84	8.41
3	9.00	9.61	10.24	10.89	11.56	12.25	12.96	13.69	14.44	15.21
4	16.00	16.81	17.64	18.49	19.36	20.25	21.16	22.09	23.04	24.01
5	25.00	26.01	27.04	28.09	29.16	30.25	31.36	32.49	33.64	34.81
6	36.00	37.21	38.44	39.69	40.96	42.25	43.56	44.89	46.24	47.61
7	49.00	50.41	51.84	53.29	54.76	56.25	57.76	59.29	60.84	62.41
8	64.00	65.61	67.24	68.89	70.56	72.25	73.96	75.69	77.44	79.21
9	81.00	82.81	84.64	86.49	88.36	90.25	92.16	94.09	96.04	98.01
10	100.00	102.01	104.04	106.09	108.16	110.25	112.36	114.49	116.64	118.81
11	121.00	123.21	125.44	127.69	129.96	132.25	134.56	136.89	139.24	141.61
12	144.00	146.41	148.84	151.29	153.76	156.25	158.76	161.29	163.84	166.41
13	169.00	171.61	174.24	176.89	179.56	182.25	184.96	187.69	190.44	193.21
14	196.00	198.81	201.64	204.49	207.36	210.25	213.16	216.09	219.04	222.01
15	225.00	228.01	231.04	234.09	237.16	240.25	243.36	246.49	249.64	252.81
16	256.00	259.21	262.44	265.69	268.96	272.25	275.56	278.89	282.24	285.61
17	289.00	292.41	295.84	299.29	302.76	306.25	309.76	313.29	316.84	320.41
18	324.00	327.61	331.24	334.89	338.56	342.25	345.96	349.69	353.44	357.21
19	361.00	364.81	368.64	372.49	376.36	380.25	384.16	388.09	392.04	396.01
20	400.00	404.01	408.04	412.09	416.16	420.25	424.36	428.49	432.64	436.81
21	441.00	445.21	449.44	453.69	457.96	462.25	466.56	470.89	475.24	479.61
22	484.00	488.41	492.84	497.29	501.76	506.25	510.76	515.29	519.84	524.41
23	529.00	533.61	538.24	542.89	547.56	552.25	556.96	561.69	566.44	571.21

	0	1	2	3	4	5	6	7	8	9
24	576.00	580.81	585.64	590.49	595.36	600.25	605.16	610.09	615.04	620.01
25	625.00	630.01	635.04	640.09	645.16	650.25	655.36	660.49	665.64	670.81
26	676.00	681.21	686.44	691.69	696.96	702.25	707.56	712.89	718.24	723.61
27	729.00	734.41	739.84	745.29	750.76	756.25	761.76	767.29	772.84	778.41
28	784.00	789.61	795.24	800.89	806.56	812.25	817.96	823.69	829.44	835.21
29	841.00	846.81	852.64	858.49	864.36	870.25	876.16	882.09	888.04	894.01
30	900.00	906.01	912.04	918.09	924.16	930.25	936.36	942.49	948.64	954.81
31	961.00	967.21	973.44	979.69	985.96	992.25	998.56	1004.89	1011.24	1017.61
32	1024.00	1030.41	1036.84	1043.29	1049.76	1056.25	1062.76	1069.29	1075.84	1082.41
33	1089.00	1095.61	1102.24	1108.89	1115.56	1122.25	1128.96	1135.69	1142.44	1149.21
34	1156.00	1162.81	1169.64	1176.49	1183.36	1190.25	1197.16	1204.09	1211.04	1218.01
35	1225.00	1232.01	1239.04	1246.09	1253.16	1260.25	1267.36	1274.49	1281.64	1288.81
36	1296.00	1303.21	1310.44	1317.69	1324.96	1332.25	1339.56	1346.89	1354.24	1361.61
37	1369.00	1376.41	1383.84	1391.29	1398.76	1406.25	1413.76	1421.29	1428.84	1436.41
38	1444.00	1451.61	1459.24	1466.89	1474.56	1482.25	1489.96	1497.69	1505.44	1513.21
39	1521.00	1528.81	1536.64	1544.49	1552.36	1560.25	1568.16	1576.09	1584.04	1592.01
40	1600.00	1608.01	1616.04	1624.09	1632.16	1640.25	1648.36	1656.49	1664.64	1672.81
41	1681.00	1689.21	1697.44	1705.69	1713.96	1722.25	1730.56	1738.89	1747.24	1755.61
42	1764.00	1772.41	1780.84	1789.29	1797.76	1806.25	1814.76	1823.29	1831.84	1840.41
43	1849.00	1857.61	1866.24	1874.89	1883.56	1892.25	1900.96	1909.69	1918.44	1927.21
44	1936.00	1944.81	1953.64	1962.49	1971.36	1980.25	1989.16	1998.09	2007.04	2016.01
45	2025.00	2034.01	2043.04	2052.09	2061.16	2070.25	2079.36	2088.49	2097.64	2106.81
46	2116.00	2125.21	2134.44	2143.69	2152.96	2162.25	2171.56	2180.89	2190.24	2199.61
47	2209.00	2218.41	2227.84	2237.29	2246.76	2256.25	2265.76	2275.29	2284.84	2294.41
48	2304.00	2313.61	2323.24	2332.89	2342.56	2352.25	2361.96	2371.69	2381.44	2391.21
49	2401.00	2410.81	2420.64	2430.49	2440.36	2450.25	2460.16	2470.09	2480.04	2490.01
50	2500.00	2510.01	2520.04	2530.09	2540.16	2550.25	2560.36	2570.49	2580.64	2590.81
51	2601.00	2611.21	2621.44	2631.69	2641.96	2652.25	2662.56	2672.89	2683.24	2693.61
52	2704.00	2714.41	2724.84	2735.29	2745.76	2756.25	2766.76	2777.29	2787.84	2798.41
53	2809.00	2819.61	2830.24	2840.89	2851.56	2862.25	2872.96	2883.69	2894.44	2905.21
54	2916.00	2926.81	2937.64	2948.49	2959.36	2970.25	2981.16	2992.09	3003.04	3014.01

TABLE B.10

Table of squares and square roots.

Part A: (cont.)

N	.0	.1	.2	.3	.4	.5	.6	.7	.8	.9
55	3025.00	3036.01	3047.04	3058.09	3069.16	3080.25	3091.36	3102.49	3113.64	3124.81
56	3136.00	3147.21	3158.44	3169.69	3180.96	3192.25	3203.56	3214.89	3226.24	3237.61
57	3249.00	3260.41	3271.84	3283.29	3294.76	3306.25	3317.76	3329.29	3340.84	3352.41
58	3364.00	3375.61	3387.24	3398.89	3410.56	3422.25	3433.96	3445.69	3457.44	3469.21
59	3481.00	3492.81	3504.64	3516.49	3528.36	3540.25	3552.16	3564.09	3576.04	3588.01
60	3600.00	3612.01	3624.04	3636.09	3648.16	3660.25	3672.36	3684.49	3696.64	3708.81
61	3721.00	3733.21	3745.44	3757.69	3769.96	3782.25	3794.56	3806.89	3819.24	3831.61
62	3844.00	3856.41	3868.84	3881.29	3893.76	3906.25	3918.76	3931.29	3943.84	3956.41
63	3969.00	3981.61	3994.24	4006.89	4019.56	4032.25	4044.96	4057.69	4070.44	4083.21
64	4096.00	4108.81	4121.64	4134.49	4147.36	4160.25	4173.16	4186.09	4199.04	4212.01
65	4225.00	4238.01	4251.04	4264.09	4277.16	4290.25	4303.36	4316.49	4329.64	4342.81
66	4356.00	4369.21	4382.44	4395.69	4408.96	4422.25	4435.56	4448.89	4462.24	4475.61
67	4489.00	4502.41	4515.84	4529.29	4542.76	4556.25	4569.76	4583.29	4596.84	4610.41
68	4624.00	4637.61	4651.24	4664.89	4678.56	4692.25	4705.96	4719.69	4733.44	4747.21
69	4761.00	4774.81	4788.64	4802.49	4816.36	4830.25	4844.16	4858.09	4872.04	4886.01
70	4900.00	4914.01	4928.04	4942.09	4956.16	4970.25	4984.36	4998.49	5012.64	5026.81
71	5041.00	5055.21	5069.44	5083.69	5097.96	5112.25	5126.56	5140.89	5155.24	5169.61
72	5184.00	5198.41	5212.84	5227.29	5241.76	5256.25	5270.76	5285.29	5299.84	5314.41
73	5329.00	5343.61	5358.24	5372.89	5387.56	5402.25	5416.96	5431.69	5446.44	5461.21
74	5476.00	5490.81	5505.64	5520.49	5535.36	5550.25	5565.16	5580.09	5595.04	5610.01
75	5625.00	5640.01	5655.04	5670.09	5685.16	5700.25	5715.36	5730.49	5745.64	5760.81
76	5776.00	5791.21	5806.44	5821.69	5836.96	5852.25	5867.56	5882.89	5898.24	5913.61
77	5929.00	5944.41	5959.84	5975.29	5990.76	6006.25	6021.76	6037.29	6052.84	6068.41
78	6084.00	6099.61	6115.24	6130.89	6146.56	6162.25	6177.96	6193.69	6209.44	6225.21

	0	1	2	3	4	5	6	7	8	9
79	6241.00	6256.81	6272.64	6288.49	6304.36	6320.25	6336.16	6352.09	6368.04	6384.01
80	6400.00	6416.01	6432.04	6448.09	6464.16	6480.25	6496.36	6512.49	6528.64	6544.81
81	6561.00	6577.21	6593.44	6609.69	6625.96	6642.25	6658.56	6674.89	6691.24	6707.61
82	6724.00	6740.41	6756.84	6773.29	6789.76	6806.25	6822.76	6839.29	6855.84	6872.41
83	6889.00	6905.61	6922.24	6938.89	6955.56	6972.25	6988.96	7005.69	7022.44	7039.21
84	7056.00	7072.81	7089.64	7106.49	7123.36	7140.25	7157.16	7174.09	7191.04	7208.01
85	7225.00	7242.01	7259.04	7276.09	7293.16	7310.25	7327.36	7344.49	7361.64	7378.81
86	7396.00	7413.21	7430.44	7447.69	7464.96	7482.25	7499.56	7516.89	7534.24	7551.61
87	7569.00	7586.41	7603.84	7621.29	7638.76	7656.25	7673.76	7691.29	7708.84	7726.41
88	7744.00	7761.61	7779.24	7796.89	7814.56	7832.25	7849.96	7867.69	7885.44	7903.21
89	7921.00	7938.81	7956.64	7974.49	7992.36	8010.25	8028.16	8046.09	8064.04	8082.01
90	8100.00	8118.01	8136.04	8154.09	8172.16	8190.25	8208.36	8226.49	8244.64	8262.81
91	8281.00	8299.21	8317.44	8335.69	8353.96	8372.25	8390.56	8408.89	8427.24	8445.61
92	8464.00	8482.41	8500.84	8519.29	8537.76	8556.25	8574.76	8593.29	8611.84	8630.41
93	8649.00	8667.61	8686.24	8704.89	8723.56	8742.25	8760.96	8779.69	8798.44	8817.21
94	8836.00	8854.81	8873.64	8892.49	8911.36	8930.25	8949.16	8968.09	8987.04	9006.01
95	9025.00	9044.01	9063.04	9082.09	9101.16	9120.25	9139.36	9158.49	9177.64	9196.81
96	9216.00	9235.21	9254.44	9273.69	9292.96	9312.25	9331.56	9350.89	9370.24	9389.61
97	9409.00	9428.41	9447.84	9467.29	9486.76	9506.25	9525.76	9545.29	9564.84	9584.41
98	9604.00	9623.61	9643.24	9662.89	9682.56	9702.25	9721.96	9741.69	9761.44	9781.21
99	9801.00	9820.81	9840.64	9860.49	9880.36	9900.25	9920.16	9940.09	9960.04	9980.01

TABLE B.10

Table of squares and square roots (cont.).

Part B: Table of $\sqrt{N}$; N = 0.0 through 99.9.

N	.0	.1	.2	.3	.4	.5	.6	.7	.8	.9
0	0.0000	0.3162	0.4472	0.5477	0.6325	0.7071	0.7746	0.8367	0.8944	0.9487
1	1.0000	1.0488	1.0954	1.1402	1.1832	1.2247	1.2649	1.3038	1.3416	1.3784
2	1.4142	1.4491	1.4832	1.5166	1.5492	1.5811	1.6125	1.6432	1.6733	1.7029
3	1.7320	1.7607	1.7889	1.8166	1.8439	1.8708	1.8974	1.9235	1.9494	1.9748
4	2.0000	2.0248	2.0494	2.0736	2.0976	2.1213	2.1448	2.1679	2.1909	2.2136
5	2.2361	2.2583	2.2803	2.3022	2.3238	2.3452	2.3664	2.3875	2.4083	2.4290
6	2.4495	2.4698	2.4900	2.5100	2.5298	2.5495	2.5690	2.5884	2.6077	2.6268
7	2.6458	2.6646	2.6833	2.7018	2.7203	2.7386	2.7568	2.7749	2.7928	2.8107
8	2.8284	2.8460	2.8636	2.8810	2.8983	2.9155	2.9326	2.9496	2.9665	2.9833
9	3.0000	3.0166	3.0331	3.0496	3.0659	3.0822	3.0984	3.1145	3.1305	3.1464
10	3.1623	3.1780	3.1937	3.2094	3.2249	3.2404	3.2558	3.2711	3.2863	3.3015
11	3.3166	3.3317	3.3466	3.3615	3.3764	3.3912	3.4059	3.4205	3.4351	3.4496
12	3.4641	3.4785	3.4928	3.5071	3.5214	3.5355	3.5496	3.5637	3.5777	3.5917
13	3.6055	3.6194	3.6332	3.6469	3.6606	3.6742	3.6878	3.7014	3.7148	3.7283
14	3.7417	3.7550	3.7683	3.7815	3.7947	3.8079	3.8210	3.8341	3.8471	3.8601
15	3.8730	3.8859	3.8987	3.9115	3.9243	3.9370	3.9497	3.9623	3.9749	3.9875
16	4.0000	4.0125	4.0249	4.0373	4.0497	4.0620	4.0743	4.0866	4.0988	4.1110
17	4.1231	4.1352	4.1473	4.1593	4.1713	4.1833	4.1952	4.2071	4.2190	4.2308
18	4.2426	4.2544	4.2661	4.2778	4.2895	4.3012	4.3128	4.3243	4.3359	4.3474
19	4.3589	4.3704	4.3818	4.3932	4.4045	4.4159	4.4272	4.4385	4.4497	4.4609
20	4.4721	4.4833	4.4944	4.5056	4.5166	4.5277	4.5387	4.5497	4.5607	4.5716
21	4.5826	4.5935	4.6043	4.6152	4.6260	4.6368	4.6476	4.6583	4.6690	4.6797
22	4.6904	4.7011	4.7117	4.7223	4.7329	4.7434	4.7539	4.7644	4.7749	4.7854
23	4.7958	4.8062	4.8166	4.8270	4.8374	4.8477	4.8580	4.8683	4.8785	4.8888

	4.9900	4.9800	4.9699	4.9598	4.9497	4.9396	4.9295	4.9193	4.9092	4.8990
24	4.9900	4.9800	4.9699	4.9598	4.9497	4.9396	4.9295	4.9193	4.9092	4.8990
25	5.0892	5.0794	5.0695	5.0596	5.0498	5.0398	5.0299	5.0200	5.0100	5.0000
26	5.1865	5.1769	5.1672	5.1575	5.1478	5.1381	5.1284	5.1186	5.1088	5.0990
27	5.2820	5.2726	5.2631	5.2536	4.2440	5.2345	5.2249	5.2154	5.2058	5.1962
28	5.3759	5.3666	5.3572	5.3479	5.3385	5.3292	5.3198	5.3104	5.3009	5.2915
29	5.4681	5.4589	5.4498	5.4406	5.4314	5.4222	5.4129	5.4037	5.3944	5.3852
30	5.5588	5.5498	5.5408	5.5317	5.5227	5.5136	5.5045	5.4955	5.4863	5.4772
31	5.6480	5.6391	5.6303	5.6214	5.6125	5.6036	5.5946	5.5857	5.5767	5.5678
32	5.7359	5.7271	5.7184	5.7096	5.7009	5.6921	5.6833	5.6745	5.6657	5.6569
33	5.8224	5.8138	5.8052	5.7965	5.7879	5.7793	5.7706	5.7619	5.7533	5.7446
34	5.9076	5.8991	5.8907	5.8822	5.8737	5.8651	5.8566	5.8481	5.8395	5.8310
35	5.9917	5.9833	5.9749	5.9666	5.9582	5.9498	5.9414	5.9330	5.9245	5.9161
36	6.0745	6.0663	6.0581	6.0498	6.0415	6.0332	6.0249	6.0166	6.0083	6.0000
37	6.1563	6.1482	6.1400	6.1319	6.1237	6.1156	6.1074	6.0992	6.0910	6.0828
38	6.2370	6.2290	6.2209	6.2129	6.2048	6.1968	6.1887	6.1806	6.1725	6.1644
39	6.3166	6.3087	6.3008	6.2929	6.2849	6.2769	6.2690	6.2610	6.2530	6.2450
40	6.3953	6.3875	6.3797	6.3718	6.3640	6.3561	6.3482	6.3403	6.3325	6.3246
41	6.4730	6.4653	6.4576	6.4498	6.4420	6.4343	6.4265	6.4187	6.4109	6.4031
42	6.5498	6.5422	6.5345	6.5269	6.5192	6.5115	6.5038	6.4962	6.4884	6.4807
43	6.6257	6.6182	6.6106	6.6030	6.5955	6.5879	6.5803	6.5727	6.5651	6.5574
44	6.7007	6.6933	6.6858	6.6783	6.6708	6.6633	6.6558	6.6483	6.6408	6.6332
45	6.7750	6.7676	6.7602	6.7528	6.7454	6.7380	6.7305	6.7231	6.7157	6.7082
46	6.8484	6.8411	6.8337	6.8264	6.8191	6.8118	6.8044	6.7971	6.7897	6.7823
47	6.9210	6.9138	6.9065	6.8993	6.8920	6.8848	6.8775	6.8702	6.8629	6.8557
48	6.9929	6.9857	6.9785	6.9714	6.9642	6.9570	6.9498	6.9426	6.9354	6.9282
49	7.0640	7.0569	7.0498	7.0427	7.0356	7.0285	7.0214	7.0143	7.0071	7.0000
50	7.1344	7.1274	7.1204	7.1134	7.1063	7.0993	7.0922	7.0852	7.0781	7.0711
51	7.2042	7.1972	7.1903	7.1833	7.1763	7.1694	7.1624	7.1554	7.1484	7.1414
52	7.2732	7.2664	7.2595	7.2526	7.2457	7.2388	7.2319	7.2250	7.2180	7.2111
53	7.3417	7.3348	7.3280	7.3212	7.3144	7.3075	7.3007	7.2938	7.2870	7.2801
54	7.4095	7.4027	7.3959	7.3892	7.3824	7.3756	7.3689	7.3621	7.3553	7.3485

TABLE B.10

Table of squares and square roots.

Part B: (cont.)

N	.0	.1	.2	.3	.4	.5	.6	.7	.8	.9
55	7.4162	7.4229	7.4297	7.4364	7.4431	7.4498	7.4565	7.4632	7.4699	7.4766
56	7.4833	7.4900	7.4967	7.5033	7.5100	7.5166	7.5233	7.5299	7.5366	7.5432
57	7.5498	7.5565	7.5631	7.5697	7.5763	7.5829	7.5895	7.5960	7.6026	7.6092
58	7.6158	7.6223	7.6289	7.6354	7.6420	7.6485	7.6551	7.6616	7.6681	7.6746
59	7.6811	7.6877	7.6942	7.7006	7.7071	7.7136	7.7201	7.7266	7.7330	7.7395
60	7.7460	7.7524	7.7589	7.7653	7.7717	7.7782	7.7846	7.7910	7.7974	7.8038
61	7.8102	7.8166	7.8230	7.8294	7.8358	7.8422	7.8486	7.8549	7.8613	7.8677
62	7.8740	7.8804	7.8867	7.8930	7.8994	7.9057	7.9120	7.9183	7.9246	7.9309
63	7.9373	7.9435	7.9498	7.9561	7.9624	7.9687	7.9750	7.9812	7.9875	7.9937
64	8.0000	8.0062	8.0125	8.0187	8.0250	8.0312	8.0374	8.0436	8.0498	8.0561
65	8.0623	8.0685	8.0746	8.0808	8.0870	8.0932	8.0994	8.1056	8.1117	8.1179
66	8.1240	8.1302	8.1363	8.1425	8.1486	8.1548	8.1609	8.1670	8.1731	8.1792
67	8.1854	8.1915	8.1976	8.2037	8.2097	8.2158	8.2219	8.2280	8.2341	8.2401
68	8.2462	8.2523	8.2583	8.2644	8.2704	8.2765	8.2825	8.2885	8.2946	8.3006
69	8.3066	8.3126	8.3187	8.3247	8.3307	8.3367	8.3427	8.3486	8.3546	8.3606
70	8.3666	8.3726	8.3785	8.3845	8.3905	8.3964	8.4024	8.4083	8.4143	8.4202
71	8.4261	8.4321	8.4380	8.4439	8.4498	8.4558	8.4617	8.4676	8.4735	8.4794
72	8.4853	8.4912	8.4971	8.5029	8.5088	8.5147	8.5206	8.5264	8.5323	8.5381
73	8.5440	8.5499	8.5557	8.5615	8.5674	8.5732	8.5790	8.5849	8.5907	8.5965
74	8.6023	8.6081	8.6139	8.6197	8.6255	8.6313	8.6371	8.6429	8.6487	8.6545
75	8.6603	8.6660	8.6718	8.6776	8.6833	8.6891	8.6948	8.7006	8.7063	8.7121
76	8.7178	8.7235	8.7293	8.7350	8.7407	8.7464	8.7521	8.7579	8.7636	8.7693
77	8.7750	8.7807	8.7864	8.7920	8.7977	8.8034	8.8091	8.8148	8.8204	8.8261
78	8.8318	8.8374	8.8431	8.8487	8.8544	8.8600	8.8657	8.8713	8.8769	8.8826

79	8.9387	8.9331	8.9275	8.9219	8.9163	8.9107	8.9051	8.8994	8.8938	8.8882
80	8.9944	8.9889	8.9833	8.9777	8.9722	8.9666	8.9610	8.9554	8.9499	8.9443
81	9.0499	9.0443	9.0388	9.0333	9.0277	9.0222	9.0166	9.0111	9.0056	9.0000
82	9.1049	9.0994	9.0940	9.0885	9.0829	9.0774	9.0719	9.0664	9.0609	9.0554
83	9.1597	9.1542	9.1488	9.1433	9.1378	9.1324	9.1269	9.1214	9.1159	9.1104
84	9.2141	9.2087	9.2033	9.1978	9.1924	9.1869	9.1815	9.1761	9.1706	9.1651
85	9.2682	9.2628	9.2574	9.2520	9.2466	9.2412	9.2358	9.2304	9.2250	9.2195
86	9.3220	9.3166	9.3113	9.3059	9.3005	9.2952	9.2898	9.2844	9.2790	9.2736
87	9.3755	9.3702	9.3648	9.3595	9.3541	9.3488	9.3434	9.3381	9.3327	9.3274
88	9.4287	9.4234	9.4181	9.4128	9.4074	9.4021	9.3968	9.3915	9.3862	9.3808
89	9.4816	9.4763	9.4710	9.4657	9.4604	9.4552	9.4499	9.4446	9.4393	9.4340
90	9.5341	9.5289	9.5237	9.5184	9.5131	9.5079	9.5026	9.4974	9.4921	9.4868
91	9.5864	9.5812	9.5760	9.5708	9.5656	9.5603	9.5551	9.5499	9.5446	9.5394
92	9.6385	9.6333	9.6281	9.6229	9.6177	9.6125	9.6073	9.6021	9.5969	9.5917
93	9.6902	9.6850	9.6799	9.6747	9.6695	9.6644	9.6592	9.6540	9.6488	9.6436
94	9.7417	9.7365	9.7314	9.7263	9.7211	9.7160	9.7108	9.7057	9.7005	9.6954
95	9.7929	9.7877	9.7826	9.7775	9.7724	9.7673	9.7622	9.7570	9.7519	9.7468
96	9.8438	9.8387	9.8336	9.8285	9.8234	9.8183	9.8133	9.8082	9.8031	9.7980
97	9.8944	9.8894	9.8843	9.8793	9.8742	9.8691	9.8641	9.8590	9.8539	9.8489
98	9.9448	9.9398	9.9348	9.9298	9.9247	9.9197	9.9146	9.9096	9.9045	9.8995
99	9.9950	9.9900	9.9850	9.9800	9.9750	9.9700	9.9649	9.9599	9.9549	9.9499

Appendix C
Answers to Even-Numbered Problems

In Appendix C, your answers may differ from the answers shown due to rounding.

Chapter 2

8. a. 6   b. 84   c. 50   d. 50
    e. 789   f. 130   g. 10   h. 10,030

10. a. $\displaystyle\sum_{i=1}^{5} X_i$ b. $\displaystyle\sum_{i=11}^{14} Y_i$

    c. $\displaystyle\sum_{i=12}^{15} gZ_i$ d. $\displaystyle\sum_{i=1}^{3} \frac{(X_i + Y_i)}{k}$

    e. $\displaystyle\sum_{i=2}^{4} g(Y_i - L_i)$ f. $\displaystyle\sum_{i=1}^{N} (X_i - G)^2$

    g. $\displaystyle\sum_{i=1}^{5} k$ h. $\displaystyle\sum_{i=1}^{m} \frac{2X_i - kY_i}{M_i}$

    i. $135 - \displaystyle\sum_{i=1}^{j} \frac{Y_i}{2c}$

Chapter 3

4.

Discrete class limits	Class midpoint
12–14	13
15–17	16
18–20	19
21–23	22
24–26	25
27–29	28
30–32	31
33–35	34

6.

Discrete class limits	Class midpoint	Continuous class limits
		.0095
.010–.019	.0145	
		.0195
.020–.029	.0245	
		.0295
.030–.039	.0345	
		.0395
.040–.049	.0445	
		.0495
.050–.059	.0545	
		.0595
.060–.069	.0645	
		.0695

Length of class interval = .010

403

8.

Continuous class limits	Class midpoint	Cum f_i More than	Cum f_i Less than
2.495		30	0
	2.62		
2.745		28	2
	2.87		
2.995		24	6
	3.12		
3.245		19	11
	3.37		
3.495		13	17
	3.62		
3.745		5	25
	3.87		
3.995		1	29
	4.12		
4.245		0	30

Length of class interval = $.25

10. (1) Lower limits not easy to visualize
    (2) Overlapping class intervals
    (3) Length of class interval (1.12) not a convenient amount
    (4) Too few classes

12.

Discrete class limits	Tally	Class frequency	Class midpoint
150–154	ⅣⅣ ⅣⅣ ⅣⅣ /	16	152
155–159	ⅣⅣ ⅣⅣ ////	14	157
160–164	ⅣⅣ ⅣⅣ	10	162
165–169	ⅣⅣ //	7	167
170–174	///	3	172
175–179	///	3	177

Continuous class limits	Cum f_i More than	Cum f_i Less than
149.5	53	0
154.5	37	16
159.5	23	30
164.5	13	40
169.5	6	47
174.5	3	50
179.5	0	53

16. Probability distributions; Sampling distributions
18. a. Normal b. 1's and 0's

22.

Discrete class limits	Relative frequency
15.0–17.9	.08
18.0–20.9	.24
21.0–23.9	.36
24.0–26.9	.20
27.0–29.9	.12
Total	1.00

24.　a.　Ordinal　　　b.　Nominal
　　c.　Quantitative　d.　Nominal
　　e.　Ordinal

26.

Opinion	Number of experts
Sharp downtrend	1
Mild downtrend	2
Retain present level	3
Small uptrend	6
Sharp uptrend	3
Total	15

Chapter 4

2.　a.　5　　b.　6　　c.　10.57
4.　60 lbs
6.　$3
8.　$\mu_w = 38.5¢$ per foot
10.　$87.40
12.　$\overline{X}_w = 14.5$
14.　11.73
16.　.280 sec
18.　42 percent a year
20.　20 percent a month
24.　a.　Arithmetic mean　　b.　Median
　　　　Geometric mean
　　c.　Arithmetic mean　　d.　Mode
　　e.　Median　　　　　　f.　Arithmetic mean
　　　　　　　　　　　　　　　Geometric mean
　　g.　Arithmetic mean　　h.　Arithmetic mean
　　i.　Arithmetic mean　　j.　Median
　　　　Geometric mean　　　　Mode
　　　　Median
　　　　Mode
　　k.　Mode

26.

	Median	Mode
a.	20	20
b.	34.9	40.1
c.	.007	.005

28.   18.95
30.   45.8
32. $Q_1 = .154$ $Q_3 = .405$
      $D_3 = .180$ $D_8 = .432$
      $P_{18} = .117$ $P_{55} = .301$
34.   Red (6)
36.   Median: opposed
      Mode: moderately opposed
38.   a.   89.7   b.   2.31 cms
40.   a.   95.16   b. $6.10

Chapter 5

2.   a.   5–18, 13   b.   100–120, 10   c.   23–34, 11
4.   400
6.   a.   64   b.   8
8.   4,400
10.   11.92
12.   (1) $\sigma = 1.23$ (2) $\sigma = 3.85$ (3) $\sigma = 4.84$
14. $\mu = 6.2$ $\sigma^2 = 10.24$ $\sigma = 3.2$
16.   a.   3–15, 12   b.   16.27   c.   4.0
18.   a.   6.8   b.   1.98   c.   3.92
20.   a.   4.88 lbs   b.   1.29 lbs   c.   1.66
22.   a.   30   b.   60
24.   40
26.   a.   24.6 percent   b.   64.2 percent
      c.   52.5 percent   d.   65.5 percent
28.   a.   60 percent   b.   59.3 percent   c.   21.1 percent
30. $z = 1.24$ (1.24 standard deviations above the mean)
32. $z_{.07} = -1.19,$ $z_{.16} = 1.63$
34.   a. $p' = .89$ or more
      b. $p = .16$ or less
      c. $p' = .44$ or more
36.   79 percent or less
38.   a.   2   b.   518,400
40.   a. $\overline{X} = 15.25,$ $s^2 = 14.79,$ $s = 3.84$
      b. $\mu = 4.88$ lbs, $\sigma^2 = 1.66,$ $\sigma = 1.29$ lbs
      c. $\overline{X} = .108$ cms, $s^2 = .001014,$ $s = .032$ cms
42.   4.18

Chapter 6

4. Element, event, sample point

8.
a.	18		f.	No; they overlap
g.	No; they overlap		h.	Yes; they do not overlap
i.	3		j.	5
k.	4		l.	1
m.	none		n.	15
o.	13		p.	17

10. a.

	Silver	Gold
Silver	·	·
Gold	·	·

   b.

	Silver	Gold
Silver		·
Gold	·	·

12. a.

		2nd card			
		1	2	3	4
1st card	1		·	·	·
	2	·		·	·
	3	·	·		·
	4	·	·	·	

   b. 8   c. 7

16. On the average, Sue will type a letter without errors 87 times out of 100.

20. 0

22. Objective probabilities

30. 3 to 2

32. a. .03   b. Definition 1
    c. .97   d.   ·   ·
                Effective   Not effective

34. Compute the proportion of automobiles involved in an accident during a period of time. Use Definition 1.

36. Make some selections with replacement and determine the proportion of red marbles obtained (Definition 1).

38. a. 3 to 1   b. 1 to 1

40. 3 to 7

42. a. .20   b. .90   c. .60   d. .10

44.
Blue	Red	Green
.50	.25	.25

46.
	Man	Woman
New	5/15	1/15
Recently employed	2/15	3/15
Oldtimers	3/15	1/15

48.   a.   *FF, FS, SF, SS*
      b.
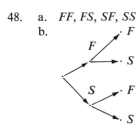

50.   120
52.   a.   blue, yellow   red, blue
           blue, red   red, yellow
           blue, white   red, white

           yellow, blue   white, blue
           yellow, red   white, yellow
           yellow, white   white, red
      b.

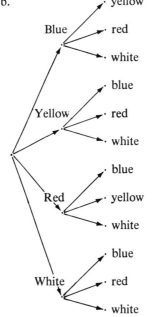

      c. $_4P_2 = 12$
54.   No,   .70 · .20 ≠ .50
56.   a.   1/6   b.   1/3   c.   1/2
58.   a.   1/2   b.   1/2   c.   7/10   d.   3/5

60.   Color   Probability
      yellow   .2
      green   .3
      blue   .1
      yellow and blue   .4

62.   Yes,   .80 · .70 = .56
64.   a.   1/2   b.   3/4   c.   1/4
66.   a.   .35   b.   .70   c.   .20
68.   a.   .65   b.   .35

70.

K	|L| A	B	C	D	Not vote	Total
A	.0625	.0750	.0500	.0375	.0250	.25
B	.0750	.0900	.0600	.0450	.0300	.30
C	.0500	.0600	.0400	.0300	.0200	.20
D	.0375	.0450	.0300	.0225	.0150	.15
Not vote	.0250	.0300	.0200	.0150	.0100	.10
Total	.25	.30	.20	.15	.10	1.00

72.   a.   .21   b.   .39   c.   .53

74.   a.   .28

    b.

	Pass	Not pass	Total
Study	.280	.420	.70
No study	.045	.255	.30
Total	.325	.675	1.00

    c.   .675

    d. $.280 \neq .70 \cdot .325 = .22750$

       Therefore, study and pass are not independent.

76.   29/60

78.   a.   1/216   b.   1/27   c.   2/27

    d.   1/8   e.   1/54   f.   1/36

80.   a.   5/66   b.   14/55   c.   1/55

    d.   8/432   e.   4/165

82.   a.   .024   b.   .144   c.   .216

84.   5/18

86.   a.   .46   b.   .12

    c.   .42   d.   (1 − part (b) answer) = probability of only one or neither buying the stock. Part (c) answer = probability of neither buying the stock.

88.   .25

90.   a.   8

    b.

	Sample space
C for	ABC
C against	ABC′
C for	AB′C
C against	AB′C′
C for	A′BC
C against	A′BC′
C for	A′B′C
C against	A′B′C′

A for — B for — C for / C against
A for — B against — C for / C against
A against — B for — C for / C against
A against — B against — C for / C against

    c.   3/4

92. a. .48   b. .07
94. a. $H \cup L$ b. $K \cap H$
     c. L' d. $(H \cup K)'$
     e. $H \cup (K \cup L)'$ f. $H \cap K \cap L'$
     g. $(L \cup H) \cap (K \cap H)$ h. $[L \cup (H \cap K)]'$

Chapter 7

10. a. 1/28   b. 1/56
12. a. 1/792   b. 1/220
16. 1/1,000
18. Four-digit numbers between 0001 and 1538 are acceptable and should be used only the first time selected.
20. The sociologist's finding could be generalized only to the city from which the sample was selected. It could not be generalized to the nationwide population, except on a judgment basis.

22. a.

$\overline{X}$	$P(\overline{X})$
2.00	.10
2.33	.10
2.67	.20
3.00	.20
3.33	.20
3.67	.10
4.00	.10

b. $\mu = 3$
    $\mu_{\overline{x}} = 3$
    $\sigma = 1.41$
    $\sigma_{\overline{x}} = .58$
c. $P(\overline{X} - \mu = 0) = .20$

26. 35
28. 495

Chapter 8

2.   a. continuous
     b. $-\infty$ to $+\infty$
     c. infinite
4. z, the variable X expressed in standard units.
6.   a. .4861   b. .9861   c. .9918
     d. .0054   e. .0168
8.   a. .01   b. .21   c. .62   d. .18
     e. .14   f. .13   g. .50
10. a. .01   b. .03   c. .98   d. .02
     e. .03   f. .17   g. .74   h. .01
12. a. 32.9 minutes   b. 41.6 minutes
18. a. $\mu = p = .40$ b. $\mu = p = .25$ c. $\mu = p = .20$
         $\sigma = .49$ $\sigma = .43$ $\sigma = .40$
20. Model A
26. a. .05   b. .11   c. .89   d. .07
28. .38
30. a. .12   b. .04   c. .77   d. .73
32. a. .02   b. .95
34. No. Model A requires a normal distribution; Model B requires a large sample.
36. .83
38. .983

40.   a. 2.70   b. 1.37   c. .30   d. .15
42.   Model D
44.   a. $\mu_X = 4$ b. $\mu_{X/n} = .40$
         $\sigma_X = 1.54$ $\sigma_{X/n} = .15$
46.   a. .35   b. .35   c. .35   d. .31
48.   a. .004   b. .21   c. .42   d. .63
50.   a. .0064   b. .9984   c. .0256   d. .4096   e. .5904
52.

$\dfrac{X}{n}$	0	.20	.40	.60	.80	1.00
$P\left(\dfrac{X}{n}\right)$	.0102	.0768	.2304	.3456	.2592	.0778

54.   a. .057   b. .341   c. .302
56.   a. .464   b. .457   c. .814   d. .053
58.   Practically zero
60.   a. .07   b. .68

Chapter 9

6.   a. 2.58   b. 2.40   c. 2.28
8.   a. $\overline{X} = 3.25$ cms   b. 3.13–3.37 cms
    c. .12 cms   d. 3.08–3.42 cms
10.   a. $7,362,500   b. $7,232,000–$7,493,000
12.   a. 170 hrs   b. 1.0 hrs   c. 167.9–172.1 hrs
14.   a. 10,200,000 hrs
    b. 60,000 hrs
    c. 10,074,000–10,326,000 hrs
16.   71.75–78.25
18.   26.6–33.8 orders
20.   a. 24,900–35,500 orders
    b. 26,600–33,800 orders
    c. 27,300–33,100 orders
22.   a. .226–.266 cms
    b. .233–.259 cms
    c. .238–.254 cms
24.   a. 30 percent   b. .26–.34
26.   a. 6,000 commuters
    b. 5,200–6,800 commuters
    c. 5,000–7,000 commuters
28.   a. .04   b. .03
30.   a. .20   b. .06
32.   .05 − .15

34.   Multiply by $\sqrt{\dfrac{N-n}{N-1}}$

38.   a. .15   b. .92   c. .026   d. .22   e. .16 − .28
40.   463 social case workers
42.   522 housewives
44.   85 units of the product

Chapter 10

4. 1.71–6.69 cms
6. $2.84–$11.16
8. a. 1.0   b. $4.16
10. 5.6–14.0
12. 5.1–7.9
14. 3.1–5.9 minutes
16. 1.42–9.02 pounds
18. .72–5.28 sec
20. .02–.12
22. .001–.099

Chapter 11

2. The population parameter involved in a test of hypothesis (mean, standard deviation, proportion).
6. a. μ, the mean test score for the population of 6th grade pupils
    b. $H_0: \mu = 100$ c. $H_1: \mu \neq 100$ d. $H_1: \mu < 100$
16. a. .01   b. .01   c. .005   d. .99
18. a. 2.33   b. 2.58
20. Two-tail test. $z = 2.40$ in region of acceptance. 12 percent is acceptable estimate.
22. Two-tail test. $z = -2.36$ in region of rejection. Reject the claim.
24. Two-tail test. $z = 2.10$ in region of acceptance. Does not need readjustment.
26. .05, because it will lead to more frequent checking and readjustment, thus, providing greater assurance of obtaining the desired result at little additional cost.
28. One-tail test. $z = 2.88$ in region of rejection. Yes, conclude that reading ability improved.
30. $t = 3.20$ in region of rejection. Reject claim.
32. $z = 5.63$ in region of rejection. Reject claim.
34. One-tail test. $t = -.92$ in region of acceptance. Accept claim.
36. Two-tail test. $z = 5.13$ in region of rejection. Reject claim.
38. a. 24.725   b. 2.088   c. 5.009, 24.736
40. One-tail test. $\chi^2 = 25.410$ in region of acceptance. Accept claim.
42. Two-tail test. $\chi^2 = 22.006$ in region of acceptance. Accept claim.
44. a. No, H_1 is not specific.
    b. Yes, H_1 is specific.
46. .94, indicates the probability of rejecting H_0 if H_1 is true.
48. $\beta = .076$
50. 43 days
52. 54 days
54. 349 subjects

Chapter 12

2. One-tail test. $z = .48$ (or $-.48$) in region of acceptance. Conclude that proportion is the same.
4. One-tail test. $z = 4.29$ (or -4.29) in region of rejection. Reject claim.
6. One-tail test. $z = 2.22$ (or -2.22) in region of acceptance. Conclude proportion is the same.

8. One-tail test. $z = 4.12$ (or -4.12) in region of rejection. Conclude western cities have a higher mean.

10. Two-tail test. $z = 2.45$ (or -2.45) in region of rejection. Conclude mean time is not the same.

12. Two-tail test. $z = 4.44$ (or -4.44) in region of rejection. Conclude that mean strength is not the same.

14. Two-tail test. $z = 7.20$ (or -7.20) in region of rejection. Weight gains, on the average, not the same.

16. Two-tail test. $t = 3.711$ (or -3.711) in region of rejection. Mean weight gains not the same.

18. Upper-tail test. $z = 7.14$ in region of rejection. Type D seed plants have a greater mean height.

20. Two-tail test. $t = .09$ (or $-.09$) in region of acceptance. Conclude that the two forms have the same mean.

22. Lower-tail test. $t = -22.87$ in region of rejection. Conclude that the film reduces the mean score.

24. Two-tail test. $F = 3.24$ in region of rejection. Variance is not the same for men and women.

Chapter 13

2. $\chi^2 = 30.482$ in region of rejection. Conclude that proportion is not the same.
4. $\chi^2 = 5.811$ in region of acceptance. Conclude that the categories are independent.
6. $\chi^2 = 5.357$ in region of acceptance. Response and sex are independent.
8. $\chi^2 = 44.667$ in region of rejection. Conclude that the distributions are not the same.
10. $\chi^2 = 4.631$ in region of acceptance. Conclude that results are consistent.
12. a. 1.32, 7.86, 23.82, 34.36, 23.63, 7.72, and 1.29
    b. $\chi^2 = 3.487$ in region of acceptance. Conclude that population distribution is approximately normal.

Chapter 14

2.

Source of variation	SS	df	MS	F
Treatments	9,236.67	3	3,078.89	51.136
Error	963.33	16	60.21	
Total	10,200.00	19		

$F = 51.14$ in region of rejection. Conclude that the different approaches are not equally effective.

4.

Source of variation	SS	df	MS	F
Treatments	25,465.45	4	6,366.36	46.301
Error	2,749.91	20	137.50	
Total	28,215.36	24		

$F = 46.301$ in region of rejection. Conclude that time required is affected by temperature.

6.

Source of variation	SS	df	MS	F
Machines	42.13	2	21.07	1.44
Experience	120.40	4	30.10	2.05
Error	117.20	8	14.65	
Total	279.73	14		

$F = 1.44$ in region of acceptance. Conclude machine used has no effect on number of units produced.

$F = 2.05$ in region of acceptance. Conclude length of experience has no effect on number of units produced.

Chapter 15

2. $z = 2.80$ in region of rejection. Conclude that film-viewing does not reduce test scores.
4. $z = \pm.98$ in region of acceptance. Accept the claim.
6. $z = .87$ in region of acceptance. Conclude that mean test score is the same for men and women.
8. $z = -2.17$ in region of rejection. Reject the claim.
10. $H = 8.913$ in region of rejection. Conclude that the mean is not the same.
12. $H = 2.448$ in region of acceptance. Conclude that the mean is the same for the four groups.
14. $z = -.66$ in region of acceptance. Conclude that a growth trend is not indicated.

Chapter 16

4. .36
6. .29
8. a. 1 to 1   b. 4 to 1
10. .40
12. a. .02, .07, .24, .40, .23, .04.   b. .09

14. a.

Number of salesclerks needed	Number of salesclerks to employ			
	5	6	7	8
5	$900	$ 880	$ 860	$ 840
6	900	1,080	1,060	1,040
7	900	1,080	1,260	1,240
8	900	1,080	1,260	1,440

b. $900, $1,040, $1,080, $1,080
c. Employ 7 or 8 clerks.

16. a.

Possible actual enrollment	Planned enrollment levels			
	30	40	50	60
30	$14,000	$13,500	$13,000	$12,500
40	14,000	16,875	16,375	15,875
50	14,000	16,875	19,625	19,195
60	14,000	16,875	19,625	22,375

b. Plan for an enrollment of 60 students.

28.   Random strategy
30.   a.   No. The maximin criterion leads to a payoff of -2 dollars to K and 1 dollar to H and these payoffs are not equal.
     b.   K should put 8 chips in a bowl with 7 chips marked A and 1 chip marked B; select a chip at random for each play and adopt A or B as noted on the chip selected.

Chapter 17

2.   b. Y = spread, X = height
         $Y_e = 1.43 + .66X$
    d.   When: $X = 4$ inches, $Y_e = 4.07$ inches
             $X = 5$ inches, $Y_e = 4.73$ inches
    e. $r = .79$ f. $r^2 = .63$
4. $r = -.42$
6.   a. $r^2 = .40$ b. $r = .63$
10.   a.   Correlation analysis   b.   Regression analysis
    c.   Correlation analysis
12.   a. Y = number of marriages
         X = number of first-born
         $Y_e = 5.42 + .82X$
    b.   When: $X = 14$, $Y_e = 16.90$
            $X = 17$, $Y_e = 19.36$
    c. $r = .93$
14.   a. Y = value of product (\$000)
         X = number of employees
         $Y_e = 7.41 + .27X$
    b.   .58   c.   When: $X = 12$, $Y_e = 10.65$
                       $X = 14$, $Y_e = 11.19$
16.   a. $Y_e = -13.32 + 2.63X$ b.   .82
18.   a. $Y_e = 120.67 - .35X$ b. $-.96$
20.   a. Y = number of letters   b.   .82
         X = months of experience
         $Y_e = 3.24 + .48X$
22.   a. Y = yield X = amount of fertilizer   b.   .62
         $Y_e = 10.65 - 2.03X$

Chapter 18

2.   a. $\log Y_e = 2.38 - 1.50 \log X$ d. $r = -.93$
4. $-.90$
6.   .19

Chapter 19

2.

	a.	b.	c.	d.	e.
(1)	1.17	.19	.94	$-.84-.22$	2.40–7.62
(2)	1.33	.19	1.03	$-.04-.88$	3.36–8.40
(3)	.89	.14	1.58	$-.05-.59$	3.78–11.04

4.   a. $t = -1.66$ (df $= 4$) is close to $t_{.20} = -1.533$, so that $b = -.31$ is not statistically significant at the .05 one-tail level of significance ($t_{.10} = -2.132$).

       b. $t = 2.21$ (df $= 6$) is not quite equal to $t_{.05} = 2.447$, so that $b = .42$ is not statistically significant at the .05 (two-tail) level of significance.

       c. $t = 1.93$ (df $= 8$) is close to $t_{.10} = 1.860$, so that $b = .27$ is statistically significant at somewhat better than the .05 (one-tail) level of significance. (In this problem b would be expected to be positive.)

6. $t = 6.31$ (df $= 18$) so that r is statistically significant at considerably better than the .001 level of significance ($t_{.001} = 3.922$).

Chapter 20

8.   a. $b_{12} = -2.4500$ $\qquad b_{13} = 1.0111$ $\qquad b_{23} = .0370$
           $b_{21} = -.2489$ $\qquad b_{31} = .2778$ $\qquad b_{32} = .1000$
       b. $r_{12} = -.78$ $\qquad r_{13} = .53$ $\qquad r_{23} = .06$
       c. $X_{1e} = 6.6037 - 2.5606X_2 + 1.1058X_3$
       d. $B_{12.3} = -.82$ $\qquad B_{13.2} = .58$
       e. $r_{12.3} = -.88$ $\qquad r_{13.2} = .77$
       f. $R_{1.23} = .9718$ $\qquad R_{1.23}{}^2 = .9444$
       g. $R'_{1.23} = .9636$

10.   a. (1) $s_{1.23} = 3.6033$
           (2) $s_{b_{12.3}} = .2881$ $\qquad s_{b_{13.2}} = .2621$
           (3) $s_{a_{1.23}} = 3.74$
           (4) $s_{X_{1e}}$ (average value) $= 1.08$
           (5) $s_{X_{1e}}$ (specific value) $= 3.76$
       b. (1) $t(b_{12.3}) = 2.140$, $t_{.05}$ (df $= 12$) $= 2.179$, $b_{12.3}$ is not statistically significant at the .05 level of significance.
               $t(b_{13.2}) = 2.387$, $t_{.05}$ (df $= 12$) $= 2.179$, $b_{13.2}$ is statistically significant at the .05 level of significance.
           (2) $t(r_{12.3}) = 2.139$, $t_{.05}$ (df $= 12$) $= 2.179$, $r_{12.3}$ is not statistically significant at the .05 level of significance.
               $t(r_{13.2}) = 2.386$, $t_{.05}$ (df $= 12$) $= 2.179$, $r_{13.2}$ is statistically significant at the .05 level of significance.
           (3) $F(R_{1.23}) = 14.71$, $F_{.05}$ (df $= 2, 12$) $= 3.89$, $R_{1.23}$ is statistically significant at the .05 level of significance.
       c. (1) $b_{12.3}$: LL $= -.011$, $\qquad$ UL $= 1.244$
               $b_{13.2}$: LL $= .055$, $\qquad$ UL $= 1.197$
           (2) $a_{1.23}$: LL $= -11.634$, $\qquad$ UL $= 4.665$
       d. $X_{1e} = 10.8$
       e. (1) X_{1e} (average value): LL $= 8.45$, $\qquad$ UL $= 13.15$
           (2) X_{1e} (specific value): LL $= 2.61$, $\qquad$ UL $= 18.99$

12.   a. (1) $s_{1.23} = 1.8933$
           (2) $s_{b_{12.3}} = .2092$ $\qquad s_{b_{13.2}} = .1534$
           (3) $s_{a_{1.23}} = 1.97$
           (4) $s_{X_{1e}}$ (average value) $= .58$
           (5) $s_{X_{1e}}$ (specific value) $= 1.98$

b. (1) $t(b_{12.3}) = 4.370$, $t_{.05}$ (df $= 9) = 2.262$, $b_{12.3}$ is statistically significant
at the .05 level.
$t(b_{13.2}) = 3.850$, $t_{.05}$ (df $= 9) = 2.262$, $b_{13.2}$ is statistically significant at the .05 level.

(2) $t(r_{12.3}) = 3.243$, $t_{.05}$ (df $= 9) = 2.262$, $r_{12.3}$ is statistically significant
at the .05 level.
$t(r_{13.2}) = 2.748$, $t_{.05}$ (df $= 9) = 2.262$, $r_{13.2}$ is statistically significant
at the .05 level.

(3) $F(R_{1.23}) = 15.90$, $F_{.05}$ (df $= 2, 9) = 4.26$, $R_{1.23}$ is statistically significant at the .05 level.

c. (1) $b_{12.3}$: LL $= .441$, UL $= 1.387$
$b_{13.2}$: LL $= .243$, UL $= .938$

(2) $a_{1.23}$: LL $= -5.706$, UL $= 3.209$

d. $X_{1e} = 10.1$

e. (1) X_{1e} (average value): LL $= 8.79$, UL $= 11.41$

(2) X_{1e} (specific value): LL $= 5.62$, UL $= 14.58$

Appendix A

2. $z' = 60$

4. a. Clerical aptitude $z' = 53$
Mechanical aptitude $z' = 75$

6. a. log $Y_e = 1.00 - .04X$ d. $Y_e = 6.0$

8. a.

1963:	7.75	1967:	13.75
1964:	7.88	1968:	15.63
1965:	8.75	1969:	22.00
1966:	11.75	1970:	32.88

10.

	a. Seasonal factors	b. Seasonally adjusted dollar volume (mil. dollars)
Jan	.590	30.5
Feb	.475	33.7
Mar	.658	30.4
Apr	.812	32.0
May	1.136	33.5
June	1.764	33.4
July	1.854	35.1
Aug	1.694	37.2
Sept	1.032	37.8
Oct	.603	36.5
Nov	.382	49.7
Dec	.998	38.1

14.

	I_p (1967 = 100)	I_q (1969 = 100)	I_p (1967– 1969 = 100)	I_v (1972 = 100)
1967	100.0	66.7	80.0	38.1
1968	125.0	83.3	100.0	59.5
1969	150.0	100.0	120.0	85.7
1970	125.0	83.3	100.0	59.5
1971	250.0	140.0	200.0	166.7
1972	175.0	100.0	140.0	100.0

16. Exact data are 14, 32, 186, and 19. Others are approximate.
18. 2(min), 1(min), 2(min), 5, 2, 2(min)
20. 74   20   978
      1   43   30
    436   504   322
22. .000694   397,000   .00640
      24.0   6.08   .466

Index

419